Making Silicon Valley

Inside Technology
edited by Wiebe E. Bijker, W. Bernard Carlson, and Trevor Pinch

A list of the series will be found on page 385.

Making Silicon Valley
Innovation and the Growth of High Tech, 1930–1970

Christophe Lécuyer

The MIT Press
Cambridge, Massachusetts
London, England

First MIT Press paperback edition, 2007
© 2006 Massachusetts Institute of Technology

Set in Baskerville by The MIT Press.

Library of Congress Cataloging-in-Publication Data

Lécuyer, Christophe.
Making Silicon Valley : innovation and the growth of high tech, 1930–1970 / Christophe Lécuyer.
    p. cm. — (Inside technology)
Includes bibliographical references and index.
ISBN-13: 978-0-262-12281-8 (hc. : alk. paper)—978-0-262-62211-0 (pb. : alk. paper)
1. High technology industries—California—Santa Clara Valley (Santa Clara County)—History—20th century. 2. Microelectronics industry—California—Santa Clara Valley (Santa Clara County)—History—20th century.
3. Entrepreneurship—California—Santa Clara Valley (Santa Clara County).
4. Military-industrial complex—California—History—20th century. 5. Santa Clara Valley (Santa Clara County, Calif.)—History—20th century. I. Title.
II. Series.
HC107.C22S395   2006
338.4'76'0979473—dc22   2005047877

to the memory of Elisabeth Henry

Contents

Acknowledgments

This book would not have been possible without the support of Timothy Lenoir and the Department of History at Stanford University; Jed Buchwald, Evelyn Simha, and the Dibner Institute for the History of Science and Technology at the Massachusetts Institute of Technology; and Michael Gorman and the Department of Science, Technology, and Society at the University of Virginia. Robert McGinn and Merritt Roe Smith provided material support at critical junctures. I would also like to thank the Charles Babbage Institute and the IEEE History Center for financial assistance.

I owe a deep debt of gratitude to Barton Bernstein for his steady support and wise counsel. I would also like to extend a special word of thanks to W. Bernard Carlson for his great help in shaping my preliminary manuscript into a book and for the many stimulating conversations we had on the history of Silicon Valley. Sheila Hochman shared with me her understanding of high-tech firms. This book would not have been completed without her friendship and her generous help. I am also deeply grateful for Judith Zakaria's encouragement and support over the course of this project. Finally, Sara Meirowitz of The MIT Press graciously shepherded a manuscript that she inherited from another editor.

This book has also greatly benefited from discussions with Amir Alexander, Babak Ashrafi, Antonio Barrera, Ross Bassett, Patsy Baudoin, Leslie Berlin, David Brock, Jack Brown, Jane Carlson, Cathryn Carson, Larry Cohen, Paul David, Margaret Graham, William Hausman, Thomas Hughes, Martin Kenney, John Krige, Stuart Leslie, Massimo Mazzotti, Bryan Pfaffenberger, Laurent Pierrot, Philippe Reymond, Michael Riordan, Nathan Rosenberg, Henry Rowen, Philip Scranton, Peter Temin, Phillip Thurtle, Takahiro Ueyama, Steven Usselman, Peter Westwick, Jameson Wetmore, and Thomas Zakaria. I would also like to acknowledge the insightful comments I received from the audience at talks at the

Dibner Institute, the Georgia Institute of Technology, the Hagley Museum and Library, the London School of Economics, the Massachusetts Institute of Technology, the National Museum of American History, the National University of Singapore, Stanford University, the University of Maryland, the University of Pennsylvania, and the University of Virginia.

Finding sources was particularly challenging. I am indebted to Henry Lowood and Maggie Kimball at Stanford and to David Farrell at the Bancroft Library for their help in locating archival materials. This project also benefited from the interest of many participants who granted me interviews. I would especially like to thank Chet Lob and Ted Taylor for sharing their understanding of the microwave tube industry and Jacques Beaudouin for introducing me to silicon technology. My thanks also extend to David Allison, Orville Baker, Roger Borovoy, John Hulme, Lionel Kattner, Eugene Kleiner, Donald Liddie, Andy Ramans, Sheldon Roberts, Arnold Wihtol, and Jack Yelverton for providing historical materials on Varian Associates, Shockley Semiconductor, Fairchild Semiconductor, Signetics, and Intersil. Finally, I would like to extend a special word of thanks to Jay Last for giving me extensive access to his papers. This was given generously and with no conditions attached.

An earlier version of chapter 4 appeared as "Fairchild Semiconductor and Its Influence" in *The Silicon Valley Edge: A Habitat for Innovation and Entrepreneurship*, edited by Chong-Moon Lee, William Miller, Marguerite Hancock, and Henry Rowen (Stanford University Press, 2000). Chapter 5 draws on the following articles: Timothy Lenoir and Christophe Lécuyer, "Instrument Makers and Discipline Builders: The Case of NMR." *Perspectives on Science* 4 (1995): 97–165; Christophe Lécuyer, "Silicon for Industry: Component Design, Mass Production, and the Move to Commercial Markets at Fairchild Semiconductor, 1960–1967," *History and Technology* 16 (1999): 179–216. I also used material first presented in the following articles: "Electronic Component Manufacturing and the Rise of Silicon Valley," *Journal of Industrial History* 5 (2002): 90–113; "High Tech Corporatism: Management-Employee Relations in U.S. Electronics Firms, 1920s–1960s," *Enterprise and Society* 4 (2003): 502–520 (by permission of Oxford University Press); "Technology and Entrepreneurship in Silicon Valley," Nobel e-Museum (www.nobelprize.org/physics/articles/lecuyer/index.html). I want to thank all those who provided me with or permitted me to reproduce illustrations, in particular Lionel Kattner, Maggie Kimball, Jay Last, Pete Nuding, the American Institute of Physics, the Bancroft Library, Fairchild Semiconductor Corporation, the Intel Corporation, the McGraw-Hill Companies, Stanford University Archives, and Varian, Inc.

Making Silicon Valley

Introduction

In 1976, two electronics hobbyists, Steve Wozniak and Steve Jobs, started a partnership, which they called Apple Computer Company, in a garage in Los Altos, California. Wozniak and Jobs were very young. They had little capital and no business experience. Their only asset was the Apple I, a small computer Wozniak had designed as a side project. Because the Apple I elicited significant interest at the Homebrew Computer Club, a local group of computer hobbyists, Jobs and Wozniak hoped to sell this machine to fellow enthusiasts and to build a small business. But their firm grew beyond their wildest dreams. Within 5 years, Apple Computer had become a major Silicon Valley corporation, with 1,500 employees and sales in the $300 million range. By 1981, more than 300,000 computers designed and manufactured by Apple were in use all over the world, and Apple had become the leading firm in a new industry, personal computing. Apple was also a major financial success. Its initial public offering in 1980 was the second largest since the listing of the Ford Motor Company on the New York Stock Exchange 25 years earlier, and Apple's founders and investors were worth hundreds of millions of dollars (Freiberger and Swaine 2000; Linzmayer 2004). How could a startup firm established by two electronics hobbyists become a household name and define a new industry?

The rise of Apple and the growth of the personal computer industry on the San Francisco Peninsula were not accidents. They were made possible by practices, skills, and competencies that had accumulated in the area for more than 40 years. Critical for Jobs' and Wozniak's success was their access to networks of engineers, entrepreneurs, and financiers in Silicon Valley. These networks gave them access to the region's unique expertise in manufacturing, product engineering, sales, and marketing. Because of these networks, Jobs and Wozniak also came to rely on the innovative techniques that local entrepreneurs and managers had devised to direct and grow electronics corporations. At the same time, Apple's founders belonged to the community of electronics hobbyists on the San Francisco Peninsula. This enabled them to hire hobbyists who were enthusiastic about personal computers and who had extensive knowledge of analog and digital techniques. In other words, the rich technical and business environment in Silicon Valley made it possible for Apple to emerge as a major corporation.

The rise of Apple was the culmination of more than 40 years of growth and innovation on the San Francisco Peninsula. Silicon Valley had humble beginnings. In the late 1920s and the early 1930s, the San Francisco Peninsula was an industrial backwater. It was home to a few radio enterprises, which had a cumulative workforce of a few hundred machinists and perhaps 80 engineers. These corporations were operating in the shadow of the Radio Corporation of America (RCA) and other large East Coast firms, which controlled most of the electronics technology at that time. Forty years later, the situation had changed dramatically. The Peninsula had become a major industrial center specializing in electronics. By the early 1970s, the Peninsula's electronics cluster had a workforce of 58,000, more than half of whom were employed by electronic component firms. Silicon Valley was, in particular, the main center for the development and production of power-grid tubes, microwave tubes, and semiconductors. Local firms manufactured about half of the microwave tubes and more than a third of the silicon transistors and integrated circuits produced in the United States. Because these components were used in virtually every advanced weapon system and in a wide range of industrial goods, the Peninsula's electronics industry became central to the US defense effort and to the political economy of US manufacturing.

In this book, I explore the formation of this industrial district and the emergence of the practices and the bodies of knowledge that made its growth possible. I analyze and interpret the expansion of this regional economy by following groups of engineers and innovator-entrepreneurs who were active in the San Francisco Peninsula's electronic component industries. I investigate the ways in which these groups developed new industries as well as innovative manufacturing and design technologies and how they competed with East Coast corporations. I also examine large-scale forces in industry, government, and the international arena that advanced their technological and business enterprise. In particular, I analyze how the military financed the Peninsula's industrial expansion and influenced the development of electronics component technologies. I also examine the role of Stanford University and other local institutions of higher education in the growth of electronic component manufacturing on the Peninsula.

In spite of its importance in the history of technology and industry in postwar America, the rise of Silicon Valley has attracted little in-depth attention. It has been the subject of a number of popular "histories" that emphasize lifestyle, tell arresting anecdotes, and eschew history. Aside from these popular treatments, historical and economic analysis of Silicon Valley's emergence has been rare. In a book and a series of pioneering articles on Stanford's Cold War research programs and the rise of Silicon Valley, Stuart Leslie, a historian of science, has argued that the Peninsula's electronics industry grew mainly out of Stanford's research and teaching programs. Leslie has asserted that Frederick Terman, Stanford's dean of engineering and later its provost, built strong research groups in electronics with military patronage in the late 1940s and the 1950s. Building on their technical achievements, Terman sought to strengthen the Peninsula's electronics industry by encouraging Stanford's engineers to establish new corporations in the area. He also persuaded General Electric and other East Coast firms to build research laboratories near the university. Like Stanford, these firms, Leslie argued, were highly dependent on military patronage and were oriented nearly exclusively toward technologies of direct interest to the Department of Defense (Leslie 1992; Kargon, Leslie, and Schoenberger 1992; Leslie 1993; Kargon and Leslie 1994; Leslie and Kargon 1994; Leslie and Kargon 1996).[1]

Whereas Leslie has emphasized military patronage and university-industry relations in the rise of Silicon Valley, AnnaLee Saxenian, a regional studies analyst, has offered a "new economy" interpretation of the industrial district, emphasizing its culture and its organizational structure. In her comparative study of Silicon Valley and Route 128 (a center of high-tech activity near Boston), she has attributed Silicon Valley's rise and Route 128's concurrent decline to the regions' respective industrial structures. Route 128, she argued, has been dominated by large, autarkic, vertically integrated firms since the postwar period. Silicon Valley, on the other hand, has been characterized since its early days by a highly fragmented and decentralized industrial structure. Silicon Valley is made up of specialized firms embedded in dense regional networks. The most successful firms are organized as autonomous and often competing teams. According to Saxenian, these different organizational structures, rooted in the regions' histories and cultures, explain the divergent fortunes of Silicon Valley and Route 128. The large and autarkic corporations on Route 128 adjusted slowly to the swift technological and market changes of the 1980s. Because Silicon Valley's decentralized industrial structure promoted collective learning and flexible adjustment, local firms were able to adapt more rapidly to change, and they thrived in the new environment (Saxenian 1994).

These interpretations of Silicon Valley's emergence and rapid expansion belong to two different intellectual traditions in the study of industrial clustering. Leslie's Stanford-centered argument resonates with Johann Christaller, August Lösch, and other German economic geographers who view industrial clusters as preordained by geographical endowments, shipment costs, and access to skilled manpower. In contrast, Saxenian's work belongs to a tradition that emphasizes the social and institutional factors of industrial clustering. She draws on Michael Piore's and Charles Sabel's "flexible specialization" argument and on the sociological literature on Italian industrial regions. In this tradition, industrial districts are made of networks of small and medium-size firms that produce small batches of output in constantly changing product and process configurations. These homogeneous and highly integrated industrial districts are superior to autarkic and mass-production-oriented firms because they foster technological innovation and adjust rapidly to shifts in technology and markets (Christaller 1966; Lösch 1967; Piore and Sabel 1984).[2]

Leslie and Saxenian and the two intellectual traditions they represent have taught us a great deal about industrial clustering and the formation of Silicon Valley. I agree with Leslie that military patronage and procurement sustained the growth of the electronics district on the Peninsula and that Silicon Valley firms often benefited from their proximity to Stanford. I partly concur with Saxenian's argument that dense relations among flexible firms in Silicon Valley gave them a substantial competitive advantage over large Eastern corporations.

But although my work shares similarities with Leslie's and Saxenian's writings, I propose a different approach to the question of Silicon Valley's formation. This approach is predicated on different theoretical foundations and relies on a different tradition in the study of industrial clustering. In *Principles of Economics*, Alfred Marshall, the father of the modern discipline of microeconomics, sketched an economic and social theory of the emergence and growth of industrial districts. This theory gave a central place to technological skills and manufacturing practices. Marshall argued that groups of skilled workers (along with their political and religious views) are critical for the formation of new industrial clusters. These districts grow and perdure because they generate specialized manufacturing knowledge and attract subsidiary trades as well as a skilled and mobile workforce. In turn, the knowledge base and the suppliers of specialized inputs bring economic advantages to local firms in the form of external economies. Michael Porter has recently explored aspects of this framework, emphasizing the role of industrial clusters in competition. According to Porter, intense rivalries within industrial clusters stimulate the formation of new businesses. They also lead to greater productivity and capacity for innovation (Marshall 1890; Porter 1990; Porter 1998a,b).

I view Silicon Valley in the period under consideration in this book as a particular kind of "Marshallian" district—one that was oriented toward the manufacture of advanced electronic components, power-grid tubes, microwave tubes, and semiconductors. Between 40 and 70 percent of Silicon Valley's electronics workforce was employed in the component sector between the early 1930s and the late 1960s. Its electronic system sector, with firms such as Hewlett-Packard, remained comparatively small during this period. Because of its specialization in components and the relative weakness of its electronic system industries, Silicon Valley depended almost exclusively on faraway markets. Firms in Silicon Valley shipped most of their output to the Department of Defense, to military electronics system firms, and later to consumer-oriented industrial sectors in the East, Midwest, and Southern California. As it grew, Silicon Valley became remarkably rich and heterogeneous. It gained many different types of firms, bodies of knowledge, skills, and manufacturing formats (such as flexible specialization and mass production). In particular, the district, which at first specialized in vacuum tubes, acquired new capabilities in the manufacture of silicon transistors and integrated circuits.

The emergence and the growth of Silicon Valley were made possible by the building of unique manufacturing, product engineering, and management competencies. Over 40 years, innovator-entrepreneurs and their engineering staffs developed new ways of manufacturing advanced electronic components. They made major innovations in manufacturing processes, and these innovations enabled local firms to introduce innovative products to the marketplace. Silicon Valley engineers and entrepreneurs also learned how to manage processes and people. They gained expertise in the control of manufacturing processes, and they learned to integrate invention, design, manufacturing, and sales logistics. These men made also important social innovations. In industries where manufacturing processes were difficult to develop and to control, they needed to attract and retain a highly skilled and motivated workforce. This led them to develop innovative forms of management-employee relations. The entrepreneurs sought to involve their professional employees in the decision-making process. They also developed unusual incentives for their employees, including profit sharing, stock ownership, and stock options.[3]

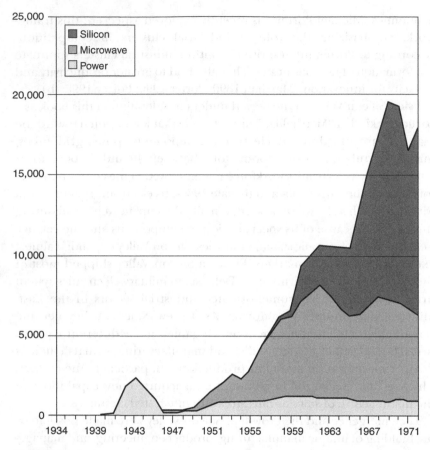

Figure I.1
Employment in manufacturing of electronic components (power-grid tubes, microwave tubes, and silicon components) in San Mateo and Santa Clara counties, 1934–1972. Source: Census of Manufactures.

Three technological and entrepreneurial groups (electronics hobbyists, microwave engineers, and semiconductor technologists) created and nurtured these technological and managerial capabilities on the San Francisco Peninsula from the 1930s to the early 1970s. These groups were indigenous to the area or moved to the Peninsula in the postwar period. Each group brought with it new technologies (often originally developed on the East Coast) as well as a distinct culture, a style of work, and a political and professional ideology. Each group also set up a new electronic component industry on the Peninsula. For example, in the 1930s and the first half of the 1940s a group of entrepreneurs and technologists with roots in the Peninsula's vibrant electronics hobbyist community established firms specialized in the manufacture of power-grid tubes. Another technical community made up of microwave engineers and physicists, who had been at the forefront of component technology on the East Coast during World War II, formed the microwave tube industry on the Peninsula in the postwar period. Similarly, scientists and technologists skilled in the design and processing of silicon components started a large number of semiconductor firms on the Peninsula in the second half of the 1950s and in the 1960s.

Although these groups were late entrants in the electronic component industries, their growing manufacturing and managerial capabilities enabled them to gain a significant competitive advantage over their East Coast competitors. Because of their technological and social innovations, firms on the Peninsula were able to capitalize on the military's growing demand for reliable high-performance electronic components during World War II and the Cold War. Several major electronics innovations were stimulated by the perceived difficulty of meeting the military's demands for reliability, by the demands of creating a broad consumer base, and by the challenges of commercial production. Thus, when the Department of Defense cut back its component expenditures and radically revised its procurement policies in the early 1960s, the rapid adaptation of a wave of innovations allowed local corporations to redirect their technologies and organizations to commercial markets. As a result, they penetrated a wide range of industrial sectors, transforming the San Francisco Peninsula into a major manufacturing center.

Thus, the rise of Silicon Valley is not a simple story about the military-industrial complex. During World War II and for much of the 1950s, the military was the largest and often the sole customer of the vacuum tubes and semiconductors made on the San Francisco Peninsula. The military services supported the development of microwave tubes and financed the construction of new factories. Because of its strict quality standards, the Department of Defense also acted as a catalyst for innovation. But the military was also a tough customer. The Department of Defense sought to control its suppliers' manufacturing lines and gain rights to their intellectual property. It also got access to their accounting books in order to reduce their bargaining powers. And the volume requirements of the military could change dramatically from one year to the next. Many firms found it so risky to work for the Department of Defense that they sought to reduce their dependence on it, and in the 1960s and the early 1970s they moved into commercial markets. In short, the military sustained the formation and growth of Silicon Valley, but at the same time it forced local firms to open up new markets for their products in the civilian sector.

The rise of Silicon Valley is not a simple story about university-industry relations either. To be sure, Stanford University trained physicists and microwave engineers who later worked for firms on the Peninsula. Stanford was also a source of important innovations in the design of microwave tubes, and these innovations were later exploited and commercialized by local corporations. But university-industry relations in Silicon Valley were much more complex than these one-way flows. In order to build active programs in electronics, Frederick Terman, an electrical engineer, and other Stanford administrators relied heavily on technologies, especially manufacturing process technologies, developed in Silicon Valley. They appropriated these technologies and built their research and teaching programs in vacuum tube and solid-state electronics on the basis of these transfers. In other words, it was because of their close relations with Silicon Valley firms that engineering groups at Stanford made technological innovations of their own and that the university was able to train a skilled engineering workforce in electronics.

To trace the district's growth and reconstruct its complex relations with Stanford and the Department of Defense, I examine the dominant technologies and the dominant firms on the San Francisco Peninsula. I focus especially on Eitel-McCullough, Litton Industries, Varian Associates, Fairchild Semiconductor, Signetics, Intel, and National Semiconductor. Eitel-McCullough (informally known as Eimac) made power-grid tubes which were used in radio transmission and high-frequency radar systems. Litton Industries and Varian Associates designed and fabricated microwave vacuum tubes—magnetrons and klystrons. These microwave tubes were the key components in advanced radar sets, radar countermeasure systems, and intercontinental communication systems. Fairchild Semiconductor and its spinoffs such as Intel and Signetics specialized in the fabrication of silicon transistors and integrated circuits, which were used in electronic products ranging from high-speed computers to television monitors and kitchen appliances. These firms were the largest and most influential electronics manufacturers on the Peninsula from the 1940s to the early 1970s. Eitel-McCullough, Litton, Varian, Fairchild, and its spinoffs also dominated the US markets for advanced tubes and semiconductors. More important, they created the main production and design innovations on the Peninsula. They were also sources of new management practices, marketing techniques, and financial arrangements. They provided a model for other electronics corporations, and they were the incubators of many other component and system ventures on the San Francisco Peninsula. I will examine Eitel-McCullough, Litton Industries, Varian Associates, Fairchild, Signetics, Intel, and National at critical periods in their history and in Silicon Valley's development.

Individual chapters tell the stories of these firms. Each chapter develops a major theme relating to how Silicon Valley evolved. In chapter 1, I examine the emergence of an indigenous power tube industry on the San Francisco Peninsula in the mid 1930s and the 1940s. Focusing on Eitel-McCullough, I argue that the power tube industry grew out of the area's vibrant amateur radio community. Established by two radio amateurs, Eitel-McCullough concentrated on the manufacture of high-quality transmitting tubes for other electronics hobbyists. The development of advanced manufacturing techniques and the very unusual performance and reliability requirements of amateur radio put Eitel-McCullough at the forefront of the tube business. Taking advantage of the enormous demand for radar tubes during World War II, Eitel-McCullough vastly expanded its activities and became one of the nation's largest manufacturers of vacuum tubes. In turn, Eitel-McCullough's rapid growth created an infrastructure for the manufacture of electronic components in the area. This infrastructure facilitated the growth of another military-related industry, microwave tubes, in the postwar period.

In chapter 2, I investigate the economic and political forces that sustained the emergence of the microwave tube industry after World War II by focusing on Litton Industries. Litton Industries was a splitoff of Litton Engineering, a supplier of tube-making equipment to Eitel-McCullough and other vacuum tube corporations. At first, the new firm (started by Charles Litton) concentrated on the design of exotic magnetrons for radar countermeasure systems. Later it benefited from the surge in the military demand for microwave tubes during the Korean War. It expanded into the manufacture of magnetrons for radar systems, and it engineered innovative manufacturing techniques for the production of these tubes. This manufacturing prowess enabled Litton Industries to make the highest-quality magnetrons in the United States and to do so at very low cost. The firm's high profits attracted the scrutiny of the federal government. As a result, its founder sold Litton Industries to a group of Southern California businessmen, who then transformed it into a major military conglomerate. The growth of Litton's magnetron business was critical for the Peninsula's manufacturing district. Litton Industries created a model of the successful startup in microwave tubes. It also helped local entrepreneurs to establish their own microwave tube businesses, thereby seeding the growth of this industry in the area.

Among the startups sponsored by Charles Litton was Varian Associates, and this company is the focus of chapter 3. Varian was established by a group of physicists and engineers who were influenced by California's utopian tradition. These men shaped Varian Associates into an engineering cooperative that specialized in the making of klystrons for the Department of Defense. In the late 1940s, Varian Associates acquired unique competence in the design and the processing of klystrons through the development of a fuse for atomic bombs. In turn, this competence enabled it to secure a large number of engineering contracts and production orders from the Department of Defense during the Korean War. In the second half of the 1950s, Varian Associates benefited from a shift in Department of Defense procurement from low-performance magnetrons to high-quality klystrons. As a result, it became the largest microwave tube corporation in the United States. The rise of Varian Associates, Litton Industries, and other microwave tube firms further enriched the Peninsula's industrial infrastructure and facilitated the emergence of another component industry: semiconductor manufacturing.

In chapter 4, I examine the ways in which Fairchild Semiconductor revolutionized the semiconductor industry and transformed the San Francisco Peninsula into the main center for the production of advanced silicon devices in the United States. Established by a group of semiconductor scientists and engineers in alliance with a New York investment bank, Fairchild adopted and refined new semiconductor manufacturing techniques recently developed at the Bell Telephone Laboratories. These techniques enabled Fairchild to be the first firm to introduce high-frequency silicon transistors to the market. These transistors met a rapidly growing demand for silicon devices in guidance and control systems for aircraft and missiles. Because Fairchild's transistors did not meet the reliability standards of the military (especially those of the Minuteman missile program), the firm's founding engineers made a second round of major process and design innovations; the planar process and the integrated circuit were the results. Capitalizing on its advances in design, processing, and manufacturing, Fairchild gained a large share of the military market for high-reliability silicon components in the late 1950s and the early 1960s. As a result, the firm emerged as the leading producer of advanced silicon devices in the United States. Fairchild also stimulated the formation of venture capital partnerships and semiconductor-equipment businesses in the region.

Chapter 5 traces how firms on the Peninsula moved into commercial markets after a major shift in defense procurement in the early 1960s. The local power-grid and microwave tube corporations were hit especially hard by military cutbacks. Eitel-McCullough and Varian Associates had to merge in order to survive. Varian also diversified into the manufacture of commercial systems; it became a producer of equipment for the manufacture of semiconductors, and it also entered the business of scientific and medical instrumentation. Fairchild, in contrast, opened up new markets for its silicon transistors in the commercial computer and consumer electronics industries. To do so, it acquired expertise in mass production from the electrical and automotive industries. It also seeded new markets for its transistors by designing electronic systems (such as television sets) around its components and by giving the designs to its customers at no cost. With these innovations in marketing and manufacturing, Fairchild developed vast markets for its transistors in the civilian sector. It also set a pattern for the later growth of the integrated circuit business on the San Francisco Peninsula.

Chapter 6 narrates the growth of the integrated circuit business on the San Francisco Peninsula in the first half of the 1960s. Because Fairchild's managers did not understand the potential of integrated circuits, the engineers who had developed the first planar microcircuit left the firm to start new semiconductor corporations, including Amelco and Signetics. These men further refined the technology of integrated circuits and helped create a small market for them. When Fairchild's managers recognized the promise of microcircuits, they resolved to crush Signetics. To do so, Fairchild's engineers copied Signetics' circuits and introduced them to the market at very low prices. They also used the innovative marketing techniques they had developed a few years earlier to create commercial markets for integrated circuits. This course of action enabled Fairchild to establish itself as the largest supplier of integrated circuits to the US civilian sector.

But starting in the late 1960s, Amelco and Fairchild spawned a number of new integrated circuit firms, including Intel, Intersil, and National Semiconductor. As chapter 7 reveals, these startups, many of them financed by local venture capitalists, relied on the rich repertoire of techniques and managerial practices that Fairchild and its spinoffs had developed since the late 1950s. The new wave of semiconductor corporations capitalized on this know-how to open up large commercial markets for microcircuits. For example, in the late 1960s National Semiconductor seeded the industrial market (automobiles, telecommunication, computer peripherals) for

integrated circuits. Taking advantage of new process and design technologies such as metal oxide semiconductors, Intel and Intersil helped
create markets for microcircuits that could be used in electronic watches
and in memory systems. In the next few years, these corporations entered
most industrial sectors, transforming silicon electronics into a ubiquitous
technology. By the early 1970s, Silicon Valley had become a major
technological and commercial center.

1

Defiant West

In 1947, Eitel-McCullough, a manufacturer of radio transmitting tubes located on the San Francisco Peninsula, sued the Radio Corporation of America and the General Electric Company, alleging patent infringement. GE and RCA, the giants of American electronics, had copied Eitel-McCullough's new line of tubes for FM radio and television broadcasting. The giants lost the lawsuit and were forced to halt production of these tubes. Exploiting its legal victory for commercial advantage, Eitel-McCullough transformed these mighty corporations into its own virtual sales force and distribution network by letting them buy its products and resell them under their own names.[1]

Such a lawsuit and its outcome would have been inconceivable 20 years earlier. RCA and GE thoroughly dominated American electronics in the late 1920s. They controlled all patents on vacuum tubes, and, along with Western Electric and Westinghouse, they dominated the manufacture of transmitting tubes in the United States. Firms that attacked their dominance were sued for patent infringement and driven into bankruptcy. How was Eitel-McCullough able to emerge as such a prominent manufacturer of transmitting or "power" tubes? How did it compete with RCA and GE and partially displace these mighty corporations in the field of power tubes? What forces at work in industry, government, and the international arena made the rise of Eitel-McCullough possible (Norberg 1976; Sturgeon 2000)?

Arthur Norberg briefly sketched the early history of power tube manufacturing on the San Francisco Peninsula in his article on the origins of the West Coast electronics industry. However, Norberg did not examine the unique social and economic context that sustained the rise of Eitel-McCullough. Nor did he explore Eitel-McCullough's history during World War II, when it emerged as a major electronics manufacturer. In this chapter, going beyond Norberg's analysis, I examine the emergence of Eitel-McCullough and of a closely allied company, Litton Engineering Laboratories, by following the careers of three innovator-entrepreneurs—William Eitel, Jack McCullough, and Charles Litton—from the early 1920s to the late 1940s.

These men started and grew the power tube industry on the San Francisco Peninsula. Eitel, Litton, and McCullough had become acquainted with the technology of power tubes through their activities in amateur ("ham") radio and their venture into tube production at local radio firms in the late 1920s and the early 1930s. In the midst of the Great Depression, these men established Eitel-McCullough and Litton Engineering Laboratories. While Litton Engineering specialized in tube-making equipment, Eitel-McCullough fabricated transmitting tubes for radio amateurs. The unusual requirements of radio amateurs and their innovative use of Litton's equipment put Eitel and McCullough at the cutting edge of the tube business. As a result, they were well positioned to supply advanced tubes for radar development programs in the late 1930s. Benefiting from the enormous demand for radar tubes during World War II, Eitel-McCullough vastly expanded the scale of its operations and became one of the largest US producers of vacuum tubes. In turn, because Eitel-McCullough and other firms heavily relied on Litton Engineering's machinery, Litton emerged as a major supplier of tube-making equipment during the war.

Training

William Eitel, Jack McCullough, and Charles Litton, unlike many subsequent electronics entrepreneurs on the San Francisco Peninsula, had deep roots in the area and came from families with a strong history of entrepreneurship. These men had been born and raised in, variously, San Francisco and the small communities of San Mateo and Santa Clara counties. Their families also shared common traits: they were middle class or lower middle class and had a strong technical and entrepreneurial bent. McCullough's parents had built a small wholesale lumber business in San Francisco. (His uncle owned a sawmill on the Peninsula.) Eitel's family had a strong mechanical orientation. His father had ventured into the design of aircraft engines in the 1910s. When the company developing his engine faltered because of a shortage of funds, he took a job managing a granite quarry. Subsequently, he ran a small stone-carving business. Eitel's uncle, E. J. Hall, had established the Hall-Scott Motor Car Company in Oakland, one of the first automotive corporations on the West Coast. One of Hall-Scott's specialties was the small-scale production of sports cars. The firm also designed and built an aircraft engine, the "Liberty engine," which was used in most American military aircraft during World War I.[2]

In addition to coming from comparable social backgrounds, Eitel, Litton, and McCullough received similar technical training in radio technology and metalworking. These men acquired a solid education in the mechanical arts by working in their families' enterprises and attending mechanically oriented educational institutions. Eitel, an energetic and resourceful youngster with limited interest in academics, gained his mechanical skills in the shop of Los Gatos High School and by working in his father's quarry as an assistant blacksmith and machine operator. One of Eitel's favorite places to visit was the shops of the Hall-Scott Motor Car Company. There he learned about machine-shop practice and the operation of complex machinery.[3]

While Eitel gained his mechanical skills mostly by doing, Litton and McCullough attended the California School of Mechanical Arts in San Francisco. One of the best technical high schools on the West Coast, it offered rigorous training in the mechanical trades and solid education in the humanities and the sciences. At the school, Litton and McCullough became excellent machinists. They also gained, as Litton later recalled, "a realistic 'feel' of materials and processes coupled with and at no sacrifice to a sound liberal arts background."[4] McCullough continued his technical education at a local junior college. Litton deepened his knowledge of mechanics and metalworking by enrolling in Stanford University's mechanical engineering department in the early 1920s. The department's curriculum at the time had a strong practical flavor. It was organized around courses in shop work and administration, machine drawing and design, and power plant engineering. It also included chemistry courses. The mechanical and chemical expertise Litton acquired at Stanford helped him greatly in his subsequent vacuum tube endeavors. Litton received a bachelor's degree in mechanical engineering in 1924.[5]

Litton, Eitel, and McCullough, like many technically minded middle-class youngsters, became interested in radio in the mid 1910s and the early 1920s. The San Francisco Bay Area was an excellent place to discover the new field of electronics. Since the turn of the century, this region had been, like Boston, one of the main centers of ham radio activity in America. By the mid 1920s, the Bay Area had more than 1,200 licensed amateurs, about 10 percent of all the radio operators in the

United States. The ham community in Northern California was remarkably dynamic. San Francisco and Oakland had radio clubs. Stanford University also had its own radio group. In addition, the Bay Area's amateur radio community generated a large share of the electronics hobbyist literature. Local clubs produced newsletters on "wireless" technology. *Radio,* one of the two magazines dedicated to amateur radio in the United States, was based in San Francisco.[6]

One might ask why an isolated and peripheral region, a continent away from important urban and industrial centers, nurtured such a large and vibrant ham radio community. A number of geographical and cultural factors seem to have played a role in the high concentration of radio amateurs in the Bay Area. Northern California had a strong maritime orientation, and San Francisco was one of the largest seaports on the West Coast. San Francisco Bay also had several military bases. The Navy and local commercial shipping firms relied heavily on radio communication to monitor their operations in the Pacific. As a result, they gave considerable visibility to the new technology—especially in the 1900s and the 1910s, when radio was used almost exclusively for ship-to-ship and ship-to-shore communication. In addition to exposing San Francisco youths to the new technology, the Navy and the shipping companies employed a significant number of radio operators, some of whom were involved in amateur radio (Pratt 1969).

The presence of a small but vital electronics industry in the area also contributed to the strength of the ham community in Northern California. The Bay Area was an active center of radio manufacturing in the 1910s and the 1920s. It was home to a number of electronics firms, including Remler (which made radio sets) and Magnavox (the leading American manufacturer of loudspeakers). Another small company, Heintz and Kaufman, designed custom radio equipment. Federal Telegraph, one of the earliest radio companies in the United States, operated a radio-telegraph system on the West Coast and produced radio transmitters in the 1910s. These firms made radio parts available to local hobbyists and hired radio amateurs (Norberg 1976; Morgan 1967).

General attitudes toward technical change in California may have reinforced the local interest in radio. Technological innovation seems to have been especially valued in California since the 1890s. California farmers, for example, mechanized their operations earlier than their counterparts in other parts of the country. In similar fashion, Californians rapidly embraced new technologies such as the automobile and the airplane in the 1900s and the 1910s (Pursell 1976).

Introduced to amateur radio through their families and their friends, Litton, Eitel, and McCullough embraced the new hobby. They rapidly found a congenial environment for their radio interests at their schools and in their communities. Eitel was tutored in the new technology by his mathematics teacher, a radio enthusiast. Eitel also learned about the new field by reading radio books and *QST*, an amateur radio magazine, in his school's library. These young men also became active in the Bay Area's radio clubs. In the process, they became acculturated to the world of amateur radio, and they soon acquired many of its values and behavior patterns.[7]

Ham radio was an unusual technical subculture in a number of ways. First, it was characterized by camaraderie and intense sociability. Hams used radio primarily as a way to make friends. In addition to communicating "over the air," radio amateurs socialized face to face in radio clubs and at conventions. They organized "hamfests" that attracted hundreds of amateurs living in the same area as well as commercial suppliers of radio equipment. By the early 1930s, an observer of the amateur community reported that radio amateurs in the United States organized more than twenty regional conventions and hundreds of "hamfests" annually. It was at one of these social gatherings that McCullough, Litton, and Eitel met (De Soto 1936; Douglas 1987). Second, the ham culture was characterized by egalitarianism and a democratic ideology. Radio amateurs gave little heed to traditional distinctions of class and educational attainment. The Santa Clara County radio club, which Eitel chaired in the mid 1920s, counted among its members farm boys, Stanford students, Federal Telegraph's technicians, and retired executives. Moreover, hams saw themselves as part of the "people." In their recurrent conflicts with the military services and with large corporations such as RCA over control of the airwaves, radio amateurs presented themselves as representatives of the citizenry against the interests of large and undemocratic organizations (Douglas 1987). Third, radio amateurs greatly valued technical innovation and resourcefulness. They were interested in extending the range and performance of radio technology. In particular, they sought to improve radio circuits and to explore new radio frequencies for long-distance communication. In short, radio amateurs built reputations among their brethren by innovating new circuitry, devising clever transmitter designs, and establishing contacts with faraway lands (Morgan 1967; De Soto 1936). Lastly, the ham culture was characterized by its mix of competitiveness and information sharing. "The predominant characteristic of the amateur," a radio hobbyist observed at the time, "is his altruism. The amateur wants every other amateur and the public to share in and benefit by his discoveries. The rivalry to accomplish something that has not been done before is intense. But it is rivalry of the friendliest sort, and no sooner does one make a new record that he wants to show all his brother amateurs not only how it was done, but how they also can do it. The slightest advance in technique, every individual discovery, any observation that promises improvement, is immediately the property of all." (De Soto 1936) Competitiveness and free sharing of information were institutionalized by the Amateur Radio Relay League (ARRL), the main association of radio amateurs in the United States. The ARRL published the most advanced amateur work in its magazine, *QST*. It also gave awards for important technical accomplishments. In 1926, for example, the ARRL established a "worked-all-continents" prize for amateurs who had established radio contact with stations in Asia, Europe, and Australia (De Soto 1936).

In tandem with their acculturation in the world of amateur radio, Eitel, Litton, and McCullough gained a solid knowledge of electronics and radio systems. They read hobbyist literature and radio textbooks and experimented relentlessly with their radio equipment. Through this intense work, these young men gained a thorough knowledge of radio circuitry. They also learned to design and build their own radio stations. Ultimately, they encountered the new field of short-wave radio.[8]

The short waves were then a largely forsaken portion of the radio spectrum. Judging the short waves (under 200 meters) to be worthless, the US Department of Commerce had given them over to radio amateurs in 1922. Waves over 200 meters, which seemed at the time to have more potential for long-distance communication, were assigned to the military, to RCA, to American Telephone and Telegraph (AT&T), and to broadcasting networks. As a result, electronics hobbyists explored the untrodden part of the radio spectrum. In 1923, radio amateurs discovered that they could use 100-meter waves to communicate with Europe. Furthermore, the radio amateurs soon found that short waves were far superior to long waves for long-distance radio transmission. Short waves made communication possible over distances greater than had ever been reached before. They also required only a fraction of the power used by the long-wave stations. This revolutionary discovery opened a new field of inquiry in which radio amateurs participated prominently. Intent upon furthering their explorations, radio amateurs experimented with ever shorter waves. They defined the characteristics of such waves and evaluated their potential for long-distance communication (De Soto 1936; Douglas 1987).[9]

Interested in joining this revolution, Eitel, Litton, McCullough, and many of their ham friends on the San Francisco Peninsula experimented with short waves. In the process, these young men made notable contributions to the art of high-frequency or short-wave radio. In 1924, Litton and a fellow member of the Stanford radio club were among the first American amateurs to establish communication in the high frequency bands with Australia and New Zealand. In 1928, Eitel pioneered the use of 10-meter waves for transcontinental communication. This important accomplishment opened the very high frequency (VHF) bands to radio communication. "I was bringing up the rear when the movement [toward the higher frequencies] started," Eitel later reminisced, "but by the time they got to 20 meters I was up with them. I beat them all in spanning the continent on 10 meters. In order to do this work, I had to build everything myself. This took an understanding of circuits and other components because most of them were marginal. In fact, they were inadequate. So I had to improvise with what was available. I worked with the circuits until I got the performance I wanted. Since I built everything myself, it was a matter of how I arranged the coils and condensers in the circuits."[10]

In conjunction with the design of transmitting stations and their innovative work in short-wave radio, Eitel, Litton, and McCullough learned about vacuum tubes, the main components of radio circuits and systems. Vacuum tubes made it possible to generate, to detect, and to amplify radio signals. A vacuum tube consisted of a glass envelope and a set of electrodes: a filament or cathode, which emitted a stream of electrons; a plate or anode, which collected them; and, finally, an open mesh grid, which controlled the flow of electrons between the filament and the plate. The grid acted as a valve, opening or closing the passage of electrons according to the voltage on it. When a vacuum tube was used as an amplifier, radio waves intercepted by an antenna came to the grid as a weak alternating current oscillating with the radio waves' frequency. The oscillating voltage thus applied to the grid modulated the flow of electrons crossing the tube to the same frequency. The electron stream then delivered an alternating current at the plate, which reproduced with great amplification the weak signal on the grid. Tubes had to be carefully evacuated of all gases to allow the flow of electrons between the cathode and the plate.

Amateur radio and especially the "exploration" of the short waves was a good school in which to learn about these complex devices. "Vacuum tubes," Eitel later recalled, "were the weak links in the chain. It was rather tricky to get them to perform properly at [high] frequencies. They were

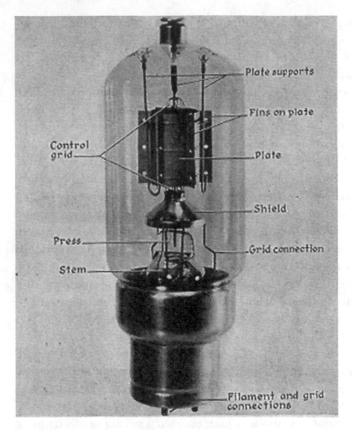

Plate supports

Fins on plate

Control grid

Plate

Shield

Press

Grid connection

Stem

Filament and grid connections

Figure 1.1
The structure of the Heintz and Kaufman 354 power-grid tube. Source: Terman 1938.

one of the components that required a sixth sense to get them to work. They had not been designed for these frequencies. They had been designed for the lower frequencies."[11] To push the tubes to their limits and get them to work at the higher frequencies, Litton, Eitel, and McCullough gained a solid knowledge of their construction and operating principles.[12]

Litton went one step further. He learned to fabricate vacuum tubes and, especially, power-grid tubes.[13] Mastering the fabrication of transmitting or "power" tubes was a remarkable achievement for an independent experimenter. These devices, which were used to generate strong radio signals for long-distance transmission, were very difficult to make. Indeed, General Electric, Westinghouse, and AT&T, which had developed high-power transmitting tubes in the early 1920s, encountered substantial difficulties in producing them in a consistent and reproducible fashion. The fabrication of power tubes required precise machining. It also required a mastery of glass blowing: transmitting tubes were made of special Pyrex glass. Their manufacture also rested on complex processing techniques. To create the high vacuum required for their operation, the tubes had to be baked at high temperatures for hours at a time in order to release the gases occluded in their metallic elements. Power tubes also required the use of exotic materials and sophisticated sealing techniques to make tight joints between the glass envelope and the metallic elements. Finally, the fabrication of high-vacuum power tubes necessitated the use of "getters"—magnesium pellets, attached to the inside portion of the tube envelope, that absorbed residual gases after the tube had been evacuated (Fagen 1975).

Although little is known about Litton's apprenticeship in vacuum tube practice, it is likely that he learned to make transmitting tubes by reading the technical literature and by playing with power-grid tubes. He also may have received technical advice from his neighbor Otis Moorhead. Moorhead, a radio amateur and a vacuum tube entrepreneur, had established a vacuum tube firm, Moorhead Laboratories, in San Francisco in 1917. Moorhead manufactured receiving tubes for radio sets until a patent-infringement lawsuit put him out of business in the early 1920s. Litton was fascinated by the complex techniques required to make power-grid tubes. In the early and mid 1920s, he experimented with materials and with tube processing techniques in his home laboratory. In parallel to this work on glass and metals, Litton mastered the design and construction of the specialized vacuum equipment required to make power tubes. He built, for instance, the vacuum pumps with which he evacuated his tubes. Litton also constructed his own ovens. By the mid 1920s, after years of trial and error, Litton was making sophisticated tubes: high-power triodes (vacuum tubes with three electrodes—cathode, plate, and grid) as well as thermionic rectifiers. He used these in his own radio transmitter and sold them to other radio amateurs on the San Francisco Peninsula (Litton and Scofield 1925; Norberg 1976; Sturgeon 2000).[14]

In 1925, to complement his training in electronics, Litton did graduate work in electrical engineering at Stanford University. At that time, Stanford's small electrical engineering department was oriented toward graduate education. Its three instructors offered a limited range of courses on electric circuits, AC machinery, and power transmission. One might speculate that in addition to these classes, Litton also took two of the few electronics-related courses that the university was offering at the time: a physics course on "ions and electrons," which covered, among other things, the theory of vacuum tubes, and a course (first offered in 1925) on "communication engineering fundamentals." The latter course included a brief treatment of electromagnetic theory and went into radio and telephony engineering in more depth. For his engineering thesis, Litton designed an instrument that recorded and helped visualize short radio waves. After graduating with an engineering degree in electrical engineering, Litton, like many West Coast engineers at that time, went East. He accepted a junior engineer position at the Bell Telephone Laboratories. There he joined a new engineering group that was developing a short-wave radio system for transatlantic telephony. Over the next 2 years, Litton designed measuring equipment and short-wave receivers. His work at the Bell Telephone Laboratories and his training in electrical engineering at Stanford transformed Litton into a professional radio engineer. But Litton never abandoned his ham radio roots and continued to play with radio transmitters and talk "over the air" for most of his life.[15]

Tube Design and Manufacture

In the late 1920s, Litton, Eitel, and McCullough found jobs with small electronics corporations on the San Francisco Peninsula. After two lonely years at the Bell Telephone Laboratories, Litton longed to return home. He had had a nervous breakdown, and he did not like the Eastern climate. In 1927 he moved back to California. With the help of Philip Scofield, a Stanford friend whom he had known through amateur radio, Litton secured a research engineer position at the Federal Telegraph Company. He negotiated a contract which stipulated that he would work only on the San Francisco Peninsula. Eitel too gained a foothold in the Peninsula's electronics industry through his ham radio connections. In 1929, Heintz and Kaufman Incorporated hired Eitel as a mechanic on the recommendation of Colonel Foster, a wealthy radio amateur who was a customer of Heintz and Kaufman. One year later, Eitel recruited McCullough to work with him at Heintz and Kaufman. Litton, Eitel, and McCullough put considerable efforts and energy into their new jobs—so much so that Litton soon became known as "Charles Vigorous Litton" among his co-workers. Eitel and McCullough, who were industrious, put in many all-nighters at Heintz and Kaufman (Southworth 1962; Millman 1984; Layton 1976).

The Federal Telegraph Company and Heintz and Kaufman Incorporated offered attractive opportunities for ambitious young men eager to prove themselves in electronics. These were the most respected electronics firms in the Bay Area. They also had active research and engineering programs in short-wave radio. Federal Telegraph, which had been formed in 1909, had pioneered continuous wave radio in the 1910s. It had also helped to develop vacuum tube technology. It was at Federal Telegraph that Lee de Forest had invented the audion oscillator and amplifier, the first vacuum tube. Exploiting these innovations for commercial advantage, Federal Telegraph became an important supplier of radio equipment to the US Navy during World War I. After these notable beginnings, Federal Telegraph declined in the 1920s, becoming a rather insignificant supplier of radio-telegraph services on the West Coast. To revive its sagging fortunes, the firm sought to gain a position in short-wave radio, as it became apparent that the new technology offered great commercial possibilities for long-distance communication.[16]

In 1927, to finance its research and development efforts in short-wave radio, Federal Telegraph secured a large contract from International Telephone and Telegraph (IT&T), a New York-based telecommunication conglomerate with operations in Europe and South America. IT&T was interested in building a global short-wave radio communication network. It contracted out the development of the required high-frequency transmitters and receivers to Federal Telegraph. Under the terms of the agreement, Federal Telegraph became the sole supplier of short-wave radio equipment to IT&T. In return, IT&T financed a large share of Federal Telegraph's research and engineering program and paid royalties to Federal Telegraph on its sales of radio communication services. With this contract, in 1927 and 1928 Federal Telegraph built a large R&D organization with more than 60 engineers and scientists working on all aspects of short-wave radio.[17]

Heintz and Kaufman was also an important player in the new field of short-wave radio. It had actually pioneered the commercial exploitation of short waves in the mid 1920s. The firm had been established in 1921 by Ralph Heintz, an inventive radio amateur and electro-mechanical engineer. At first, Heintz had repaired scientific instruments and produced radio sets and broadcasting transmitters. Sensing the future commercial importance of short-wave radio, Heintz re-oriented the firm toward the design and manufacture of high-frequency radio equipment in 1924. The corporation produced high-frequency transmitters and receivers on a custom basis for a wide variety of users. Among these were the Army, the Navy, the Boeing Airplane Company, and wealthy radio amateurs. Heintz and Kaufman's transmitters were also used in various expeditions to Antarctica and the North Pole. In addition to these custom jobs, Heintz and Kaufman secured a large procurement contract from the Dollar Steamship Company. Dollar, a large shipping company based in San Francisco, asked Heintz and Kaufman in 1929 to build an extensive short-wave radio network in the Pacific. These transmitters would connect its fleet with shore stations in Hawaii, Guam, China, and the Philippines as well as major ports in the United States. In conjunction with this large contract, Dollar also acquired a controlling interest in Heintz and Kaufman, transforming it, in effect, into its in-house supplier of radio equipment (Olson and Jones 1996; Niven 1987).[18]

Working for Federal Telegraph and Heintz and Kaufman, Litton, Eitel, and McCullough soon gravitated toward the design and production of power-grid tubes. In less than a year, Litton and Eitel respectively became the heads of Federal Telegraph's and Heintz and Kaufman's tube shops.[19] The manufacture of transmitting tubes was a new activity for these corporations. Until then, Federal Telegraph and Heintz and Kaufman had specialized in the operation of radio-telegraph systems and the manufacture of radio transmitters and receivers for long-distance transmission. It was only in the late 1920s that they moved into power tube manufacturing. They did so because they could not procure transmitting tubes on the open market. RCA, GE, Western Electric, and Westinghouse—the sole producers and distributors of power-grid tubes in the United States—refused to sell these tubes to Federal Telegraph or Heintz and Kaufman. RCA went one step further and threatened to sue Federal Telegraph and Heintz and Kaufman for patent infringement in case they procured transmitting tubes from European suppliers.[20] The reasons for this refusal were clear. RCA, which had been set up in 1919 at the instigation of GE and the Navy to ensure American predominance in radio, controlled ship-to-shore and transoceanic communication in the United States. It considered Federal Telegraph and Heintz and Kaufman threats to its domination of long-distance radio communications. Allowing Federal Telegraph and Heintz and Kaufman to buy power-grid tubes would permit them to establish transoceanic radio circuits for IT&T and the Dollar Steamship Company in direct competition with RCA. RCA could deny the sale of transmitting tubes to Federal Telegraph and Heintz and Kaufman because of its control of radio technology. RCA, which was partially owned by GE and Western Electric (AT&T's manufacturing arm), had signed a series of exclusive cross-licensing agreements with AT&T, GE, and Westinghouse. These cross-licensing agreements gave RCA control of more than 2,000 patents in the field of radio, including all the important vacuum tube patents. Making the most of these patents, RCA aggressively sued firms that infringed on its intellectual property rights and put them out of business.[21]

RCA's monopolistic practices forced Federal Telegraph and Heintz and Kaufman to manufacture their own power-grid tubes. Litton's and Eitel's job was to produce power tubes and to make them sufficiently different from General Electric's, Western Electric's, and Westinghouse's devices so that they would not fall under RCA's patents. As Litton and Eitel soon realized, designing and making transmitting tubes that did not infringe on the patents of the radio monopoly was an enormously difficult task. RCA had a seemingly impregnable patent position. It controlled more than 250 patents which covered all aspects of tube design and manufacture. Furthermore, RCA held the fundamental patents on device structures and tube elements such as cathodes, getters, and glass-to-metal seals. Circumventing these patents was highly risky. In the late 1920s, large electronics firms such as Sylvania tried to manufacture transmitting tubes that would bypass RCA's intellectual property rights, but they failed to circumvent some of RCA's key patents. As a result, RCA sued Sylvania and forced the firm to stop manufacturing power-grid tubes (Stokes 1982).

Litton and Eitel also had to confront some challenges that were specific to their location on the West Coast. Unlike Sylvania, which was located in Massachusetts and had ready access to a workforce skilled in vacuum tube manufacture, Federal Telegraph and Heintz and Kaufman operated in an industrial backwater. Though Litton and Eitel could find good mechanics in the Bay Area, they lacked access to a workforce skilled in vacuum tube practice. There were few operators with a knowledge of vacuum techniques and chemical handling in the Bay Area. Glass blowers too were rare, and Eitel considered the local ones incompetent. Furthermore, most suppliers of the special materials required for the fabrication of power tubes were located on the East Coast. For instance, the Corning Glass Works, which produced the Pyrex glass used in tube envelopes, had its main plant in New York State. Ordering materials required long and expensive trips to the East and entailed high shipping costs. In other words, the Bay Area's industrial infrastructure was inadequate for the manufacture of complex electronic devices such as power-grid tubes.[22]

But Litton, Eitel, and McCullough had access to significant technical and financial resources. Because the production of transmitting tubes was essential for the development of short radio communication systems, the managers of Federal Telegraph and Heintz and Kaufman allocated significant resources to the tube laboratories. Litton, for instance, directed a group made of ten college-trained engineers and scientists as well as a number of draftsmen and machinists. He had an ample budget, which allowed scouting trips to the East Coast to identify potential suppliers. Litton received considerable support from IT&T's legal department and the engineering groups of its European subsidiaries. A group of French engineers, for example, was dispatched from IT&T's research laboratory in Paris to aid with Federal Telegraph's tube-development efforts in 1930. Although the Dollar Steamship Company did not have IT&T's financial resources, Eitel was able to build a team of a dozen mechanics, glass blowers, and radio amateurs at Heintz and Kaufman. He could also rely on the counsel of patent lawyers and on the inventive mind of Ralph Heintz, who participated in the design and construction of transmitting tubes.[23]

The two groups at Federal Telegraph and Heintz and Kaufman also collaborated with each other in the late 1920s and the early 1930s. Their tight collaboration was predicated on common interests. They had to solve similar legal and design problems and solve them fast. They also had to make power-grid tubes, a difficult undertaking. The collaboration was also facilitated by the fact that they did not compete directly with each other and by the fact that they had a common enemy: RCA and the Eastern radio monopoly. One can also speculate that the cooperation between Litton, Eitel, and Heintz was also shaped by the friendships they had built and the values they had acquired through amateur radio. In the late 1920s and the very early 1930s, these men shared substantial information on tube design and production. Litton, who had more experience with transmitting tubes than his counterparts at Heintz and Kaufman, taught them the fundamentals of tube processing and manufacture. He also gave them production blueprints of tube-making machinery and detailed information on material suppliers. As Eitel, Heintz, and McCullough became more proficient in the tube art, the two groups frequently discussed the difficulties they were facing. "We learned from each other," Heintz later reminisced. "We went through the same agonies of decisions on how to do this and on how to do that. [Litton's] mind and mine were running pretty parallel."[24] The cooperation was so close that Federal Telegraph and Heintz and Kaufman jointly ordered their glass blanks from Corning.[25]

The development and manufacture of power-grid tubes was as much a legal endeavor as a technical one. To engineer transmitting tubes that would not infringe on RCA's patents, Litton, Eitel, and Heintz worked closely with their patent attorneys. "The patent department," recalled a former engineering manager at Federal Telegraph, "would point out that the elements of a proposed high-power tube would have to be designed to avoid infringing upon RCA's patents. The actual method we followed was to start with a group conference with two or three engineers and one or two men from the patent department. All present would discuss tube problems."[26] At these meetings, Litton would propose ways of circumventing RCA's patents, which the patent experts would then discuss. Based on their response, Litton would then work on the tube's detailed design. Eitel and Heintz proceeded in the same way at Heintz and Kaufman.[27]

These legal constraints guided the design of transmitting tubes. Litton, for example, devised a clever tube design that bypassed an important structure patent controlled by RCA. The patent covered a tube with a grid that "surrounded" the cathode. Taking advantage of the patent's phrasing, Litton devised a grid that, instead of encircling the filament, went 179° around it. Because of the grid's unusual shape, Litton decided to support both the grid and the plate from the side rather than at the end of the tube envelope. He also attached the filament to the tube's extremities. The resulting tube (nicknamed "the crying pig" for its odd shape) was less efficient than Bell's and General Electric's products. It was also much more difficult to make. But it did a reasonable job at the high frequencies, and it made possible the building of IT&T's short-wave radio communication system. Similarly, Eitel and Heintz made use of an old tube design that had fallen into the public domain—a design with a filament and two plates instead of the standard filament, grid, and plate. Experimenting with this old design, Heintz, Eitel, and McCullough engineered the gammatron, a rugged power tube that worked well at high frequencies.[28]

Because of the complexity of these designs, Litton and Eitel were forced to use new materials and develop new manufacturing techniques. They also had to circumvent manufacturing-process patents held by RCA. That corporation had a solid portfolio of patents on tube-manufacturing processes and evacuation techniques. For example, it had patents on the manufacture of special metal-to-glass seals intended to withstand high thermal stresses. RCA also controlled the use of getters. To circumvent these patents, Litton, Heintz, Eitel, and McCullough used new materials and developed novel techniques. To replace getters, Eitel and Heintz made tube plates of tantalum, a rare and exotic metal. When pre-heated at very high temperatures, tantalum acted as a getter and absorbed the gases released by the tube elements. Litton invented and patented a new technique for making shock-resistant seals. The use of tantalum and Litton's seals were important innovations. They made it possible to create a high vacuum in the tubes' envelopes. Because the quality of the vacuum was closely related to tube reliability, these new techniques allowed the fabrication of transmitting tubes that were more reliable than those distributed by RCA.[29]

Litton also made innovations in tooling. Relying on his mechanical and glass blowing expertise, he invented the glass lathe, an apparatus that mechanized tube-making operations such as assembly, glass blowing, and sealing. Litton developed this new machine to overcome the production and manpower difficulties he was encountering at Federal Telegraph. The tube he had designed to bypass RCA's patents was hard to make: because the grid and the plate were attached to the side of the glass envelope rather than to its extremities, the tube required complex glass work. Most glass blowers in the Bay Area did not have the advanced skills required. Furthermore, they were not amenable to strict industrial discipline. Glass blowers had the habits of pre-industrial craftsmen: they worked irregular hours and got drunk in the shop. As Litton and Eitel attempted to discipline them, some turned violent. In 1929 an irate and drunken glass blower at Heintz and Kaufman destroyed Heintz's new automobile and ransacked the shop.[30]

To rid himself of rebellious glass blowers and make complex tubes in large quantities, Litton invented the glass lathe, an ingenious machine that made it possible to simultaneously form a complex glass envelope and seal it to the tube's elements. A glass lathe's two heads rotated in synchronism and supported the glass blank as well as the tube's filament, grid, and plate. The machine operator would fabricate the tube envelope by applying a fire to the glass blank and blowing gas into it. At the same time he would join the metallic elements to the glass and seal them into the tube envelope. The glass lathe enabled the production of high-precision tubes. It was soon adopted at Heintz and Kaufman and later became one of the most important pieces of manufacturing equipment in the power-grid tube industry.[31]

In the process of designing and producing power tubes and tube-making machinery at Federal Telegraph, Eitel, Litton, and McCullough gained expertise in product engineering and vacuum tube manufacturing. They learned about the importance of process technology for the engineering of high-quality, high-precision transmitting tubes. They also

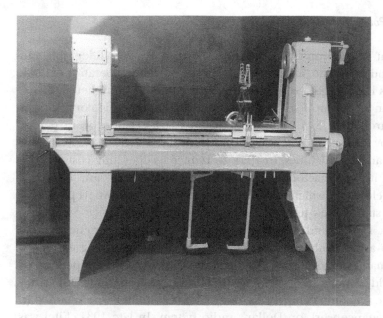

Figure 1.2
Litton's glassworking lathe, late 1930s. Courtesy of Bancroft Library, University of
California, Berkeley.

gained intuitive knowledge of materials and a deep understanding of the
functioning of power-grid tubes. For example, Eitel and McCullough
gained the ability to visualize the complex interrelationships that gov-
erned the design of vacuum tubes. Litton became an expert in materials
and manufacturing processes. One of Litton's associates later reported
that at Federal Telegraph "Litton [gained] a fantastic feel for Mother
Nature. He knew exactly what he could do with what Mother Nature gave
him in the way of physical materials: the elements, tungsten, copper,
glass, and so forth; just how far he could push Mother Nature to where
she finally cried 'uncle' and gave up. He developed this at Federal
Telegraph building the tubes, how to get a high vacuum."[32] This manufacturing expertise shaped much of their latter careers in electronics and
informed their approach to the vacuum tube business. At Federal
Telegraph and Heintz and Kaufman, Litton, Eitel, and McCullough also
acquired solid management skills. They learned to direct engineering
projects, to oversee the production of transmitting tubes, and to handle
personnel relations. These supervisory skills, along with their technological competence and the development of new manufacturing processes,
helped them greatly in their subsequent entrepreneurial activities.[33]

Building Power Tube and Tube Machinery Businesses

The Great Depression had a severe impact on Federal Telegraph and Heintz and Kaufman and their power tube-making operations. Bank credit was hard to get. Sales of manufactured goods plummeted. IT&T, the Dollar Steamship Company, and the small electronic corporations they controlled in San Mateo and Santa Clara counties were deeply affected by these economic conditions. In the early 1930s, IT&T ran into severe financial difficulties. In 1931, to reduce its operating costs, it consolidated its manufacturing operations on the East Coast. It forced Federal Telegraph, which it had recently acquired, to move its plant and technical staff to New Jersey at this time. Similarly, the Dollar Steamship Company, Heintz and Kaufman's parent company, experienced huge losses. To avoid bankruptcy, Dollar sharply cut its operating expenses. Dollar's management forced Heintz and Kaufman to fire 75 percent of its fifty-odd workers in September 1930. Four months later, Dollar dismissed the remaining employees. Only Heintz and Eitel were retained to do maintenance work on Dollar's radio system. In late 1931, Eitel was allowed to hire back McCullough and a few other employees to repair transmitting tubes. But their position remained precarious, as it was dependent on the evolution of the trans-Pacific trade and Dollar's shipping business.[34]

The dire economic conditions of the early 1930s compelled Litton, Eitel, and McCullough to start new businesses. When Federal Telegraph moved to New Jersey, Litton, who had no interest in living in the East, decided to stay in California. In 1932, with $6,000 in savings, he established a small proprietorship, Litton Engineering Laboratories. He also built a vacuum tube shop on his parents' property in Redwood City. Around the same time, Eitel and McCullough built a new commercial business at Heintz and Kaufman. Their primary incentive was to create new revenue streams and thereby safeguard their jobs. To generate these monies, they commercialized a new transmitting tube, which they had developed for their own use in amateur radio, under the Heintz and Kaufman brand name. Dissatisfied with the power-grid tubes that were then on the market, Eitel and McCullough set out to engineer a ham radio tube for their own use in early 1932. They wanted to make a power tube that would be more reliable than the one marketed by RCA. The RCA tube had the added disadvantages of operating poorly at the very high frequencies and working only at low voltages. These were serious limitations for ham radio use. At the time, radio amateurs applied high voltages to their tubes in order to get a high output. By doubling the voltage applied to the tube, they could increase the power of their radio frequency signal by a factor of 4.[35]

To make a better tube, Eitel and McCullough used the unique tools and processes they had developed for their first tube project at Heintz and Kaufman. They assembled and sealed the tube directly on a glass lathe. They also used tantalum for the grid and the plates. But Eitel and McCullough also appropriated the latest vacuum tube innovations developed on the East Coast. They used a new tube structure, developed at RCA, that enabled the electrons to better focus on the plate. They also took advantage of the development of new cathodes at General Electric. In the early 1930s, engineers at GE devised new filaments made of thoriated tungsten. Thoriated filaments emitted more electrons than conventional cathodes. They also could last a long time, and they were resistant to high voltages. As a result of these design and processing choices, Eitel's and McCullough's new power tube lasted longer and withstood higher voltages than RCA's products. Unlike its East Coast counterparts, it could also operate efficiently at high and very high frequencies. In other words, it was an excellent amateur radio tube, as Eitel and McCullough soon verified by using in their own radio transmitters.[36]

Eitel and McCullough thought this tube would sell well in the ham radio market. They also felt that the time was ripe for commercializing power-grid tubes. RCA, GE, Western Electric, and Westinghouse had lost some of their control of vacuum tube technology with the expiration of several of the most important tube patents in the early 1930s. And in 1930 the Department of Commerce filed an antitrust lawsuit charging RCA, GE, Western Electric, and Westinghouse with violations of the Sherman Antitrust Act. In its brief to the court, the government claimed that these corporations had conspired to restrain competition. According to the Department of Commerce, they had "continuously refused except on terms prescribed by them to grant licenses to any individuals, firms, or corporations for the purpose of enabling the latter to engage in radio communication, radio broadcasting, or interstate commerce in radio apparatus, independently or in competition with the defendants." After 18 months of negotiations, RCA and the other corporations accepted to sign a consent decree by which all the cross-licensing agreements were made non-exclusive. GE and Western Electric also divested themselves of all their RCA holdings. In addition, RCA became an independent company manufacturing its own power tubes and radio equipment. Though RCA, Westinghouse, GE, and Western Electric remained major players in the vacuum tube business, they had to use more restraint when dealing with their smaller competitors. They could not put them out of business as easily as before (Sobel 1986; Maclaurin 1949).

Taking advantage of this change in the legal environment, Eitel and McCullough convinced the Dollar Steamship Company to allow them to sell their tube on the open market under the Heintz and Kaufman name. They advertised their product in ham radio magazines and rapidly built a small tube business. But Eitel and McCullough soon met substantial resistance from Dollar when they sought to expand the scope of their tube activities and introduce more products to the market. Dollar had no interest in building a substantial component business, which lay outside of its core activities. It may also have been concerned about a possible lawsuit from RCA. Dollar's lack of interest in the ham radio tube business led Eitel and McCullough to consider starting their own tube operation. Their determination to form a new firm was reinforced by Dollar's decision to lay off some of their subordinates in the tube shop in 1934. At this time, the Dollar Steamship Company suffered from a general strike on the San Francisco waterfront. This strike and the financial crisis that it brought about led Dollar's managers to lay off 25 percent of Heintz and Kaufman's workforce in the spring of 1934. In spite of its profitability, the tube shop saw its manpower reduced by one-fourth. "McCullough and I," Eitel later reminisced, "figured out that if that was the way [Dollar] operated, there was not very much future there. We decided we were wasting our time trying to develop a complete line of tubes and market them."[37]

In September 1934, Eitel and McCullough left Heintz and Kaufman to set up their own transmitting tube corporation, Eitel-McCullough Inc., with the financial support of two small businessmen, Walter Preddey and Bradshaw Harrison. Harrison was a real-estate agent in San Bruno; Preddey operated a chain of movie theaters in San Francisco. According to the terms of the agreement, Harrison and Preddey invested $2,500 each in the partnership, and Eitel and McCullough brought their know-how to the table. The first profits were to be shared equally between the two investors until they had reached $2,500; Eitel and McCullough then would become equal partners. Preddey was the president, Eitel a vice-president.[38]

Establishing a power tube business in the midst of the Great Depression was risky if not foolhardy. The market for transmitting tubes shrank in the early 1930s. Furthermore, in spite of the partial breakup of the radio monopoly, RCA, GE, and Westinghouse thoroughly dominated the main markets for power tubes used in commercial broadcasting and long-distance radio communication. To survive in this inauspicious environment, Litton, Eitel, and McCullough focused on niche markets, which large East Coast firms did not fully control. To compete in these markets, Eitel, Litton, and McCullough emphasized quality, customer service, and technological innovation (especially through the development of new manufacturing processes). The entrepreneurs also introduced a series of products, which met the multifaceted needs of their customers. They adjusted flexibly to new business opportunities and exploited them aggressively. Following the practices they had started at Federal Telegraph and Heintz and Kaufman, Litton, Eitel, and McCullough also cooperated closely. Litton helped Eitel and McCullough set up their own vacuum tube shop by giving them the castings and engineering blueprints of his glass lathe. This gift enabled Eitel and McCullough to construct their own high-quality glass lathe at low cost. In the next few years, Litton, Eitel, and McCullough freely exchanged technical and commercial information in order to reduce the many risks, including the manufacturing risks, associated with the running of small tube-related businesses.[39]

Litton Engineering Laboratories had a difficult start and nearly collapsed in the early 1930s. "Litton," an employee recalled, "was struggling very desperately. That was at the depth of the Depression. He struggled, making various tubes, doing a little bit of research work and development work, and so forth, some for RCA and some for Federal Telegraph. He had to scrounge around and look for business."[40] Litton's situation started to improve in 1935. At this time, he discovered that there was a demand among East Coast tube manufacturers for the glassworking apparatus he had invented at Federal Telegraph. Federal Telegraph, for instance, asked him to produce precision glassworking equipment for its New Jersey plant. Similarly, RCA and Westinghouse ordered glass lathes from Litton Engineering. As a result, Litton re-oriented his small shop toward the design and production of glass lathes and other pieces of machinery used in tube manufacturing.[41]

To meet the demand for precision glassworking equipment, Litton designed four different types of glass lathes in close consultation with East Coast manufacturing concerns. Each lathe was developed to make transmitting tubes of a specific size. Litton also developed a machine that could seal irregularly shaped tubes. A manager at Litton Engineering later reported that "each year, as new tubes were developed, changes in the design of the glass working lathes were made by Litton Engineering

Laboratories. New type glass working lathes were designed and manufactured and consultations with leading tube manufacturers were held to obtain their reactions to proposed design changes."[42] To meet the requirements of its customers, Litton produced high-quality glass lathes. These machines were carefully designed and produced with the utmost precision. To enable the fabrication of advanced tubes, the lathes had tolerances on the order of 0.001 inch—very unusual in machine-shop practice at the time. These machines were also characterized by their near-perfect alignments.[43]

Litton later diversified into the manufacture of vacuum pumps. In 1938, he designed and constructed a new pump, which used low-vapor-pressure oil as its evacuating medium. Until then, most vacuum pumps relied on mercury. Mercury pumps required, among other disadvantages, that the mercury vapor traps be cooled by liquid air. This made them bulky and ineffective. Unlike its mercury-based counterparts, Litton's vapor oil pump was compact. It also operated at higher speed and made possible the attainment of higher vacuum. Because low-vapor-pressure oils were not available on the market, Litton invented a new distillation apparatus and produced his own pumping oil out of a commercial brand of motor oil. Exploiting these inventions, Litton built a small manufacturing equipment business. By the late 1930s, Litton Engineering devoted 90 percent of its activity to equipment manufacture. The other 10 percent was devoted to tube development and consulting, notably for Federal Telegraph. By 1939, Litton Engineering had $25,000 in sales and employed five machinists.[44]

While Litton concentrated on the design and production of glass lathes and vacuum pumps, Eitel and McCullough oriented their new business toward the production of transmitting tubes. Their objective was to manufacture high-quality tubes for radio amateurs. Eitel and McCullough viewed product reliability and performance as key to survival in the ham radio business—for a number of reasons. Eitel-McCullough had to compete with the tube they had developed themselves at Heintz and Kaufman. After Eitel's and McCullough's departure, the managers of the Dollar Steamship Company actively commercialized the tube the pair had recently designed for radio amateurs. Eitel and McCullough also faced direct competition in the radio-amateur market from RCA, from GE, from Raytheon, and from Taylor Tubes (a new Chicago-based venture). In addition to their brand names, financial resources, and intellectual property rights, these firms had a definite cost advantage over Eitel-McCullough: they were located close to their markets and material suppliers and, as a result, had lower shipping costs. "We had real handicaps to overcome in building [a company on the San Francisco Peninsula]," Eitel later reflected. "We had to ship our goods further, to the big centers of use; there we had to pay disproportionately high costs for many of the commodities we used for production. We had to learn to offer something better to the world."[45]

In addition to these competitive pressures, Eitel and McCullough had a further incentive to produce high-quality products. Radio amateurs were the most demanding users of power tubes in the mid 1930s. They applied very high voltages to their components to increase the power output of their transmitters. Radio amateurs also required tubes, which operated efficiently in the short-wave portion of the radio spectrum. In 1936, 82 percent of all radio amateurs in the United States used high frequencies. Another 10 percent were active in the very-high-frequency (VHF) band.[46]

To compete with RCA and Heintz and Kaufman and to meet the reliability and performance requirements of radio amateurs, Eitel and McCullough concentrated their efforts on improving manufacturing. They perfected the processing techniques they had developed at Heintz and Kaufman and devised new ones. In particular, they developed a novel sealing and assembly technique, which relied heavily on the use of Litton's glass lathe. The new assembly procedure worked as follows: using their glass lathe, Eitel and McCullough first sealed the plate to the top of the glass envelope. They would then hold the filament and grid on one head of the glass lathe while attaching the glass envelope and plate assembly to the other. The grid was aligned with the plate by carefully melting the glass stem to which the filament and grid were attached. In the final step of the process the two heads were joined together, which allowed the insertion of the grid at the center of the plate. This was an important technique and one of Eitel-McCullough's most closely guarded secrets. This technique enabled the close spacing of the tube elements. Because the spacing of the cathode, grid, and plate was closely related to tube performance, this process made possible the fabrication of VHF transmitting tubes. The close alignment of the tube elements also enabled radio amateurs to operate the tube at high voltages.[47]

Paralleling this new assembly technique, Eitel and McCullough designed a highly efficient system to evacuate transmitting tubes. Their system enabled them to outgas the power tubes thoroughly and thereby create a very high vacuum. At Heintz and Kaufman, Eitel and McCullough had removed the occluded gases in the tube's metallic parts by shooting electrons at the grid and the plate. Electrons emitted by the filament would heat a tube's elements to remove the occluded gases (gases that were contained in the tube's metallic parts before its operation). This technique, however, did not drive all the gases out of the grid. The grid would receive fewer electrons than the plate and, as a result, would be heated to lower temperatures. To attain identical temperatures for both the plate and the grid, Eitel and McCullough devised a new technique. They heated the grid and the plate separately. They then alternately bombarded the grid and the plate with electrons, thereby effecting the independent but concurrent heating of the plate and the grid. As a result, both elements were maintained at very high temperatures, and all occluded gases were eliminated from the tube. This new technique, soon patented by Eitel and McCullough, was a major process innovation. It decreased the time and cost of manufacture. The new procedure also made it possible to evacuate tubes thoroughly for greater reliability.[48]

These new manufacturing techniques enabled Eitel and McCullough to design a series of high-quality power triodes (tubes with three electrodes) for radio amateurs. They first produced the 150T, an improved version of the amateur-market tube they had developed at Heintz and Kaufman.[49] They also developed both a small and a large version of this tube to fill differing needs of radio amateurs. Because of their unique processing, Eitel-McCullough's tubes were substantially more reliable and had better electrical characteristics than products then on the market (including the radio tube they had designed at Heintz and Kaufman). Eitel-McCullough's tubes operated efficiently at the high frequencies. They could resist tremendous overloads and were characterized by long lifetimes. The average life of a power tube was then between 600 and 1,000 hours. Eitel-McCullough's tubes could last as long as 20,000 hours. In 1936 the firm introduced more powerful tubes for airline radio transmission. Eitel and McCullough distributed their tubes through manufacturing representatives and ham radio shops. They also actively advertised their products in *QST*, the journal of the Amateur Radio Relay League, and in *Radio*.[50]

Eitel-McCullough's tubes soon gained wide acceptance among radio amateurs and small manufacturers of aircraft radio equipment. By 1937, the firm's sales reached $100,000, half to airlines and half to radio amateurs. It was also highly profitable, which enabled Eitel and McCullough to finally become full partners. To meet the growing demand for their products, Eitel and McCullough gradually enlarged their workforce.

Figure 1.3
An advertisement featuring Eitel-McCullough's first tube, the 150T (1936).
Courtesy of Varian, Inc.

The two entrepreneurs had started the firm with just one mechanic in 1934. Three years later, they had 15 employees. To fill the new positions, Eitel and McCullough relied almost exclusively on the electronics hobbyists they had met at radio clubs on the San Francisco Peninsula. Radio amateurs had the skills Eitel and McCullough needed: familiarity with transmitting tubes and expertise in the design of radio systems. Furthermore, they had an intimate knowledge of Eitel-McCullough's ham radio market. As their new hires had no prior knowledge of vacuum tube practice, Eitel and McCullough trained them on the job in glass blowing, assembly, evacuation, and sealing. As a result of the founders' employment and training practices, Eitel-McCullough was an unusual firm. Most of the employees were in their early twenties. The culture of amateur radio was influential. Technical resourcefulness and innovation were valued highly. Other important characteristics were camaraderie, competitiveness, and a democratic ideology.[51]

Wartime Expansion

In the late 1930s, because of growing threats to international peace from Japan and Germany, President Franklin D. Roosevelt and his administration rebuilt American military and naval power, expanding the Army and the Navy and procuring new airplanes, cruisers, and aircraft carriers. A significant aspect of this rearmament effort was the development and deployment of an entirely new electronic system: radio detection and ranging (radar). This new system came out of secret research programs in short-wave radio at the Naval Research Laboratory (NRL) and the Signal Corps Engineering Laboratories (SCEL) at Fort Monmouth, New Jersey. In the late 1920s, radio engineers at these military laboratories discovered that ships and airplanes reflected high-frequency radio signals. This finding had great military potential: it promised the detection and location of enemy ships and airplanes at great distances. Building on their strength in short-wave radio, engineering groups at these laboratories developed experimental radar systems in the mid 1930s. These systems used VHF radio pulses to detect approaching airplanes. Though these radar sets helped identify incoming aircraft, they could only do so at close range (Van Keuren 1994; Allison 1981; Gebhard 1979; Page 1988; Zahl 1972).[52]

To extend the reach of their radar systems to 100 miles or more, the engineering groups at NRL and SCEL needed transmitting tubes that could function at high voltages. The transmitting tubes for radar had to operate with momentary voltages many thousands of volts higher than the normal voltage applied to radio communication tubes. The radar tubes also had to work efficiently at very high frequencies. None of the tubes manufactured by RCA, Westinghouse, Western Electric, or Raytheon met these requirements. Only the new Eitel-McCullough tube, which the firm introduced to the amateur market in 1937, had the desired performance and reliability characteristics. Engineers at NRL and SCEL, who knew about the tube through their ham radio activities, procured it from electronics parts dealers on the East Coast and soon incorporated it into their experimental radar systems (Van Keuren 1994; Allison 1981; Gebhard 1979; Page 1988; Zahl 1972).

In December 1937, NRL and SCEL engineers asked Eitel-McCullough to adapt its tubes to the specific requirements of their radar systems. They wanted Eitel and McCullough to make modifications to their transmitting tubes so that these tubes would better fit the electrical characteristics of their radar circuits. To better understand the radar systems, Eitel and McCullough visited the military laboratories and conferred about their requirements. Out of these discussions came two different versions of Eitel-McCullough's amateur tube. Because the SCEL engineers wanted tubes with short leads, Eitel and McCullough changed the tube's shape and lead arrangement. The new tube had a rectangular envelope. Its leads came from each side of the glass envelope rather than from its extremities. Eitel and McCullough also developed another version of the same tube for the Navy.[53]

When hostilities began in Europe, the Army and the Navy decided to bring their prototype radar systems to production and opened bidding for manufacture of the radar systems that had been developed at NRL and at SCEL. The military services selected RCA and Western Electric to do the job. The award of these production contracts to RCA and Western Electric created both an opportunity and a challenge for Eitel-McCullough. It opened a large potential market for the firm's radar tubes. But Eitel-McCullough would have to convince RCA and Western Electric to use its tubes in their radar systems. RCA and Western Electric, which produced their own power-grid tubes, had no interest in buying transmitting tubes from Eitel-McCullough. They wanted to use their own tubes in their radar transmitters. Only the steadfast support of radar engineers at NRL and SCEL helped Eitel and McCullough secure large subcontracts from RCA and Western Electric in the summer of 1940. The tube orders they received from RCA and Western Electric were significant indeed for a firm as small as Eitel-McCullough. Western Electric, for example, ordered 10,000 tubes for $500,000—five times the company's annual sales in 1939.[54]

These large orders created considerable dissension between Eitel and McCullough and their financial backers. Walter Preddey, who had helped finance the firm's formation and who was its president, opposed its going into high-quantity production for the military services. He worried that these large military contracts would make the firm too dependent on a few customers. He also thought the firm was not ready to execute such large contracts. On the other hand, Eitel, McCullough, and the other investor, Bradshaw Harrison, were eager to transform Eitel-McCullough into a larger operation. They forced Preddey to resign from the presidency of Eitel-McCullough in December 1939. Eitel replaced him as president, and McCullough became the firm's vice-president. In May 1941, Preddey sold his shares to Eitel and McCullough for $57,500. Eitel and McCullough were now in full control of their business.[55]

In 1940 and the first half of 1941, to meet RCA's and Western Electric's large orders for radar tubes, Eitel and McCullough converted their firm to volume production. With financing from the Bank of America, they constructed a new plant in San Bruno. Eitel and McCullough also greatly expanded their workforce, from 17 employees in July 1940 to 125 in May 1941 and to 170 in July 1941. The entrepreneurs hired local radio amateurs and machinists. (There were many precision machinists in the Bay Area, many of Swiss or German origin.) As Eitel and McCullough soon exhausted the supply of radio amateurs on the Peninsula, they increasingly hired women for delicate assembly operations such as the making of grids, plates, and filaments. To train and manage the fast-growing workforce, Eitel and McCullough relied heavily on the crew of radio amateurs they had assembled in the 1930s. These men instructed the new hires in the complex techniques of power tube production. They also built large departments around specific manufacturing processes such as pumping, glass working, and assembly.[56]

In 1939 and 1940, labor unions based in San Francisco sought to organize the plant. The Bay Area was the largest and most active center of trade unionism in the western United States. Its labor unions were particularly powerful and militant. They were also eager to extend their sway to the electronics industry on the Peninsula. Unwilling to relinquish control of the shop floor to union organizers, Eitel and McCullough fought vigorously against the unions. To thwart these organization efforts, they adopted managerial techniques that had been developed in the 1930s at Eastman Kodak, at Sears, Roebuck, at Thomson Products, and at other large corporations. These techniques were corporatist in nature. They sought to define the world of work not in terms of a sharp divide between employer and employee, but in terms of cooperation and mutual obligations between managers and workers in the same firm. To do so, these corporations gave pensions and job security to their employees. They also established profit-sharing programs, whereby a portion of the company's profits would be distributed to its employees. Inspired by these corporatist programs, Eitel and McCullough set up a medical unit, and a cafeteria that offered subsidized meals. In December 1939 they instituted a profit-sharing program that transferred one-third of each year's profits to the employees. These policies enabled Eitel and McCullough to keep the labor unions out of the plant in the 1940s.[57]

In conjunction with the hiring of a much larger workforce and the development of new personnel practices, Eitel and McCullough reengineered their radar tubes and their manufacturing processes. Producing radar tubes in quantity, as Eitel and McCullough soon discovered, was particularly difficult. The tubes regularly failed after 50 hours of operation. The high peak powers required by radar systems heated up the tube elements to unprecedented temperatures. As a result, the thorium in the filaments would evaporate and deposit itself on the grids. In turn, the grids would emit electrons, which led to uncontrolled current flows to the plates. To solve this problem, Eitel and McCullough made the grids out of platinum, a material known as a poor emitter of electrons. The use of platinum eliminated electron emission, but the new grids lacked rigidity. They would short the filaments and thereby ruin the tubes. Eitel and McCullough also discovered that their seals would fail at high temperatures. Finally, when the firm started to make thoriated tungsten filaments in large quantities, it had difficulty in maintaining uniformity in the process and in producing filaments with comparable electrical characteristics.[58]

To tackle these difficulties, Eitel and McCullough expanded the firm's research laboratory, which they had established in 1938 to develop new products. They also hired some chemistry graduates from the University of California, and some inventive radio amateurs. These men concentrated their efforts on the reliability problems of radar tubes. To develop non-emitting grids, the laboratory's chemists developed a new grid-making process. They coated molybdenum wires with carbon, platinum, and zirconium, and later they sintered these elements into the wires in a high-temperature furnace. Grids fabricated with this process were mechanically strong and emitted fewer electrons than the standard grids.

Eitel-McCullough chemists also perfected the filament-making process and worked out a series of procedures that could be followed by inexperienced operators. Finally, these men developed a new material, Pyrovac, to replace tantalum in the tubes' plates. Pyrovac, which was made out of zirconium and carbon, absorbed gases much better than tantalum and, as a result, made possible the fabrication of tubes with much longer lives. These important process innovations were carefully patented. The new manufacturing processes enabled the firm to produce reliable radar tubes in quantity and to obtain good manufacturing yields. (The yield is the proportion of good products coming out of the manufacturing line.) In turn, these tubes enabled the Army and the Navy to deploy a significant number of radar systems in late 1940 and the first half of 1941.[59]

The rapid expansion of Eitel-McCullough and the growing military demand for transmitting tubes encouraged others to enter the power-grid tube business. In May 1941, Charles Litton, Philip Scofield, and Ralph Shermund established a new power tube corporation, Industrial and Commercial Electronics (ICE). At Stanford, Litton had befriended Scofield and Shermund through their common interest in amateur radio. In the second half of the 1930s, Shermund, who had graduated from Stanford with a bachelor's degree in bacteriology, had gained substantial experience in tube manufacturing. Because of a recommendation from Litton, Shermund found a job at Raytheon, a Massachusetts-based tube maker. He later directed Heintz and Kaufman's tube shop after Eitel's and McCullough's departure in 1934. After several years at Heintz and Kaufman, Shermund, like his predecessors, ran into difficulties with Dollar's management. He sought to convince Dollar to spin out the tube shop and sell it to him—to no avail. Seeing the great commercial potential of power-grid tubes, Shermund decided to leave Heintz and Kaufman and start a company of his own. Litton soon joined the project. He was keenly interested in the power-grid tube business, but he knew that he could not produce power-grid tubes in his own shop and under his own name. Litton Engineering supplied glass lathes and other pieces of manufacturing equipment to most makers of transmitting tubes. Producing power tubes under the Litton label would make him compete directly with his own customers and would soon lead to the downfall of his equipment business. Investing in ICE and assisting Shermund discreetly on the manufacturing side of the business would permit Litton to participate in the financial rewards of tube manufacture without losing his own glass lathe business. Litton, Shermund, and Scofield each owned a third of the new corporation.[60]

In the summer of 1941, Shermund and Litton incorporated the new organization and built a tube-making shop (all the manufacturing equipment came from Litton Engineering). They also set up a profit-sharing program for their employees—mostly as a way to avoid labor unrest. (Litton Engineering Laboratories also had a profit-sharing plan, which gave half of the profits to the employees.) Thanks to Litton's reputation and his wide contacts in the US electronics industry, he and Shermund soon received contracts from Bendix and the Navy for the manufacture of vacuum relays and power-grid tubes. Litton worked at ICE two days a week, supervising tube production and working on manufacturing processes. He and Shermund had a business running by the end of the year.[61]

The attack on Pearl Harbor led to expansion of the transmitting tube companies on the San Francisco Peninsula. The Army and the Navy procured and deployed tens of thousands of high-frequency radar systems during the first years of the war. They also built a worldwide network of radio communication stations. These systems required millions of transmitting tubes every year. Eitel-McCullough, ICE, and Heintz and Kaufman, with their competence in the manufacture of reliable tubes, benefited from the enormous growth in the military demand for transmitting tubes. They were inundated by tube orders from the Navy and the Army, and also from prime military contractors such as RCA, Bendix, GE, Hallicrafters, and Wilcox Electric. For example, ICE received a large number of production orders for power-grid tubes from the Navy and the Army. These tubes had often been designed elsewhere, including at Eitel-McCullough. Because Litton became the manager of Federal Telegraph's vacuum tube division in November 1942, it was Shermund who ran ICE during much of the war.[62] Under Shermund's direction, ICE expanded rapidly. Its sales grew from $51,142 in 1942 to $1,333,693 in 1944. By that time, ICE had several hundred employees. It was also enormously profitable. With sales of $817,000 in 1943, ICE had a net profit of $305,693. Similarly, in spite of recurrent managerial infighting, Heintz and Kaufman expanded substantially during the war. By January 1943, Heintz and Kaufman had 300 employees.[63]

But it was Eitel-McCullough that benefited the most from the exploding demand for transmitting tubes during the war. After Pearl Harbor, Eitel-McCullough received very large orders for its transmitting tubes. The firm also secured second-source production contracts for components designed by General Electric and other East Coast manufacturers (including receiving tubes that could operate at very high frequencies). These huge orders led Eitel and McCullough to expand their workforce .

by a factor of 20 between Pearl Harbor and mid 1943. By the summer of 1943 they employed 3,600 operators and technicians. Extensive training programs for supervisors, foremen, and operators were needed. In conjunction with the rapid growth of its workforce, Eitel-McCullough hastily expanded its plant in San Bruno. It also opened a new factory in Salt Lake City. The primary impetus for the building of this new plant came from the military services. The services were concerned about Eitel-McCullough's proximity to the Pacific Coast, which made it vulnerable to a Japanese attack. In the spring of 1942, Eitel and McCullough chose the site of the new factory. The plant was operational by August 1942. The Salt Lake City plant was financed by the Defense Plant Corporation (a federal agency recently established to fund the construction of manufacturing plants for the war effort) and was owned by the federal government.[64]

To handle mass production of power-grid tubes, Eitel, McCullough, and their associates thoroughly transformed their manufacturing methods. In particular, they reinforced the production-control function. They set up a traffic department to schedule and expedite the flow of materials, tube elements, and semi-assembled tubes throughout the plant. In an effort to better control manufacturing, Eitel-McCullough's management also split up large production departments into smaller ones. For example, in 1943 the assembly department was divided into three divisions: punch press, grid making, and plate assembly.[65]

Eitel and McCullough also mechanized the manufacture of power-grid tubes. Until that time, transmitting tube manufacture had been, to a large degree, a craft-based activity requiring highly skilled workers. To mechanize the production of power tubes, Eitel and McCullough hired mechanical engineers with solid experience in machine-tool design. Many had worked in local shops producing machine tools for canneries and other Bay Area industrial establishments. Among the production bottlenecks they mechanized were the exhaust process and the fabrication of grids. The latter was a very labor-intensive operation. Each grid had to be delicately wound and spot welded. Because the production of a single grid required tens of operations, Eitel-McCullough's grid department could not meet the demand for its products. At the peak of production, the firm used more than 250,000 grids per month. The mechanical engineers developed a machine that could automatically produce grids of remarkable uniformity in huge quantities.[66]

Eitel, McCullough, and the head of the pump department designed an ingenious machine to evacuate and de-gas transmitting tubes. "We used stand pumps," Eitel later recalled, "when our volume was small. [The pumps] were arranged in rows and one operator could man four sections, which at most meant sixteen tubes. This was a bottleneck because skilled operators were required to constantly monitor and adjust the current to the tubes. There was no way we could have attained the volume necessary to meet our commitments with that system."[67] To solve this problem, Eitel-McCullough's founders invented a rotary exhaust machine. This machine was made of sixteen exhaust setups attached to a rotary wheel. Each setup had five vacuum pumps. The rotary machine could evacuate 768 tubes in 24 hours and could be tended by a relatively unskilled worker. Because the exhaust schedules were pre-programmed, the operator's only task was to seal the tubes on the exhaust setup and seal them off when the wheel had made its revolution.[68]

As a result of these and other innovations, Eitel, McCullough, and their associates were able to raise production rates from a few thousand tubes per month in mid 1941 to 150,000 tubes per month in 1943. At the peak of production, in late 1943, Eitel-McCullough had sales revenues of about $2 million a month. By the end of World War II, Eitel-McCullough, having manufactured more than 3 million transmitting tubes for military applications, was one of the largest US manufacturers of vacuum tubes and by far the largest electronics firm on the San Francisco Peninsula.[69]

Because of large orders from Eitel-McCullough and other transmitting tube firms, Litton Engineering grew substantially—from a few machinists in 1939 to 85 employees in 1944.[70] In early 1942, to meet the growing demand for its lathes, Litton built a new plant in Redwood City and tooled up for larger-scale production. Litton also transformed the organization of production. Whereas lathes had been built one at a time in the 1930s, the firm produced batches of standard machines during the war. As a result, its output increased substantially. Before 1940, Litton Engineering fabricated fewer than 10 glass lathes a year; in 1943, it produced 222. These lathes were allocated by the War Production Board to Eitel-McCullough, ICE, RCA, Westinghouse, Raytheon, Heintz and Kaufman, Western Electric, GE, Sperry Gyroscope, and Federal Telegraph. By making possible a dramatic surge in the production of power tubes, they played an important part in the war effort.[71]

Postwar Crisis and Renewal

After the enormous boom brought about by the war, the Peninsula firms experienced a difficult transition to peacetime production. In the immediate postwar period, Litton Engineering saw its orders for tube machinery decline. But the power-grid tube firms were even harder hit than Litton Engineering. Starting in March 1944, the Signal Corps, which had large inventories of power-grid tubes, canceled many of the production contracts it had placed with Eitel-McCullough, ICE, and Heintz and Kaufman. As a direct result of these contract cancellations, Eitel and McCullough laid off 1,100 workers and closed their Salt Lake City plant. With the capitulation of Japan, the military services canceled most of their remaining contracts with tube manufacturers. As a result, vacuum tube corporations drastically reduced their workforce. By December 1945, Eitel-McCullough had only 390 employees, a far cry from the 3,600 it had had in mid 1943. ICE and Heintz and Kaufman also slashed their workforces.

But the worst was still to come. Starting in late 1945, the military dumped its enormous inventories of surplus vacuum tubes on the market. Radio amateurs and manufacturers of electronics equipment could now buy advanced vacuum tubes for roughly 10 percent of their original price. As a result, Eitel-McCullough, ICE, and Heintz and Kaufman saw their tube sales decline to almost nothing in late 1945 and 1946. The military essentially killed the market. ICE and Heintz and Kaufman never recovered. ICE, also weakened by fights between Shermund and Litton over stock ownership, went bankrupt in 1949. After years of anemic sales, Heintz and Kaufman closed down in 1953.[72]

Eitel-McCullough survived and prospered by developing a new line of power-grid tube products. These new tubes made the components produced during the war obsolete. For example, Eitel and McCullough introduced new triodes to the market in late 1945 and 1946. They also commercialized a family of power tetrodes that could operate at very high frequencies. A tetrode (a tube with four electrodes) had the usual plate, cathode, and control grid. In addition, it had a screen grid, which helped screen or isolate the control grid from the plate. With this additional grid, a tetrode had lower capacitance than a triode. The screen grid also had an electron-accelerating effect and increased the gain dramatically. In other words, tetrodes amplified signals better than triodes. Eitel-McCullough engineers had designed their first four-electrode tube in 1941. But after Pearl Harbor, they had shelved this design for the duration of the war. When it became imperative to introduce new products to the market, Eitel-McCullough engineers resurrected this tube design. In 1945 and 1946, these men also engineered more powerful tetrodes.[73]

The more powerful tetrodes found a ready market. FM (frequency modulation) radio broadcasting had been developed in the 1930s by Edwin Armstrong, an independent inventor. In spite of its advantages over AM (less static, less station interference, more faithful reproduction of a wide range of tones), FM radio had little commercial success in the late 1930s. The war years, however, brought a surge of interest. Numerous business groups applied to the Federal Communications Commission for licenses to set up FM radio broadcasting stations after the war. In the last years of the war, the growing demand for FM equipment led many electronics firms to develop FM radio transmitters and the vacuum tubes they required. But in June 1945, in a surprise decision, the FCC decided to change the frequency band it had allocated for FM radio from 42–50 megacycles to 88–108 megacycles. It was widely suspected at the time that the FCC made this decision at the request of AM radio stations and the Radio Corporation of America, which wanted to thwart or at least slow down the adoption of FM radio. The FCC decision had the immediate effect of making the FM transmitters that had been developed during the war obsolete. Electronics firms had to design new FM transmitters capable of operating at very high frequencies. They also needed new power-grid tubes for them.[74]

The decision of the FCC had the unanticipated consequence of creating a large market for Eitel-McCullough's new line of power tetrodes. Eitel-McCullough's products were among the rare vacuum tubes on the market that had the frequency and power characteristics needed for the new FM radio transmitters. Eitel-McCullough also benefited from the reputation it had acquired during the war among many electronic system designers for making high-quality products. As a result, many system firms chose to design their FM radio transmitters around Eitel-McCullough's tetrodes. When these corporations moved their new transmitter designs to volume production (in 1946 and 1947), Eitel and McCullough received large orders for their power-grid tubes. Not surprisingly, the commercial success of Eitel-McCullough's tetrodes encouraged RCA, GE, and other companies to produce similar tubes. To protect their sales, Eitel and McCullough sued these corporations for patent infringement. Because of the strong patent position Eitel-McCullough had acquired during the war, Eitel and McCullough won the patent lawsuits and forced RCA and GE to halt the production of tetrodes. Exploiting their victory for commercial advantage, the entrepreneurs let RCA and GE buy their products and resell them under their own brand names. RCA and GE had become Eitel-McCullough's de facto distributors. These commercial and legal victories made Eitel-McCullough the largest American manufacturer of transmitting tubes. In 1947, with sales of about $1.5 million, it controlled more than 40 percent of the US market for power-grid tubes.[75]

Conclusion

Reflecting on Eitel-McCullough's rise to prominence in the 1930s and the 1940s, Jack McCullough attributed its success to its ability to meet the component needs of new electronics systems: "We were the ones that were able to supply the missing link. For instance, we made the first tubes used in radar and in FM radio." Eitel-McCullough's ability to supply reliable high-performance power-grid tubes for radar sets and other advanced systems (including FM radio transmitters) was predicated on its unique assemblage of competencies. Because of their training in amateur radio, Eitel, McCullough, and their employees had solid expertise in electronic circuit and system design. They also acquired competence in high-vacuum processing and electron tube manufacturing. Eitel and McCullough developed some of these processing methods at Heintz and Kaufman in order to bypass RCA's patents. They then improved upon them to meet the requirements of radio amateurs, and later to meet the requirements of the radar-development groups at the Naval Research Laboratory and at the Signal Corps Engineering Laboratories. They also made innovative use of Litton's advanced tube-making equipment.

Eitel-McCullough's manufacturing processes had no counterparts on the East Coast. East Coast engineers who had visited Eitel-McCullough's San Bruno factory during the war had been struck by the unusual production methods. "Engineers from Westinghouse," Eitel later recalled, "came to our plant to familiarize themselves with our production methods. After they had gone through our plant and saw how we made our tubes, we met with them to see if they had any questions on our methods. One engineer spoke up and commented that we could not make tubes the way we were making them! Everything he saw in our plant was foreign to him and he was unable to comprehend our approach to tube making."[76] With its innovations in tube manufacturing, its system expertise, and its leadership in processes and in products, Eitel-McCullough rose to prominence in the power tube industry and outcompeted GE and RCA.[77]

Eitel and McCullough, in collaboration with Litton, not only built a strong power tube industry on the Peninsula but also drove much of the area's subsequent growth in the engineering and manufacturing of electronic components. In the late 1930s and the first half of the 1940s, Litton helped make Stanford University an important player in vacuum tube technology. In the mid 1930s, Frederick Terman, a young and ambitious professor of radio engineering in Stanford's electrical engineering department, became interested in setting up a teaching and research program in vacuum tube engineering. Terman knew Litton well through amateur radio and had become acquainted with Eitel and McCullough by consulting at Heintz and Kaufman in the late 1920s. Terman closely followed Litton's, Eitel's, and McCullough's work and their development of new power tube designs and high-vacuum processing techniques. Terman reasoned that their presence on the Peninsula offered a wonderful opportunity to build a program in vacuum tube engineering at Stanford. The acquisition of tube-making techniques from local firms would enable him to establish a research program on electron tubes and transform vacuum tube electronics into an academic discipline. His goals were to create new courses, to write new textbooks, and to establish a well-equipped research and teaching laboratory.[78]

Terman enticed Litton to join Stanford's teaching staff. In 1936 he appointed Litton a lecturer in the electrical engineering department. In the late 1930s and the early 1940s, and again after World War II, Litton lectured regularly on vacuum tubes and their processing techniques to electrical engineering students. He also shared his knowledge of vacuum tube making with faculty members at Stanford. For example, Litton trained Karl Spangenberg, a young instructor whom Terman had recently hired, in the fabrication of vacuum tubes. He also helped Spangenberg establish a vacuum tube laboratory on campus. This new laboratory and his knowledge of vacuum tube production techniques enabled Spangenberg to initiate and conduct research projects on electronic phenomena and vacuum tube design in the late 1930s and the early 1940s. These projects were funded by IT&T and by Sperry Gyroscope, a military contractor based on the East Coast.[79]

Litton also supported the tube program in the electrical engineering department by giving a $1,000 grant to Terman in 1938.[80] With this grant, Terman brought one of his favorite students, David Packard, back to the university for further studies. During his year at Stanford, Packard worked with Litton on vacuum tubes and established a close friendship with him. With another Stanford student, William Hewlett, Packard also established Hewlett-Packard, an electronic instrumentation company, at this time. Litton was their business mentor. In parallel with his collaboration with Terman and his group of electrical engineers, Litton helped another research group in the physics department develop the klystron, a revolutionary vacuum tube that could operate at extremely high frequencies. These innovative research projects and the writing of highly respected textbooks on vacuum tubes enabled Stanford University to rise to prominence in vacuum tube electronics after World War II.[81]

In addition to sharing their production expertise with Stanford University, Litton, Eitel, and McCullough also applied their unique tube-making skills to the development and manufacture of an entirely new class of vacuum tubes. Vacuum tubes capable of generating microwaves were even more difficult to make than power-grid tubes. They required a higher vacuum, tighter tolerances, more complex processing procedures, and a much higher level of cleanliness than the transmitting tubes for radio transmitters and high-frequency radar sets that Litton, Eitel, and McCullough had fabricated in the 1930s and during World War II. Litton, Eitel, and McCullough pioneered the microwave tube industry on the San Francisco Peninsula. In the late 1930s and the early 1940s, Litton diversified into the development of klystrons for IT&T. He later established Litton Industries, which specialized in the manufacture of magnetrons (another type of microwave tube used primarily in radar transmitters). Similarly, Eitel and McCullough conducted a few magnetron-development projects during World II. In 1951, they branched out into the production of microwave devices. They applied their firm's manufacturing competence to the production of klystrons for long-distance communication and television broadcasting. By the late 1950s, Eitel-McCullough and Litton Industries were among the largest manufacturers of microwave tubes in the United States.[82]

Litton, Eitel, and McCullough created an infrastructure and a predisposition for electronic component manufacturing on the San Francisco Peninsula. They attracted suppliers of specialized inputs, and they trained a workforce skilled in vacuum techniques and chemical handling. They also helped develop a culture of collaboration in the region. In the postwar period, this infrastructure and this modus operandi facilitated the formation of other corporations in San Mateo and Santa Clara Counties. Litton, Eitel, and McCullough also showed that it was possible to build successful electronic component businesses in the area. Local firms, in order to establish themselves in industries pioneered by large East Coast firms, had to concentrate on the development and constant improvement of manufacturing processes. They also had to commercialize high-quality products. These lessons were not lost on other innovator-entrepreneurs in the area. In the late 1940s and the 1950s, product quality and a commitment to manufacturing processes became the hallmarks of the San Francisco Peninsula's electronic component industries.

2

Diversification

In 1953 the British military released a report that rated the magnetrons produced by Litton Industries the best in the Western world. Litton's magnetrons were more reliable and had better electrical characteristics than those made by European firms. The British military also considered these magnetrons better than the tubes produced by large US corporations such as Raytheon, Sylvania, General Electric, and Western Electric (Byrne 1993). This ranking was surprising because Litton Industries was a relatively new entrant in the magnetron business. It was only in 1944 that Charles Litton, a manufacturer of tube-making equipment, had diversified into the design of advanced magnetrons. In 1951 Litton began producing these tubes in volume. By this time, magnetrons were well-established products. They had been mass produced by Raytheon, Western Electric, General Electric, Westinghouse, and Sylvania during World War II and were in the product lines of many US electronics firms. How did Litton diversify into the development and manufacture of magnetrons? How did he establish himself in the magnetron business and partially displace Raytheon, Western Electric, and Sylvania, which had dominated the production of magnetrons during World War II? What did Litton owe to Stanford University and the military?

The emergence of Litton Industries and the formation of the microwave tube business on the San Francisco Peninsula have attracted little historical attention. In a book and a series of pioneering articles on Stanford's Cold War research programs and the emergence of Silicon Valley, Stuart Leslie has argued that the Peninsula's microwave tube industry grew out of Stanford University's research and teaching programs. Frederick Terman, the university's dean of engineering and later its provost, the argument goes, built strong research groups in microwave tube engineering with military patronage in the immediate postwar era. Building on their technical achievements, Terman sought to strengthen the Peninsula's electronics industry by encouraging Stanford's engineers to establish new microwave tube corporations in the area. Like Stanford, these firms, Leslie argued, were highly dependent on military patronage and were oriented nearly exclusively toward technologies of direct interest to the Department of Defense (Leslie 1992, 1993; Kargon et al. 1992; Kargon and Leslie 1994; Leslie and Kargon 1994).

Though Leslie has argued convincingly that the local microwave tube industry was closely connected to Stanford University and that its expansion was financed by military funding, the rise of Litton Industries cannot be explained solely by military patronage and university research. Critical to Litton Industries' emergence and expansion were Charles Litton's innovations in magnetron engineering and manufacturing. Litton, one of the foremost experts in tube-manufacturing processes in the United States, approached the design of magnetrons in a novel way. He took into account production techniques in the design of his tubes and engineered them in such a way that they would maintain a very high degree of vacuum (high vacuum was essential to tube reliability). Litton also developed new processing techniques to make magnetrons and innovated in the manufacture of these devices by emphasizing cleanliness in tube processing and establishing effective quality-control procedures. These innovations enabled Litton to produce the highest-quality magnetrons on the market. Litton's relations with Stanford were important for Litton's success in this business. But these relations were more complex than is often argued. They were not unidirectional. Instead they involved the two-way flow of people, ideas, and processing techniques between Litton's venture and the university. Charles Litton learned about microwave tube design from a Stanford group that developed the klystron in the late 1930s, and he recruited Stanford students and faculty members to his magnetron business in the postwar period. But Litton also gave engineering assistance to Stanford's microwave tube research groups and granted subcontracts to the university. In short, Litton's relations with Stanford benefited the university as much as they benefited Litton.

Litton's emergence as an important manufacturer of magnetrons was further shaped by the evolution of international relations and the changing strategies of large electronics firms in the East. In the late 1930s, Litton first entered the microwave tube business by designing klystrons for the French subsidiary of International Telephone and Telegraph. Because of Germany's growing military threat, IT&T's subsidiary in Paris urgently needed microwave tubes for its radar systems. It was on the basis of this experience that in 1944 Litton obtained a large contract from the National Defense Research Committee to develop an advanced magnetron. In the immediate postwar period, Litton expanded his fledgling magnetron-engineering business. It was a propitious time to do so. Electronics corporations in the East, that had dominated the microwave tube business during World War II, drastically curtailed their activities in this area in the second half of the 1940s. This opened a window of opportunity for Litton. To exploit these opportunities, Litton set up a new corporation, Litton Industries, that concentrated on the design and manufacture of magnetrons. (His old firm, Litton Engineering Laboratories, continued to produce tube-making machinery.) With military research contracts, Litton designed a family of continuous-wave magnetrons for electronic warfare in the second half of the 1940s. The Korean War enabled Litton to transform his small magnetron engineering business into a first-class manufacturing operation. At the Navy's request, Litton manufactured pulse magnetrons that other corporations could not fabricate effectively. Using innovative processes and management techniques, Litton Industries produced these tubes at very high yields and therefore very low cost. At the end of the Korean War, Litton Industries' enormous profits attracted the wrath of the Renegotiation Board. This led Litton to sell his company to a group of businessmen who proceeded to use the profits of his magnetron business to build a large conglomerate under the Litton name in Southern California.

Pioneering

Charles Litton first became involved with the new technology of microwave tubes because of his association with Russell and Sigurd Varian, the two inventors who developed the klystron at Stanford University in 1937. Litton was a friend of the Varian brothers and often discussed engineering with them. Russell Varian, like Litton, was a Stanford graduate. After obtaining a master's degree in physics from the university in 1927, Russell Varian had worked on the design of television tubes and circuits at the Television Laboratory in San Francisco. This laboratory, established by Philo Farnsworth, pioneered the development of television in the mid 1920s and the early 1930s. Russell Varian did consulting work for Heintz and Kaufman and for other Bay Area electronics firms. Sigurd Varian, Russell's adventurous and charismatic younger brother, was a pilot as well as a mechanic. A barnstorming pilot in the early and mid 1920s, he later worked for Pan American Airways where he supervised the airline's Mexican operations. Sigurd Varian had also opened new routes into Mexico and South America. By repairing his own airplanes, Sigurd Varian had gained an excellent knowledge of the mechanical arts (Varian 1983; Everson 1949; Farnsworth 1989).

In 1935, Russell Varian and his brother Sigurd set up their own laboratory at Halcyon, the socialist-theosophist community in which they had grown up. At Halcyon, located on the central California coast, they worked on inventions of their own: a process for reducing iron ore to iron and a ruling engine for diffraction gratings. This engine was an extremely precise machine used for the making of optical devices for spectroscopic research. The Varian brothers also became increasingly interested in developing a radar system that would make flying safer and detect enemy bombers at great distances. They were motivated to start this project by their deep concerns about Germany's increasingly aggressive foreign policy and its swift buildup of a strong air force. The Varians, who were not aware of the radar programs at the Naval Research Laboratory and the Signal Corps Engineering Laboratories or of related work on tube development at Eitel-McCullough, decided to use microwaves (under 1 meter in wavelength) to detect approaching aircraft. It was only by using these frequencies, they reasoned, that they would be able to precisely determine the location and direction of an attacking airplane.[1]

Their problem was how to generate these very short waves with sufficient power. No vacuum tube then on the market had the power and the wavelength required to drive a radar transmitter. To build microwave radar systems, the Varians would have to develop a radically new vacuum tube. Russell Varian, who had kept in close touch with the work of William Hansen, a friend of his and a young physics professor at Stanford, realized that a new device invented by Hansen, the rhumbatron, could help generate very short waves. The rhumbatron can be best described as a cavity enclosed by a metallic shell. The rhumbatron resonated to waves of a certain frequency, so that an oscillating electric field was created within the cavity. When the rhumbatron was of the right size, it could resonate at microwave frequency. Interested in collaborating with Hansen and getting access to Stanford's physics laboratory, the Varian brothers entered into an unusual partnership with the university in April 1937. Under the terms of the agreement, Stanford gave the Varian brothers access to the physics laboratory and the right to consult with Hansen and with David Webster, the chairman of the physics department. The university also provided $100 for materials and supplies. The Varians supplied their services without charge and brought most of their shop equipment to Stanford. Financial returns from the group's inventions were to be divided equally between the university and the Varian brothers. Russell and Sigurd Varian moved to Stanford in May 1937 (Hansen 1938; Bloch 1952; Galison et al. 1992; Hevly 1994).

In order to obtain the microwave signals they needed for airplane detection and landing, Hansen and the Varians at first thought of combining the rhumbatron with the multiplier vacuum tube that Philo Farnsworth had developed for his television system. The multiplier tube, which Russell Varian knew well through his work with Farnsworth, used secondary emission as a way to generate electrons. (When a metal is bombarded with electrons, electrons may be emitted from the surface of the material. The bombarding electrons are called primary and the emitted electrons are designated secondary.) Because of this mode of operation, the multiplier tube was an excellent amplifier, which made it an interesting candidate for the Varians' microwave generator. To build their combination of multiplier tube and rhumbatron, the Varians relied, as they often did, on Litton's technical expertise.[2] Russell Varian visited Litton at Litton Engineering Laboratories to discuss the patent situation regarding Farnsworth's vacuum tube. Sigurd Varian, who was new to the vacuum tube art, also conferred with Litton several times about vacuum techniques, vacuum equipment, and the design of cathodes. From these free consultations, Sigurd Varian learned a great deal about vacuums and vacuum techniques. At this time, Sigurd Varian later reported, "I learned firsthand the wonders of the vacuum, the main wonder to me at that time being that anyone ever got a vacuum at all." With Litton's advice, Sigurd Varian built a vacuum system and tested cathode structures for the multiplier tube.[3]

Before Sigurd Varian started building the multiplier tube and rhumbatron combination, his brother found a more promising way of generating microwaves. In June 1937, Russell Varian conceived of the klystron. In the next few weeks, he refined this idea with Hansen's theoretical and mathematical assistance. (Russell Varian had considerable physical intuition, but his knowledge of mathematics was lacking.) In July and August 1937, using his newly acquired knowledge of vacuum tube practice and without direct input from Litton, Sigurd Varian fabricated the first klystron. The klystron was significantly different from conventional gridded tubes of the sort produced by Eitel-McCullough.

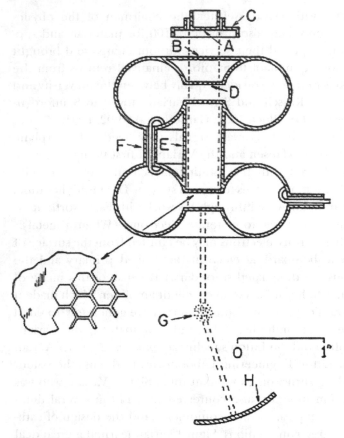

Figure 2.1
Scale drawing of essential parts of a typical klystron oscillator. A and C: cathode
parts. B: tungsten grid. D: honeycomb grid. E: drifting tube connecting two rhumbatrons. G and H: magnetic field and fluorescent screen used by the Varians to
detect oscillation in their first tube. This drawing, published in Varian and Varian
1939, is reproduced with the permission of the American Institute of Physics.

It was both a vacuum tube and an electronic circuit. The klystron consisted of a cathode, several grids, a collector, and two rhumbatrons
connected by a drift tube. The klystron in its amplifier configuration
operated in the following way: At one end the cathode, as in any vacuum tube, fed a stream of electrons into the tube. The electrons were
pulled toward the opposite end by the positive charge on the collector.
The electrons and the oscillating electric field of the first rhumbatron
interacted as the electrons passed through the cavity. As seen by the

electrons, the oscillating field of the first cavity alternately presented a negative and a positive charge. As a result, some electrons were slowed down and others were sped up. In the next stage of their journey through the drift tube, the faster electrons caught up with the slower ones and bunches were formed. When the bunching became greatest, the electron groups reached the second rhumbatron. If this cavity was tuned to the frequency of their arrival, the electron bunches excited electrical oscillations in it, just as hitting a pendulum with periodic strokes timed to its natural frequency would keep it swinging. The electrical energy delivered at the second rhumbatron was taken off by a transmission line. To transform this amplifying klystron into a klystron that generated microwaves, one had to feed part of the power from the second rhumbatron back to the first. The klystron developed by Hansen and the Varian brothers operated at a wavelength of 13 centimeters. The klystron was a major innovation. The first electron device that could operate in the microwave range, it was highly efficient and had good frequency stability (Varian and Varian 1939).

In 1938, Litton became involved in the klystron project as a supplier of tube-making equipment.[4] At this time, Hansen and the Varian brothers secured a large contract from Sperry Gyroscope, a manufacturer of airplane instruments located on Long Island, to patent and further develop their klystron invention. In return, Sperry got an exclusive license on Stanford's klystron patents. Sperry's patronage enabled Hansen and the Varians to expand their project significantly. They recruited expert machinists from Heintz and Kaufman and doctoral students from Stanford's electrical engineering department. With Sperry's monies, Hansen and the Varians also procured tube-making equipment from Litton Engineering. They purchased Charles Litton's glassworking lathe, his spot welder, and the vacuum pumps he had recently designed. This new equipment enabled the Stanford group to fabricate a number of experimental klystrons in 1938 and 1939. By helping the Varian group use his equipment, and by advising them on high-vacuum techniques, Litton quickly became familiar with klystrons and their operation. He also learned about microwave theory, which Hansen and Webster were then developing. But Litton did not get access to the most important design information: the klystrons' physical dimensions and the ways in which these dimensions were calculated.[5]

In September 1939, Litton began making klystrons for International Telephone and Telegraph. IT&T was keen to get access to klystron technology. Its French subsidiary had an urgent need for it. The impending war with Germany led IT&T's engineers in Paris to start a microwave radar development program. But they did not have a suitable microwave generator to power their radar transmitter. This urgent requirement led IT&T vice-president Harold Buttner to contact Sperry Gyroscope. Buttner proposed that Sperry sign a cross-licensing agreement and supply detailed information on klystron fabrication techniques to IT&T. Sperry turned these requests down. (Two years later, Sperry did cross-license its patents with IT&T's.) When this attempt failed, Buttner asked Litton to make klystrons for the French radar system. As Buttner well knew, Litton had a good understanding of klystrons. Litton was also closely connected to IT&T. Litton had directed the tube laboratory of Federal Telegraph, an IT&T subsidiary, several years earlier. By the late 1930s, Litton had a consulting arrangement with IT&T. He had also signed a patent agreement whereby he assigned all his tube and radio inventions to IT&T. Acceding to Buttner's request, Litton agreed to design and produce klystrons for the French. He also consented to Buttner's demand that he limit his contacts with the Varian group to a minimum. Buttner and IT&T's patent lawyers feared that Litton's klystron ideas would be patented by Sperry. For this work, IT&T paid Litton $3.50 an hour—a good hourly rate for the time. In 1940, Litton received $1,000 in consulting fees from International Telephone and Telegraph.[6]

Designing and fabricating good klystrons was a complex undertaking. Klystron technology was still experimental. The tubes made in the physics department at Stanford were temperamental and unreliable. They were also attached to a vacuum pump, which made them difficult to use in the field. Litton had to transform these laboratory devices into manufacturable products. The tubes also had to be sealed off and hold a very high vacuum. (Klystrons required a much higher vacuum than power-grid tubes to operate properly.) Litton also had to overcome significant construction and materials difficulties. Klystrons were complex in structure—more complex than power-grid tubes. Klystrons were also made of a wider range of materials, and their fabrication required much higher tolerances. To solve these problems, Litton relied on the expert knowledge of materials and vacuum techniques he had acquired at Federal Telegraph and in his own tube shop at Litton Engineering. In less than a year, Litton, working on his own, engineered three different klystrons and supplied them in small quantities to IT&T. Among these products were klystrons for the transmitter and the receiver of the

French radar system. The klystrons were completed too late to be of any use to the French war effort, as the French government signed an armistice with Germany in July 1940. But, according to Buttner, Litton's klystrons were first-rate pieces of engineering. They had better performance and reliability characteristics than the klystrons developed at Stanford and Sperry Gyroscope at the same time.[7]

In 1941 and 1942, Litton received more microwave tube development jobs from IT&T. By this time, Litton Engineering had become de facto the advanced tube-development laboratory of International Telephone and Telegraph in the United States. Litton did all the research and engineering work on ultra-high-frequency and microwave-frequency tubes for IT&T. In contrast, the firm's laboratories on the East Coast concentrated on the design of lower-frequency power-grid tubes. With these new projects, Litton's vacuum tube development business expanded substantially. In 1941, he received $7,700 in consulting fees from IT&T. The next year, IT&T gave Litton an annual development budget of $30,000. Litton hired a few technologists, including Robert Helm, a recent graduate of the electrical engineering department at Stanford, to help him with vacuum tube development. With this group, Litton worked on a variety of tube projects. He designed several continuous-wave klystrons, and a few VHF triodes for radar applications. He also worked on a high-power tunable cavity magnetron. In addition to these tube-engineering projects, Litton produced small quantities of microwave tubes designed by IT&T's British subsidiary.[8]

In parallel with the development and the small-volume production of klystrons and other vacuum tubes, Litton made a series of inventions designed to improve the manufacturability and applicability of klystrons. For example, he devised ways to increase their power output. At this time, the power output of klystrons was low. This characteristic limited their use as transmitting tubes in radar systems. Litton also came up with ideas on how to increase the energy efficiency of these tubes and extend their tuning range. In addition, Litton devised ways of reducing the volume that had to be evacuated within the klystron envelope. (The size of this volume presented a substantial manufacturing challenge.) Litton also conceived of a klystron with external cavities—namely, cavities which were external to the tube's vacuum envelope. In the postwar period this idea was exploited commercially by several firms on the San Francisco Peninsula. IT&T, which wanted to build a patent position in microwave tubes, carefully patented these inventions.[9]

Figure 2.2
A triumphant Charles Litton in front of his home-made car in the High Sierras, late 1930s or early 1940s. Every year, Litton sought to be the first motorist to cross the Sierras at the end of the winter. Courtesy of Bancroft Library, University of California, Berkeley.

Turnaround at Federal Telegraph

In 1943 and early 1944, Litton abandoned his work on klystrons and concentrated on the production of radar triodes and cavity magnetrons at Federal Telegraph.[10] In the summer of 1942, Federal Telegraph had difficulty meeting its contractual obligations to the military services. It had signed several contracts with the Navy for the production of a high-frequency triode. This tube, originally developed at Eitel-McCullough, was used in airborne radar systems and was important for military operations in the Pacific. While Eitel-McCullough succeeded in manu-facturing this high-frequency triode in volume, Federal Telegraph's manufacturing yield (the proportion of good tubes coming out of the production line) was nil. The Navy, which had an urgent need for the tube, put enormous pressure on Federal Telegraph to meet its deliv-ery deadlines. This led Harold Buttner, IT&T's vice-president, to fire the manager of Federal Telegraph's vacuum tube division. To replace him,

Buttner turned again to Charles Litton. He asked Litton to run the division and solve Federal Telegraph's production difficulties. He also gave him the mandate of building a new tube plant for the firm. Litton accepted Buttner's offer on the condition that the job would last only 18 months and that Federal Telegraph would send a manager to run his glass lathe business in Redwood City during his absence. Litton moved to the East Coast in November 1942.[11]

At first, Litton concentrated on producing the triode for the Navy. When Litton joined Federal Telegraph, the company's small tube shop specialized in power-grid tubes for radio broadcasting and communication. Its annual sales amounted to $500,000. This shop was utterly unprepared to produce high-precision radar tubes in large quantity. To manufacture the high-frequency triode, Litton decided to set up a separate shop. This production line focused exclusively on radar tubes, while the old shop manufactured conventional tubes. Litton supervised the tooling of the new line and trained more than 100 operators in the production of radar tubes. He also devised a new manufacturing process, using techniques he had developed at Litton Engineering before the war. Litton's efforts soon bore fruit. In 1943, Federal Telegraph produced the triode in significant quantity. This rapid "ramp up" enabled the company to meet the Navy's demand for the tubes and to restore its reputation with the military services. Federal Telegraph's vacuum tube division also grew substantially during this period. Its monthly sales increased from $47,000 in December 1942 to $600,000 in December 1943. The division did $3.6 million in business in 1943.[12]

In parallel with the production of power-grid tubes, Litton also designed and built a new vacuum tube manufacturing plant for Federal Telegraph. His goals were twofold: he wanted to increase Federal Telegraph's manufacturing capacity to meet the wartime demand for radar vacuum tubes. But he was also interested in designing a state-of-the-art production facility that would enable the firm to fabricate new tubes after the war. Litton's factory was innovative in its design. All power supplies, vacuum pumps, and maintenance facilities were located in the basement. This permitted efficient use of the manufacturing space on the first floor and also enabled management to rapidly reconfigure the plant's layout for the production of new tubes. In addition, Litton strongly emphasized cleanliness in the new plant. (His focus on cleanliness later became the hallmark of his approach to tube production.) In Litton's view, cleanliness was essential for the production of advanced vacuum tubes. Cleanliness reduced tube contamination. In turn, this enabled the plant

a

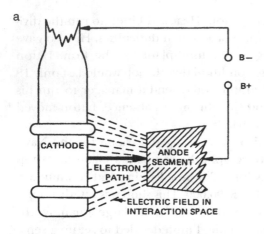

b

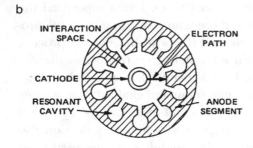

Figure 2.3
Two diagrams of a magnetron without a magnetic field applied. (a) Electron path.
(b) Electron path looking into cavity.

to obtain high manufacturing yields and to produce tubes that were reliable and had long lifetimes. Litton also gave considerable attention to the quality of the materials used in the production process. For example, he installed new gas-purifying machines. The new factory, which opened in January 1944, substantially enlarged Federal Telegraph's production capacity and enabled the firm, which until then had manufactured only power-grid tubes, to move into the production of magnetrons and other advanced products.[13]

Litton put Federal Telegraph in the magnetron business in 1943. By that time, cavity magnetrons, which had been invented in England in 1940, had become the main vacuum tubes used in microwave radar systems in the United States. Magnetrons, like klystrons, generated microwaves. But they had a distinct structure and generated these waves in a

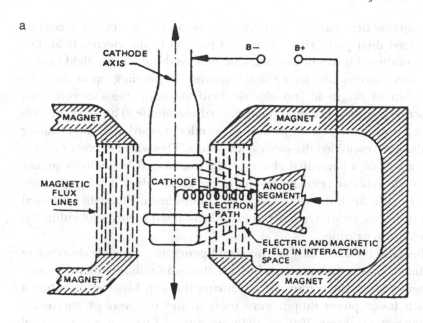

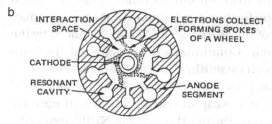

Figure 2.4
Operation of a magnetron with a magnetic field applied. (a) Electron path. (b) Electrons rotating synchronously.

different way than klystrons. Magnetrons were made of a cylindrical cathode, a cylindrical anode block, and a magnet. The anode block contained a number of equally spaced cavity resonators. These resonators or cavities had slots along the anode surface adjacent to the cathode. Cavity magnetrons operated in a complex way. When a voltage was applied between the anode and the cathode without the presence of a magnetic field, electrons leaving the cathode surface would be accelerated by the anode voltage and move directly across the interaction space to strike the anode. (The interaction space was the area situated between the anode and the cathode.) The course taken by the electrons was straight and parallel to the lines of force produced by the electric field.[14] When a magnetic field was added to this cathode-anode system, the electrons changed their path. They still moved parallel to the electric field. But the combined forces of the magnetic field and the electric field caused the electrons to take on a curved motion. Depending upon the relationship of magnetic and electric field strength, the electrons also described a spiral-like movement toward the anode. The electrons collected with other electrons and proceeded toward the anode along paths that resembled the spokes of a wheel. This wheel of electrons was the result of a tangential electric field established by radio-frequency oscillations within resonant cavities in the anode block. Passing through the electric field, the electrons were slowed down. Thus, they released energy to the radio-frequency field and sustained oscillations within the magnetron's cavities.

Because the cavity magnetron could generate extremely short waves in bursts of very high power, it became the main radio-frequency source for microwave radar transmitters during the war. Klystrons that had a much lower power output were used in the receivers of microwave radar systems. From 1940 to 1945, enormous financial and technical resources were devoted to the further development of magnetron tubes in the United States. Much of this work was guided and financed by the National Defense Research Committee. The National Defense Research Committee had been established in June 1940, at the instigation of Carnegie Institution president Vannevar Bush, to foster and finance the development of new weapon systems at research universities and industrial corporations. During the war, the NDRC devoted a large share of its budget to the development of microwave radar systems, radar countermeasure devices, and the magnetrons that these systems required.

With funding from the NDRC, the Massachusetts Institute of Technology established the Radiation Laboratory, a large lab that focused on the development of cavity magnetrons and radar systems. The Bell Telephone Laboratories, GE, Westinghouse, Sylvania, Raytheon, and RCA also set up large magnetron research and development programs during the war. These organizations pushed magnetrons to higher and higher frequencies and higher and higher power. They also developed ways to produce these tubes in large volumes. Like many other firms, Federal Telegraph entered the fray in the early 1940s. Although the historical record is largely silent on Federal Telegraph's magnetron activities, it is clear that Litton initiated a few magnetron development projects at Federal Telegraph in the winter and spring of 1943. In early 1944,

Litton supervised the transfer of these tubes to production. Federal Telegraph manufactured tens of thousands of highly precise magnetrons in the last 2 years of World War II.[15]

Litton's short stint as manager of Federal Telegraph's vacuum tube division was important to his later career in microwave tube manufacturing. It reinforced his interest in the design and manufacture of advanced magnetrons. Running Federal Telegraph's tube division also gave him unique experience in the volume production of vacuum tubes and in the management of a large manufacturing organization. Litton's stay in the East helped him build a network of contacts in the electronics industry. In 1942, Litton became a member of the New York-based Vacuum Tube Development Committee, which monitored and coordinated all vacuum tube development projects in the United States. The most prominent vacuum tube engineers and managers in the United States sat on this committee. While at Federal Telegraph, Litton also established close ties with the Navy and its Bureau of Ships. These associations later proved invaluable for his building of a magnetron engineering and manufacturing business in California.

The NDRC Contract

Upon his return to the San Francisco Peninsula, Litton received an important contract from the National Defense Research Committee for the development of a continuous-wave magnetron. This ultra-secret project had the highest priority in the NDRC's vacuum tube development program. In early 1944, the US Navy discovered that Japan had recently deployed a microwave radar. This was the first time during the war that the military services had detected an enemy radar operating at microwave frequency. To counter this threat, the NDRC set up a crash R&D program at Harvard University's Radio Research Laboratory. The Radio Research Laboratory, an offshoot of the MIT Radiation Laboratory, specialized in the design of radar jamming equipment. The goal of the new NDRC program was to design a shipborne system that would jam Japan's radio detection equipment and other microwave radars.[16]

To engineer this electronic countermeasure system, the Radio Research Laboratory needed a new family of magnetron products. Unlike radar magnetrons, which generated short pulses, the jamming magnetrons had to produce continuous waves. In other words, they had to emit continuous high-power signals. The magnetrons also had to be tunable in order to jam a variety of radar systems. GE's research laboratory had recently developed prototypes of continuous-wave magnetrons, but the NDRC strongly doubted GE's ability to engineer these tubes and bring them to production fast enough to meet the Japanese radar threat. Since the beginning of the war, GE had been consistently late in its vacuum tube engineering projects. To improve its chances of getting jamming magnetrons rapidly, the NDRC gave development contracts for the same electron tubes to Raytheon and Sylvania. But the main contract went to Litton's firm, Litton Engineering Laboratories. The NDRC asked Litton to develop a magnetron that would operate at 10 centimeters, the approximate wavelength of the new Japanese radar system. In July 1944, Litton received a cost-plus contract which stipulated that the NDRC would reimburse Litton for his expenses ($70,000) and would pay him an overhead fee of 65 percent.[17]

Some were surprised at first by the award of this contract to a small firm on the West Coast. But Litton was well known in the vacuum tube community for his ability to turn experimental tubes into manufacturable products. Litton also had strong supporters at the Radio Research Laboratory and the National Defense Research Committee. The head of the Radio Research Laboratory was his old friend Frederick Terman. Terman had left Stanford in 1942 to head the US radar countermeasure efforts at the RRL. Terman, whom Litton had helped build a vacuum tube laboratory at Stanford before the war, was keen to help his friend obtain this important magnetron contract. The US Navy, which was to use the microwave jamming transmitter on its ships, also strongly approved the granting of the magnetron contract to Litton.[18]

Litton, who thrived in competitive and fast-paced environments, made a go of this tube-engineering project. He was well aware of the military significance of the continuous-wave magnetron. But he also expected that this contract would continue after the war and as a result would protect him and his firm from the decline in the demand for glass lathes that he expected in the immediate postwar period. Once Litton received the NDRC contract, he immediately constituted a team of experienced vacuum tube engineers and technicians. For example, he recruited Winfield Wagener, an electrical engineering graduate of the University of California who had designed a variety of power-grid tubes at RCA and Federal Telegraph. Wagener had most recently served as chief engineer at Heintz and Kaufman. Helm, an electrical engineer, and Roy Woenne, a machinist and shop-trained mechanical engineer who had worked with Litton since the mid 1930s, also joined the project.

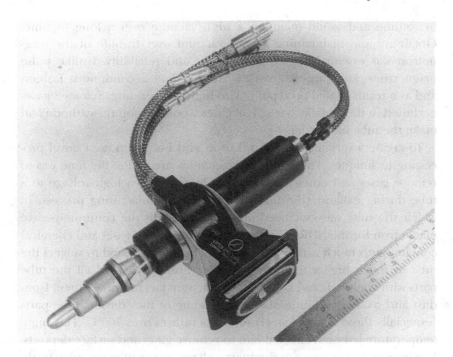

Figure 2.5
Litton's continuous-wave magnetron, 1945. Courtesy of Bancroft Library, University of California, Berkeley.

This was a well-balanced team with experience in designing and fabricating advanced tubes. In July and August these men visited GE and the Radio Research Laboratory in order to learn about GE's tube prototype and the preliminary design of the radar countermeasure system. They also visited the Radiation Laboratory at MIT, where they discussed the magnetron's electrical design.[19]

These men approached the engineering of the continuous-wave magnetron in an innovative fashion. At the time, most university laboratories and electronics firms developed magnetrons and other microwave tubes in several stages: first research, then tube modeling, then detailed engineering, and then pilot production. Litton and his group simultaneously worked on the magnetron's mechanical and electrical design, developed new fabrication methods for it, and did research. This concurrent approach enabled them to speed up the development schedule. It also enabled them to take production issues into account early in the design process. Indeed, manufacturing considerations were paramount for Litton. He aimed at engineering a tube that could be manufactured in volume and would maintain a high vacuum over a long lifetime. Obtaining and maintaining a high vacuum over the life of the magnetron was essential to its performance and reliability. Unlike pulse magnetrons, jamming magnetrons operated in a continuous fashion and as a result heated up rapidly. The high temperatures released gases occluded in the metallic parts. The gases would poison the cathode and make the tube inoperable.[20]

To create a very high vacuum, Litton and his group used novel processing techniques. The standard evacuation method at the time was to remove gases and other contaminants by applying a high voltage to a tube during exhaust (the final step of the manufacturing process, in which the tube was evacuated and sealed). With the continuous-wave magnetron for the NDRC, Litton decided to remove gases and chemical contaminants much earlier in the production process and to subject the tube to much higher temperatures. He carefully cleaned all the tube parts with chemicals and kept them in vacuum jars to protect them from dust and oxidation. Litton also heated some of the "dirtier" tube parts (especially those made of steel) to temperatures over 700°C. This high-temperature bake-out enabled him to remove gases and surface deposits. In addition, Litton gave a preliminary exhaust to the tube assembly (without the cathode and the glass stem) at very high temperatures. To drive out any remaining foreign matter, Litton exhausted the whole tube for 12 hours before sealing it off.[21]

Litton's unusual processing choices shaped the design of the continuous-wave magnetron. They required that the group select materials for the tube carefully. For example, because Litton and his engineers cleaned the tube parts with chemicals, they could not use porous materials such as porcelain for the tube's insulators. The tube elements were also subjected to very high temperatures, which led Litton to use copper and copper-gold alloys. The high-temperature exhaust procedures also led to considerable thermal stresses, as the different materials in the tube had different expansion coefficients. Litton had to take these stresses into account when he designed the magnetron's constructional features. His solutions to these mechanical problems impinged upon the tube's electrical operation. As a result of these processing and design choices, Litton's magnetron differed greatly from the magnetron prototype developed at GE. Only the dimensions of the cathode and resonating cavities were similar. But the magnetron's structural features and methods of assembly had been radically redesigned. The electrical design of the tube was also significantly different.[22]

Litton and his group completed the continuous-wave magnetron project in May 1945. It had taken them less than 10 months to solve a complex problem in tube engineering. The project's outcome was a remarkable vacuum tube, an engineering tour de force. The magnetron operated at 10 centimeters, and its average output was in excess of 1,000 watts—the highest output ever achieved at this frequency. The tube could also operate for more than 500 hours, whereas most pulse magnetrons at the time failed after 50 hours of operation. The continuous-wave magnetron could also be manufactured easily. This achievement was made all the more remarkable by GE's failure to transform the same experimental tube into a manufacturable product. By February 1945, GE was so far behind schedule that the NDRC canceled the contract and asked the firm to reallocate its engineering resources to other tube-development projects.[23]

Litton's successful execution of the NDRC contract gave him a solid foothold in the continuous-wave magnetron business. In April 1945, the Navy granted Litton a $200,000 pilot production contract to further refine his magnetron design and to fabricate 100 units. This was a fixed-price contract. In other words, Litton committed himself to produce 100 magnetrons at the rate of $2,000 per tube. Should his costs be higher than that, he would have to assume the project's financial losses. With the end of the war, the primary responsibility for the microwave jammer and continuous-wave magnetron development program moved from the NDRC and the RRL to the military services. The NDRC and the RRL were closed in the fall of 1945. The Navy and the Army took over their most promising R&D projects. The Navy, which had a strong interest in the microwave jammer and wanted to jam microwave radars operating at many different frequencies, decided to finance the development of more continuous-wave magnetrons. Since Litton Engineering had been the only firm to succeed in productizing one of GE's jamming magnetrons during the war, the Navy asked Litton to engineer more tubes in this magnetron family. The Army Air Force, which wanted to gain a competence in microwave jammers and continuous-wave tubes, also approached Litton regarding the development of continuous-wave magnetrons.[24]

Litton was interested in taking advantage of these business opportunities. In the fall of 1945, to prepare himself for more magnetron development and production contracts, he upgraded his vacuum tube laboratory and invested in new tube-making equipment and measuring instruments. He also planned to split his proprietorship, Litton Engineering Laboratories, into two independent firms. One would make glass lathes and vacuum pumps, while the other would design and produce magnetrons and other advanced vacuum tubes. The primary reason for this move was that these product lines had very different customers. Litton sold glass lathes and vacuum pumps to civilian users, while the military was the only buyer of highly specialized magnetrons. As a result, the two businesses had different accounting procedures and different capital requirements. This was especially important since certain military contracts had price-redetermination clauses attached to them and gave rights to military auditors to examine the accounting books of military suppliers. In addition, Litton was keenly aware that by expanding his tube business he would start competing with his own customers of tube-making equipment. He needed to separate these two businesses more clearly to avoid alienating part of his customer base. At the same time, a new tax law made it advantageous to incorporate proprietorships and partnerships. Litton was not the only entrepreneur to think along these lines at this time. David Packard, Bill Hewlett, and other electronics entrepreneurs on the San Francisco Peninsula also incorporated their businesses in 1945 and 1946.[25]

Building a Magnetron Engineering Business

A major fire in Litton's plant in November 1945 set back his plans of building a large magnetron business for several years. The fire destroyed the tube-development laboratory and the factory where Litton produced glass lathes and other vacuum tube-making equipment. "The laboratory burned to the ground," Litton despondently wrote to a friend a few weeks later, "together with an accumulation of 25 years of work. All laboratory equipment disappeared in thin smoke or became encased in 8000 molten nonex and lead glass bulbs—such a mess you never saw."[26] The financial blow was significant as well. Litton later estimated that he lost $250,000 in the fire. He later recovered only half of this sum from his insurance company. To ride out the storm, Litton decided to retrench. In December 1945, he laid off 45 of his 80 employees. He disbanded his vacuum tube engineering group, with the exception of Helm and Woenne. With the remaining employees, Litton rebuilt the machine tools that he could reclaim from the burned-out plant. He also set up temporary production facilities. Later a new factory was constructed in San Carlos. With these new facilities, Litton met his customers' orders for glass lathes. He also continued making continuous-wave magnetrons for the Navy—although at a much reduced pace. The Navy helped Litton to produce these tubes by commandeering government-owned equipment at GE and sending it to Litton's plant. To bring in funds, Litton actively consulted for Federal Telegraph and other East Coast firms.[27]

Now that he had resumed production, Litton decided to slowly re-grow his glass lathe and vacuum tube business. He reinvested his profits and consulting fees in the rebuilding of his tube laboratory and his manufacturing facilities. Because his business was now too small to split into two corporations, Litton resorted to a temporary expedient. In August 1946, he incorporated a new firm, Litton Industries, the charter of which was to develop and manufacture advanced vacuum tubes. Litton also kept alive his proprietorship, Litton Engineering Laboratories. Under the new arrangement, Litton Engineering focused exclusively on the development and sale of glass lathes and other specialized tube-making machinery. Litton, the proprietor of Litton Engineering, also gave Litton, the sole owner of Litton Industries, a 10-year contract for the manufacture of glass lathes for Litton Engineering. These legal maneuvers enabled Litton to substantially reduce his tax outlays and at the same time make the two businesses look separate to potential customers and military auditors (it did not work out with the military).[28]

To expand his vacuum tube business, Litton actively looked for military development and pilot production contracts. It was a propitious time to do so. After the surrender of Germany, the uneasy alliance between the United States and the Soviet Union began to unravel. In the next few years, the victors of World War II clashed over the fate of Germany, free elections in Eastern Europe, and the international control of atomic energy. The Truman administration adopted a policy of containment of the Soviet Union and established the Marshall Plan to prop up European democracies against the perceived Communist threat. These tensions also led the United States to rearm. When in 1946 the military services detected the building of a network of microwave radars in the Soviet Union, they substantially expanded their programs dedicated to the development of electronic countermeasures.

In 1946, the Aircraft Radio Laboratory of the Army Air Force opened the development of high-power microwave jammers for bids. The Navy gave research and engineering contracts to General Electric and a small electronics startup, the Airborne Instruments Laboratory. The Army Air Force and the Navy also actively supported the development of new microwave tubes for electronics countermeasures. In particular, the Aircraft Radio Laboratory and the Navy's Bureau of Ships funded the reengineering of the family of continuous-wave magnetrons that GE had designed during the war. These tubes were to act as the radio-frequency source of high-power radar-jamming transmitters (Price 1984; Harris et al. 1978). But few firms bid on these projects. In the immediate postwar period, the Bell Telephone Laboratories, Western Electric, GE, Westinghouse, and RCA, which had developed klystrons and magnetrons during the war, reconverted to peacetime production. They shifted most of their engineering and technical resources from microwave tube and radar projects to the development and production of communication equipment, radio sets, and television monitors. They also established large research groups to develop color television. As a result, they had little interest in devoting their research and engineering resources to military projects. Only Raytheon, Sylvania, Federal Telegraph, and Sperry Gyroscope maintained active research programs in microwave tubes in the immediate postwar period. They were the only large Eastern corporations to pursue military contracts for the development of microwave tubes during that period.[29]

Increased military funding and the withdrawal of most East Coast corporations from the microwave tube field opened a window of opportunity for small and rather peripheral establishments such as Litton Industries. Exploiting these opportunities, Litton bid on many military development contracts in the second half of the 1940s. He focused especially on continuous-wave magnetrons. Litton reasoned that his expertise in jamming magnetrons would give him a cost advantage over Raytheon and Sylvania. He was also interested in building a whole family of continuous-wave magnetron products, and in doing it at the government's expense. From 1946 to 1949, Litton received eight cost-plus contracts for the development of microwave tubes from the Navy, the Air Force, and the Signal Corps. Most of these contracts were for continuous-wave magnetrons. Applying the same design methodology to these tube projects as to the NDRC magnetron, Litton engineered a family of high-quality continuous-wave magnetrons. These electron tubes operated at either higher or lower frequencies than Litton's first continuous-wave magnetron. In late 1947, Litton Industries also received a $500,000 fixed-price contract from the Navy to produce 600 magnetrons of the kind that Litton had designed during the war.[30]

These military development and production contracts enabled Litton to rebuild what he had lost during the fire. They indirectly financed the purchase of new production equipment and measuring instruments. The military monies also enabled Litton to rebuild his engineering team. Starting in late 1946, Litton recruited experienced microwave tube engineers from East Coast firms to work with him at Litton Industries. For example, he hired Robert Schmidt, a magnetron tube specialist from Raytheon. He also brought in Paul Crapuchettes, a graduate of the University of California at Berkeley who had worked on microwave tubes at GE. Crapuchettes, who like most Western-born engineers wanted to go back to the West after the war, accepted a position at Litton Industries in 1946. Two years later, Litton hired Norman Moore, an entrepreneurial physics PhD from MIT. Moore had worked at the Radiation Laboratory and at Dalmo-Victor, an antenna manufacturer based on the San Francisco Peninsula. He had also held a teaching appointment in the electrical engineering department at Stanford. In the late 1940s, Litton trained these men in his unique processing and design techniques and indoctrinated them in the importance of cleanliness and high-vacuum processing in electron tube manufacturing. Litton, who desired to spend more and more of his time in the Sierras, also groomed Moore in the management of a small and highly technical business. In addition to attracting and training these engineers and managers, Litton educated a cadre of technicians and skilled machinists in the production of high-precision magnetrons.[31]

Litton actively supported the resumption of microwave tube development at Stanford in the immediate postwar period. In late 1945 and early 1946, Stanford reentered the microwave tube field after a lull during World War II when the group around William Hansen and the Varian brothers had left the university and moved to Sperry Gyroscope and the Radiation Laboratory. Critical for Stanford's new push into microwave tubes was Frederick Terman, the former head of the Radio Research Laboratory. Terman, who had recently become Stanford's dean of engineering, viewed microwave tube technology as central to the postwar growth of the electronics industry and wanted the university to establish itself as a major player in this new field. These ambitions led Terman to create new faculty appointments in the electrical engineering department for microwave tube technologists. Terman also encouraged his faculty members to look for military R&D contracts and pushed them to focus their research on microwave tubes for electronic countermeasures applications. At the same time, another microwave tube group formed in the physics department around William Hansen and Edward Ginzton. Ginzton had worked on the klystron project at Stanford in the late 1930s before moving with Hansen and the Varian brothers to Sperry Gyroscope. In the immediate postwar period, Hansen and Ginzton concentrated on the development of linear accelerators for physics research, and of powerful klystrons to drive them.[32]

Litton became involved with both of these groups in the late 1940s. In particular, he became a patron of Terman's microwave tube program. In 1947, he gave a subcontract to the electrical engineering department for the design of a traveling-wave tube (a new kind of microwave tube that had been recently developed at the Bell Telephone Laboratories). A year earlier, he had received an exploratory contract from the Navy to investigate the potential use of traveling-wave tubes as receiving tubes in electronic countermeasure systems. But Robert Helm, one of Litton's engineers, had been stricken by cancer. As a result, Litton Industries had not had much success developing this project. At the Navy's suggestion, Litton subcontracted this work to Stanford and especially to Lester Field, a new faculty member in the electrical engineering department. This subcontract enabled Field to make important innovations in traveling-wave tubes and strengthened Stanford's electronic countermeasures research program.[33]

Litton also designed klystrons for physics research. In 1948, Edward Ginzton and Marvin Chodorow (a Sperry physicist whom Ginzton had recently invited to join the Stanford faculty) set out to develop a very powerful klystron. Their goal was to scale up klystron power from the 30 kilowatts achieved by some British klystrons during World War II to 30 megawatts. This was an extraordinarily ambitious objective. Usually, most engineers tend to scale up devices and machines in small increments. To attain this unusual goal, Ginzton's group did considerable theoretical analysis on electron dynamics and beam focusing. They also completely redesigned the cathode and the collector. By 1949, Ginzton and Chodorow had obtained a klystron prototype that could generate 14 megawatts. But this tube was too unreliable to be used in the linear accelerator. This led Ginzton and Chodorow to ask Litton to redesign the high-power klystron for them. Litton, Ginzton later remembered, "was excited by such a major engineering challenge and agreed to try. He said that, in his experience, engineers were overly timid in designing tubes. He thought he could design a klystron which would work the first time." (Ginzton and Cottrell 1995) In 1949, drawing on his expertise in microwave design and processing, Litton designed a klystron capable of delivering 30 megawatts at peak power. After a few trials, it worked. The megawatt klystron, a remarkable advance, made Stanford the center of advanced klystron work in the United States.[34]

By the end of 1949, Litton had helped transform Stanford into a major player in microwave tubes. But he also had built two solid businesses—one in magnetron engineering and one in glass lathe manufacturing. His tube and glass lathe business had now fully recovered from the fire and regained the size that it had at the end of World War II. In late 1949, Litton had 80 employees. He also did $600,000 in business a year—half in tubes and half in glass lathes and other specialized tube-making equipment. Litton now had a strong basis for the expansion of his tube business. He had state-of-the-art facilities for advanced magnetron engineering and production. He also had an excellent team of engineers and managers and a group of skilled technicians and operators who could handle large military engineering and manufacturing contracts. Litton Industries was ready for growth. And grow it did in the early 1950s.[35]

Magnetron Production

The Korean War made Litton Industries a much larger enterprise. In June 1950, the army of North Korea, with the support of the Soviet Union, invaded the Republic of South Korea. The Truman administration committed significant American forces to the defense of South Korea. The long and uncertain war that ensued led the US government to launch a massive rearmament effort and increase its defense spending significantly. National defense expenditures increased from 4 percent of the gross national product in 1948 to 13 percent by 1953. The microwave tube industry was a major beneficiary of this surge in military spending. It saw its sales increase from $4.6 million in 1947 to $58.2 million in 1954. Magnetron sales increased from $3 million in 1947 to $49.6 million in 1954. In the mid 1950s, magnetrons represented 88 percent of the US market for microwave tubes. In conjunction with the large-scale procurement of magnetrons, the military greatly enlarged its support of microwave tube research and engineering during the Korean War. For the sake of expediency, the military services also transformed the procurement process during the Korean War. They gave more and more cost-plus or cost-reimbursement contracts to manufacturers of vacuum tubes. Conversely, the share of fixed-price contracts declined at this time. The Navy, the Army, and the Air Force also granted more sole-source contracts (contracts were awarded to only one contractor) for vacuum tubes. These changes enabled the military to speed up the development and production of microwave tubes. But at the same time, these changes made it less risky and more profitable for corporations to do business with the military.[36]

Because of these growing opportunities in microwave tubes, large electronics firms such as RCA and GE, which had nearly abandoned the microwave tube field in the immediate postwar period, reactivated and greatly enlarged their microwave tube operations during the Korean War. But it was the organizations active in microwave tube research and production in the late 1940s that benefited the most from the military's growing demand for microwave tubes. Raytheon captured nearly half of the US market for magnetrons during the Korean War. Sylvania and the microwave tube research laboratories at Stanford received large contracts during the war. Litton Industries also benefited greatly from these rearmament efforts. The increasing military demand for microwave tubes enabled the firm to expand into the volume production of continuous-wave magnetrons and pulse magnetrons for radar transmitters. To transform Litton Industries into a manufacturing company, Litton focused most of his financial and technical resources on production. He diverted much of the skilled personnel who produced glass lathes and other tube-making equipment to the magnetron business. Moore, Woenne, and Crapuchettes, who constituted Litton's product engineering team, also extended their activities into manufacturing and oversaw the move into volume production. At the same time, Litton de-emphasized contract engineering activities—with the exceptions of a few subcontracts for the development of radar klystrons and traveling-wave tubes which he received from Stanford.[37]

In spite of Litton Industries' reputation as a developer of magnetron tubes, it took nearly 6 months after the invasion of South Korea for Litton and Moore to receive a magnetron production contract. Breaking into the tight world of military contracting proved difficult. In 1950, Litton and Moore bid on a large subcontract for pulse magnetrons from Hughes Aircraft. Hughes Aircraft, a corporation based in Southern California, had entered the military avionics business in the immediate postwar period. It developed and manufactured electronics systems used in aviation and aeronautics. In the late 1940s, Hughes Aircraft developed a sophisticated fire-control system for jet aircraft. This fire-control system enabled the crew of a fighter jet to lock on a target plane, extrapolate its course, and hit its target. When the Korean War broke out, Hughes emerged as the sole source of fire-control systems for the Air Force and military orders for its products exploded. Because these fire-control systems required pulse magnetrons, Hughes put out a request for bids for these tubes (the magnetrons helped guide the aircraft missiles to their target). Litton Industries bid on this subcontract 'and quoted a price of $3 million. This quote was equal to 5 times Litton Industries' revenues in the previous year. To investigate this proposal, Hughes sent a team of engineers and managers to Litton Industries. These men wrote a favorable report on Litton and advised that Hughes place its subcontract with the firm. But the Air Force decided that Litton Industries did not have enough of a track record in magnetron manufacturing to execute a production contract of this size and importance. It vetoed Hughes' choice and decided instead to give the subcontract to Raytheon.[38]

This setback only strengthened Litton's resolution to move into the magnetron manufacturing business. To establish Litton Industries as a maker of pulse magnetrons, Litton and Moore turned to the Navy, their oldest and largest military customer, and asked for advice on how to break into this business. The Navy's procurement bureau in the Midwest suggested that they place a bid for a magnetron, the 4J50, which was already in production at Raytheon and Sylvania. This pulse magnetron, which had been designed at the Bell Telephone Laboratories during the war and had similarities with the one that Litton had recently bid on, was used by the Navy in some of its advanced radar systems. Because Raytheon and Sylvania did not produce the magnetron in sufficient quality and quantity, the Navy was interested in creating a new supply source for this tube. When Litton and Moore declared their interest in this job, the Navy's procurement bureau gave them the magnetron's specifications as well as a few inoperable tubes made by Raytheon and Sylvania. After analyzing these tubes, Litton and Moore offered to produce 5,000 magnetrons at the price of $400 per tube. This quote was substantially lower than Raytheon's and Sylvania's prices for the same tube. But Litton and Moore were confident that they would make a profit on this contract. After taking a $700,000 loan from a local bank to finance the tooling for the pulse magnetron, these men obtained a $2 million fixed-price contract for the production of 5,000 tubes from the Navy in December 1950. This was by far the largest contract that Litton Industries had ever received.[39]

Now that the corporation had received this contract, many more contracts for the same magnetron came from military contractors. In June 1951, Hughes Aircraft, which had not received Raytheon's tube on time, decided to use the 4J50 magnetron in its fire-control system and ordered 1,000 units of this tube from Litton. (The Air Force did not object to this subcontract this time.) Other orders for the same tube from the Navy, Hughes, RCA, Westinghouse, GE, and Sperry Gyroscope followed in the next 2 years. By April 1953, Litton Industries had contracted for the production of 13,302 4J50 magnetrons for a total price of $4,114,000. In addition, Litton and Moore produced another magnetron, the 4J52, for the Signal Corps. This magnetron had also been designed at the Bell

Telephone Laboratories and the Radiation Laboratory in 1943 and 1944. Because Sylvania, Amperex, and four other firms could not meet their delivery deadlines for this tube, the Signal Corps gave a large contract to Litton in 1951 to make 3,750 4J52 magnetrons. In parallel with these pulse magnetron contracts, starting in 1951 Litton Industries also received a growing number of sole production contracts for the continuous-wave magnetrons it had designed in the second half of the 1940s.[40]

To make these tubes, Litton and his group developed an innovative approach to magnetron manufacturing.[41] Their goal was to produce high-quality products at low cost. "We are trying to do two things," Litton wrote to a Navy procurement officer in 1952. "(1) Make the best [magnetrons] by as wide a margin as possible—by this I mean long life—good shelf life and no aging. (2) Be the low cost producer in the field."[42] To fabricate highly reliable magnetrons at low cost, Litton made the most of his long experience with vacuum tubes. He used many of the lessons he had learned at Federal Telegraph during World War II. He also relied on the magnetron fabrication techniques he had developed since the mid 1940s. The end result was a unique way of manufacturing magnetron tubes in volume.[43]

One of the most salient characteristics of Litton's manufacturing operation was its focus on cleanliness and the removal of tube contaminants. Operators systematically baked all tube parts at very high temperatures in order to remove gases and other contaminants and obtain a very high vacuum. At the end of the process, the magnetrons were exhausted for 11 hours at temperatures oscillating between 440°C and 650°C. To prevent the contamination of the tube parts during the manufacturing process, Litton's workers avoided direct physical contact with them and cleaned them at each step in the production process. "At Litton, tweezers and gloves are mandatory," a visitor to the plant reported in 1951. "Each part is cleaned with ether copiously and often, by each operator, and at each bench."[44] Litton also asked his employees to leave the tube assemblies and their components as little as possible in outside air (the metallic parts would oxidize and dust would fall on them). His rule was that no processed part should be in the air for more than three hours and that no part should be produced more than three days in advance of its use. Tube parts, which were not used immediately, were stored in small vacuum jars in order to protect them from outside contaminants. Plant cleanliness was also important for Litton. Every Friday, Litton asked his operators to thoroughly clean their workstations (whether it was in the machine shop or in the tube-assembly part of the plant). Every Friday afternoon, workers at Litton cleaned their tools. They also scrubbed the floor and waxed it.[45]

To produce components with very high precision and to obtain good yields, Litton and his engineers also tightly controlled the manufacturing process by imposing a tight discipline on the workforce. As a first step, they developed process sheets, which described in great detail the various steps in the production process. They also asked the operators to follow these procedures very carefully. No deviations were allowed. "Ninety percent of the success in tube manufacturing," Litton often reminded his engineer-managers, "depends upon the exact procedures of processing and fabrication of parts and it is in this know how that we will sink or swim. The one sign that has to be painted on the door is 'Don't change anything.' For this is one great cause of trouble, especially if the change is not subject to discussion."[46] Indeed, this process discipline was essential. Uncontrolled and undocumented process changes often had catastrophic results. They led to significant yield losses and decreases in product quality. They also made it much more difficult to identify the causes of tube failures and fix them.[47]

In conjunction with these moves, Litton also established an unusual quality-control system at Litton Industries in order to obtain highly reliable high-performance tubes. This system was based on worker self-inspection. At Litton, each worker inspected his own work and was responsible for its quality. At other microwave tube plants, inspection was distinct from production and was performed by a different group of people. In 1952, speaking to some Air Force officers, Moore described Litton Industries' approach to quality control as follows:

Our whole company philosophy and policy revolves around the fact that the person who performs the act, whatever it may be, in the manufacture of our products is responsible for that act and, as doer of the deed, is most qualified to judge whether or not it has been done according to directions. Each worker is thus his own inspector. Each worker is expected to determine that the part is correct both as it arrives and leaves his station and this system of push-pull gives in essence a double check at each station by the people most qualified to know, the actual workers on the part. The system presumes that there is no pressure on any employee to make more parts possible in a given time or to sacrifice anything for quality, and no such pressure is ever exerted. No piece work basis is ever used, actually or implied.[48]

This unusual "quality-control" system had a number of benefits. It dispensed with the need to hire tube inspectors and the need to organize an independent inspection department. As a result, it reduced the firm's overhead expenses. It also showed that management trusted operators— which in turn had an excellent effect on worker morale, pride, and productivity. This system of self-inspection also contributed to high man- ufacturing yields, as bad tube parts were identified at the earliest possible time in the manufacturing process.[49]

Because product quality and manufacturing yields depended to a large degree on his employees' skills and motivation, Litton needed a cooper- ative workforce. To motivate his workforce, Litton relied on managerial techniques developed at General Radio, a manufacturer of electronics measurement apparatus located in Cambridge, Massachusetts. General Radio was an engineer-owned corporation. It also had liberal personnel policies. In an effort to bridge the gap between capital and labor and to give financial incentives to its workforce, General Radio set up a plan under which its employees received a share of the company's profit every year. Inspired by General Radio's example, Litton shared half of Litton Industries' profits with his workforce. He also attended to his employees' recreation needs. In 1949, Litton bought a cabin and a large tract of land surrounding Jackson Lake in the High Sierras. He later built a camp- ground for his employees and invited them to spend their weekends and summer vacations there. In the summer of 1950, Litton posted this notice in his plant:

I wish to extend to all of our employees and to their families a welcome to make this location their headquarters for the vacation period, and to their families for as long as they wish to stay, both before and after. There will be three large tents available but there is ample room to pitch a tent of your own. It is our intention to set aside a sizeable area permanently for your use and to have it equipped for food storage, power, laundry facilities, etc., but probably these will not proceed very far this season. Feel free to come and go as you wish throughout the summer or on week-ends.[50]

In addition to providing a service to his employees, Litton's summer campground enabled him to get to know them individually and establish personal relations with them. Litton expected that these close ties would enable him to identify discontented workers and to thwart unionization.[51]

Because of these managerial and technological innovations, Litton had one of the best, if not the best plant in the microwave tube industry in the United States. It produced high-quality magnetrons at high yields. In the early 1950s, the Litton name became a synonym for quality in the microwave community. Magnetrons made by Litton Industries had good electrical characteristics and, more important, a very "hard" vacuum.

This high vacuum made them highly reliable and long lasting. At the same time, Litton obtained remarkable manufacturing yields. Because of his advanced processes and unusual quality-control system, Litton's manufacturing yields on pulse magnetrons were in the order of 80 percent in early 1952. And the yields kept climbing up. As the plant's engineers carefully improved manufacturing procedures, yields reached a high of 95 percent in 1953. This was an extraordinary yield for the microwave tube industry. The usual proportion of good tubes coming out of the line at this time was around 30–40 percent. It could often be much lower. Litton Industries' remarkable yields enabled it to produce magnetrons at low cost. Taking advantage of these low costs, Litton steadily reduced his prices on pulse magnetrons. He lowered the price of the 4J50 from $400 in 1950 to $238 in 1953. The 4J52, which sold for $270 in 1951, could be purchased for $165 in 1953. These price reductions enabled Litton to drive his competitors out of the 4J50 and 4J52 business. By the end of the Korean War, Litton Industries was the sole producer of these pulse magnetrons for the military services. At the same time, the plant's yields were so high that Litton garnered enormous profits. In 1953, Litton Industries made $1.2 million in profits over sales of $3 million.[52]

The Sale of Litton Industries

Litton was almost too successful. The firm's rapid growth and its enormous profitability had the paradoxical effect of increasing its owner's financial risks. During the Korean War, Litton Industries' growth rate became so great that Litton could not finance the firm's expansion out of his own funds and had to rely on bank loans. Litton also could not safely use the firm's profits to finance its expansion either. Because Litton Industries was overwhelmingly a military business, a large share of its profits was subject to reviews from the Renegotiation Board. The Renegotiation Board had been organized by Congress to reclaim excess profits from military contractors. Starting in 1951, Litton became increasingly worried about the Renegotiation Board's future rulings regarding his company. Litton Industries' profits far exceeded the profits usually allowed by the Renegotiation Board (10 percent or less). Litton knew well that he might face a large financial liability in the next few years. At the same time, Litton was becoming more risk averse. Now 49 and remarried, he had three small children. (He had divorced his first wife, Gertrude, in 1944.) Litton was increasingly worried about what would happen to his family and to his company in the event of his death. He wanted to create an estate for his wife and children and put his company on a more stable financial basis and ensure that his company would survive him.[53]

The growing financial risks attached to the magnetron business and his own desire to provide for his family led Litton to decide to sell Litton Industries in 1952. But Litton, who had no desire to retire and wanted to continue working on vacuum tube manufacturing techniques, chose to keep his proprietorship, Litton Engineering Laboratories. (By then, Litton Engineering had annual sales of about $300,000 and was one-tenth the size of Litton Industries.) To prepare for the sale of his magnetron business, in August 1952 Litton more clearly separated Litton Industries from Litton Engineering. He canceled the 10-year agreement whereby Litton Industries manufactured Litton Engineering's glass lathes and vacuum pumps. He also saw to it that the two firms had separate workforces and production facilities. A few months later, Litton incorporated Litton Engineering Laboratories. This corporation specialized in tube-making equipment. It was owned by Litton and his wife and by a trust fund in their children's names.[54]

Now that Litton Industries was clearly separated from Litton Engineering, Litton looked for a buyer for his microwave tube company. In early 1953 Litton approached Hughes Aircraft, one of his main customers, and proposed that Hughes buy his magnetron business. When Hughes' management declined Litton's offer, he let it be known that his company was for sale. No less than nine firms expressed an interest in buying it. Among them were Sylvania, Clevite, Kaiser, General Electric, and Federal Telegraph. Litton Industries had become a major player in the magnetron business and many of its competitors were interested in acquiring it in order to strengthen their own position in the fast growing microwave tube industry. Because Sylvania gave him the best offer, Litton decided to sell Litton Industries to this firm. But the Navy, which was Litton Industries' largest customer, strongly opposed the sale on the grounds that it would reduce competition in the magnetron business. The Navy also signified its opposition to Litton Industries' sale to General Electric, Federal Telegraph, and other makers of microwave tubes. Litton had to find a buyer outside the microwave tube industry.[55] This led him to consider a proposal from a group of Hughes managers who were leaving the firm to start their own business, the Electro Dynamics Corporation. This group was headed by Tex Thornton and Roy Ash, two experienced managers who had figured in Hughes Aircraft's stellar performance of the late 1940s and the early 1950s. These men were interested in building a new type of corporation (now called a conglomerate).

This firm would take advantage of loopholes in corporate income tax laws by investing money, which would otherwise go to the federal government, into the acquisition of small and medium-size firms and the establishment of new research and development programs in unrelated technological fields. "Thornton's plan," Litton later reflected, "was to grow by acquisition rather than by the new creation of wealth. By judicious selection of candidates, money that might have gone for taxes went into effective acquisition."[56] Thornton and Ash also planned to manage these new acquisitions through a detailed analysis of their financial results. Thornton and Ash, who knew Litton Industries well through their previous association with Hughes, were interested in buying Litton's company. Litton Industries had a solid competitive position. Its unusual manufacturing processes were difficult to duplicate. It had solid managers in Moore, Woenne, and Crapuchettes. But more important, Litton Industries was enormously profitable. It would generate the cash flow that Thornton needed to purchase other businesses. In other words, Litton Industries was the perfect nucleus around which to build a conglomerate.[57]

In November 1953, Thornton and his team convinced Litton to sell them his company. To finance the deal, they raised $1.5 million from Lehman Brothers, the investment bank in New York. With these monies, Thornton and Ash bought Litton Industries, Litton's plant in San Carlos, as well as rights to Litton's microwave patents for $1.25 million. They also hired Litton as a consultant at the rate of $75,000 a year for 10 years. In return, Litton agreed not to compete with Electro Dynamics in the microwave tube business for this period. In addition, Thornton proposed to Litton that he purchase a significant interest in Electro Dynamics—a deal which Litton turned down. The agreement between Litton and Thornton also covered Litton's main lieutenants at Litton Industries, Moore, Woenne, and Crapuchettes. Litton, who wanted his men to profit from the transaction, obtained that each of them get 5,000 options on Electro Dynamics stock. Stock options were a new form of financial incentive. They offered the right to buy stock at a predetermined price at a future date, thereby enabling the optionees to make substantial capital gains if the stock increased in value. This was an important social innovation, as Electro Dynamics became the first firm on the West Coast to systematically use stock options to motivate and retain its top engineers and managers. In addition, there is substantial evidence that Litton shared some of the sale's proceeds with Moore, Woenne, and Crapuchettes and that in turn these men shared Litton's gift with their own subordinates.[58]

With the monies that he kept for himself, Litton established a trust fund for his family. He also invested in his equipment business, Litton Engineering Laboratories. In 1954, Litton, who had for a long time wanted to live close to the Sierras, relocated Litton Engineering from San Carlos to Grass Valley, a small town in the Sierra foothills. In Grass Valley, Litton, ever the eccentric, bought an abandoned hospital building. He refurbished it and transformed it into his residence. He also converted part of the building into a well equipped machine shop and vacuum tube development laboratory. Over the next 15 years, Litton conducted his glass lathe and tube-making machinery business from Grass Valley. He kept the corporation relatively small. In the second half of the 1950s and in the 1960s, he employed between 25 and 80 technicians and machinists. Much of the firm's business was in glass lathes. In particular, Litton introduced new glass lathes designed especially for the manufacture of microwave tubes. But Litton also designed microwave tubes for Litton Industries and consulted on vacuum tubes with Hughes Aircraft and Federal Telegraph. He also made hermetic transistor packages for Fairchild Semiconductor and other semiconductor firms on the San Francisco Peninsula. In addition to running his business, Litton attracted small electronics enterprises to Grass Valley and built a small electronics cluster in the area.[59]

The sale of Litton Industries also led to the emergence of a large conglomerate in Southern California. Thornton's and Ash's new firm benefited from Litton's excellent reputation in the Navy and the Air Force. His reputation was so good that in the summer of 1954 Thornton decided to drop the name of Electro Dynamics and call his corporation Litton Industries, Inc. (The magnetron company was renamed Litton Industries of California at this time.) Critical to the rise of the conglomerate was also the profit stream coming out of San Carlos. Under Moore's, Woenne's, and Crapuchettes' direction, the magnetron business flourished in the mid and late 1950s. By 1957, the microwave tube division in San Carlos had 700 employees and did roughly $12 million in business. It was the second-largest maker of magnetrons (behind Raytheon) in the United States. The microwave tube division also was highly profitable. After a hearing with Moore, Litton, and Ash in January 1956, the Renegotiation Board awarded the microwave tube division a 27.4 percent profit on military business (the division received such a high profit allowance because its magnetron prices were much lower than those of its competitors). These profits went straight to Litton Industries' headquarters in Beverly Hills. But the actual profits diverted from the magnetron business and reinvested by Litton Industries, Inc. were even higher. Woenne later recalled that the group in San Carlos "was putting money into Litton Industries, Inc. in all ways that looked like expenses and money was pouring out of San Carlos into the main office." The microwave tube division had become a cash cow and Thornton bled it of most of its cash.[60]

The millions of dollars generated by the microwave tube division financed the building of Thornton's conglomerate. With the profits made on magnetrons, Thornton and Ash acquired numerous corporations (they purchased or merged with 27 corporations between 1954 and 1959). These firms tended to be small or medium sized (with a few exceptions such as the Monroe Calculating Machine Company, a maker of mechanical calculators, which Litton merged with in 1957). Most of Litton Industries' acquisitions manufactured specialty products for the military sector. For example, Aircraft Radio, Inc. specialized in military communication equipment. The Ahrendt Instrument Company in College Park, Maryland produced servomechanisms and differential analyzers. Other companies manufactured resistors, transformers, and navigation equipment. Thornton and Ash bought these firms at a low price. Most of them had aging founders who were interested in retiring and transforming their corporations into cash.[61]

In parallel with these acquisitions, Thornton and Ash invested the profits of the magnetron business in the development of new military avionics systems, especially airborne computers and inertial guidance systems. In 1954, they recruited Henry Singleton and Henry Kozmetsky, two alumni of Hughes Aircraft, to build these businesses. As an additional incentive, Thornton and Ash gave them liberal stock options. At Litton Industries, Singleton established and directed an inertial guidance laboratory. This laboratory developed a small and light inertial guidance system that was later used in many military aircraft. Kozmetsky started a digital computer design group. The computer laboratory developed airborne systems that automated many of the tasks associated with the surveillance and control of a given airspace. By 1960, the electronic equipment division headed by Singleton and Kozmetsky had 5,000 employees and $80 million in sales.[62]

As a result of Thornton's numerous acquisitions and the rapid expansion of the electronic system division, Litton Industries became very large indeed in the late 1950s. The corporation, which had $8.5 million in sales in 1955, had revenues in the order of $180 million in 1960. This expansion enabled Thornton and Ash to bring their company public. In 1956,

Litton's stock was listed on the American stock exchange, before appearing on the board of the New York stock exchange the next year. Because Litton Industries' stock gained considerable value in the second half of the 1950s, Thornton, Ash, and the other early investors in Electro Dynamics made a killing. Moore, Woenne, and Crapuchettes, who rapidly vested their stock options, also became multi-millionaires at this time. Litton, who had turned down Thornton's offer of buying Electro Dynamics' stock, became the laughing stock of the business press in the late 1950s and the 1960s. This was indeed a curious fate for the innovator-entrepreneur who had pioneered microwave tube manufacturing and had played a vital part in the creation of an electronics district on the San Francisco Peninsula.[63]

Conclusion

Litton's diversification into microwave tubes was a long and checkered process. It took Litton almost 14 years to move from klystron design to full-scale production of magnetrons. Entering this new business and establishing himself in it did not prove easy for Litton. Some of the chief obstacles that he faced were the large capital requirements of magnetron manufacturing and the fact that the military services tended to give research and, even more so, procurement contracts to large and established corporations. To build a magnetron business, Litton mobilized a variety of resources. He used his close relations with Stanford researchers and faculty members. Through Stanford, Litton gained access to design and theoretical knowledge. (He also relied on the theoretical expertise of the Radiation Laboratory and the Bell Telephone Laboratories.) In 1944, Litton relied on Terman to get an important development contract from the National Defense Research Committee. Litton also used public capital to finance his magnetron business. Research and development contracts from the NDRC and the military services enabled him to form new engineering teams and develop a line of continuous-wave magnetrons. Similarly, the large profits that Litton made on fixed-price production contracts financed the building of his manufacturing plant in San Carlos.

But the most important resource that Litton mobilized to establish himself in the microwave tube business was the manufacturing process and product engineering expertise he had acquired since the late 1920s. This expertise, which he shared with various research groups at Stanford, allowed him to devise new and innovative ways of engineering and fabricating advanced magnetrons. In the US microwave tube industry, Litton was unique in identifying high vacuum as the primary objective of his tube designs. To obtain this very high vacuum, Litton used only certain materials and baked his tube parts to very high temperatures. These processing techniques guided the magnetrons' mechanical and electrical designs. Litton also innovated by developing new, highly clean manufacturing processes and adopting self-inspection as his primary tool for quality control. Because of these innovative engineering and manufacturing methods, Litton received a series of development contracts from the military services in the second half of the 1940s and later established himself as a major supplier of high-quality magnetrons to the Department of Defense.

Litton's successful entry into the microwave tube business opened a path for other local electronic entrepreneurs to follow. Litton provided these men with a "recipe" on how to build a microwave tube firm with little initial capital: build engineering teams and a product line with military research and development contracts, then move to the production of these tubes, again with military funding. Most microwave tube entrepreneurs followed this path in late 1940s and the 1950s. They also emphasized the development of new manufacturing processes and the design of high-quality products. Litton also trained some of the new entrepreneurs in tube-making technologies and helped others establish their firms. It was Litton who trained Raymond Stewart, an early microwave tube entrepreneur, in high-precision machining and vacuum tube fabrication techniques. Stewart established Stewart Engineering in 1952. At first, Stewart focused on the production of furnaces used in the manufacture of vacuum tubes. He later diversified into the manufacture of microwave tubes. Litton also assisted other men to start their own enterprises. For example, he acted as an advisor to R. Huggins, the founder of Huggins Laboratories, and invested in his tube venture. In 1948, Litton helped Russell Varian, Sigurd Varian, and Edward Ginzton (the klystron specialists with whom he had collaborated since the late 1930s) to start their own microwave tube corporation. It was because of Litton's recommendation that Ginzton and the Varian brothers received their first microwave tube contract in the summer of 1948. This was a contract from the Airborne Instruments Laboratory (AIL) for the fabrication of traveling-wave tubes. This contract brought much-needed cash to the new venture. In the 1950s, Varian Associates became the largest maker of microwave tubes on the San Francisco Peninsula.[64]

3

Military Cooperative

When the Varian brothers, Edward Ginzton, and Myrl Stearns opened their new klystron plant in the Stanford Industrial Park in the fall of 1957, Robert Williams, the undersecretary of commerce in the Eisenhower administration, declared at the factory's opening ceremonies that "this plant [was] one of the most important facilities supporting [the] national [defense] effort and was a strong factor in making the Stanford University area the 'Microwave Capital' of America."[1] By 1957, Varian was indeed a major supplier of advanced electronic components to the Department of Defense and the American defense industry. Its klystrons were important to US air defense and to the development of ballistic missiles and other weapons. Varian Associates was also the main producer of microwave tubes on the San Francisco Peninsula. It created many of the local industry's technical and social innovations and controlled more than one-third of its sales. Varian also had a major impact on other microwave tube firms in the area. It provided a model for other tube startups such as Watkins-Johnson. Varian also emerged as a major source of engineering talent and process and design knowledge for these firms as well as more established corporations such as Eitel-McCullough and Litton Industries. In short, Varian Associates became the dominant microwave tube firm on the Peninsula and the largest manufacturer of such tubes in the United States.

How did Varian become such a powerful force in the growing electronics district on the San Francisco Peninsula? How can one explain its rapid rise to prominence in the microwave tube business? Like Charles Litton, the founder of Litton Industries, Russell Varian, Sigurd Varian, Edward Ginzton, and Myrl Stearns benefited from the growing demand for microwave tubes during the Cold War. But these men also came to electronics entrepreneurship with an unusual background and a rich set of military connections that gave them a significant competitive advantage in the postwar period. The Varian brothers, who were influenced by the utopian socialist tradition in California, shaped their venture into a communal and employee-owned laboratory, where engineer-owners had a stake in the financial well-being of their firm. The founding group of Varian Associates also included a few Stanford faculty members, which gave the firm direct access to the university's best students and research findings. Perhaps more important, the founders had substantial prior experience in the microwave tube industry on the East Coast. They had worked during World War II at Sperry Gyroscope, the military system firm based on Long Island. At Sperry, they had developed new klystron process and design technologies and built close contacts with the military services.

The Varians, Stearns, and Ginzton made the most of these resources when they received an important contract from the Diamond Ordnance Fuse Laboratory in 1948 to develop an exotic klystron for the fuse of atomic bombs. This contract helped Varian gain a unique competence in the design and ultra-clean processing of advanced klystrons. It also enabled the firm to secure a large number of engineering contracts and production orders during the Korean War. Because the military shifted its procurement of microwave tubes from low performance magnetrons to high-quality klystrons in the second half of the 1950s, Varian Associates expanded into volume production of microwave tubes to meet these requirements. By doing so, the firm transformed itself. It gradually opened itself to outside investors in order to muster the necessary financial resources that it needed to expand. Varian's founding group also built a flexible manufacturing organization, which met the military demand for complex tubes in short production runs. This flexible production capability, the firm's unique engineering competence, and its close coupling with user needs helped Varian become the largest microwave tube firm in the United States and displace established corporations such as GE, Raytheon, and RCA in the late 1950s and the early 1960s.

Progressive Politics and the Formation of Varian Associates

Varian Associates was established in 1948 by Russell and Sigurd Varian, Edward Ginzton, Myrl Stearns, Frederick Salisbury, and Donald Snow. These men were an unusual group, different from the entrepreneurs who had established Eitel-McCullough and Litton Industries. Unlike Bill Eitel, Charles Litton, and Jack McCullough who were from the middle class, the founders of Varian Associates came from families of modest

Figure 3.1
Founders and close associates of Varian Associates in front of the firm's first build-
ing, early 1950s. Left to right: Russell Varian, Sigurd Varian, Marvin Chodorow,
Dorothy Varian, Richard Leonard, Esther Salisbury, Edward Ginzton, Frederick
Salisbury, Donald Snow, Myrl Stearns. Courtesy of Varian, Inc. and Stanford
University Archives.

means and progressive politics. This was important because these backgrounds led them to shape their firm in a different way than Litton, Eitel,
and McCullough. Snow, Salisbury, and Stearns were of working-class origin. Stearns' father was a lumberman and belonged to the Industrial
Workers of the World (IWW), the most radical labor union in the inter-
war period. His older brother worked as an IWW organizer. Similarly,
Ginzton came from a socialist family. His mother had participated in the
revolution of 1905 in Russia. After the Bolshevik revolution, the family
had lived in the Ukraine, in Manchuria, and in China before migrating
to California in 1929.[2]

The Varian brothers, however, had the most unusual background.
They came from an impecunious Theosophist family and had been
brought up in Halcyon, a utopian community on the central California
coast. Their father had a small and unprofitable chiropractic practice
and wrote socialist tracts and religious poetry for Theosophist journals.

Halcyon, one of many utopian communities in California, had been established in 1905 by a group of Theosophists. Their goal was to "spiritualize material conditions" and prepare for the coming of a messiah.[3] Halcyon's founders also aimed at constructing a socialist community "wherein all the land [would] be owned all of the time by all of the people, where all the means of production and distribution, tools, machinery and natural resources, [would] be owned by the people and where Capital and Labor [would] meet on equal terms with no privileges to either."[4] The community permitted individuals to own houses and other personal objects. But the pottery factory and the farm were owned and operated collectively.[5]

As a result of their unusual upbringing, most of the founders of Varian Associates had progressive politics. Stearns was a staunch New Deal Democrat. Ginzton was an ultra-liberal who associated himself with longshoremen, labor organizers, and communist sympathizers in San Francisco. Russell Varian and, to a lesser degree, Sigurd Varian had strong socialist leanings. The Varian brothers were deeply distrustful of big business and critical of the distinction between capital and labor which alienated workers from their work. They longed for a system where workers would be in control of their work and would share in the ownership of the means of production. Russell Varian was also a friend of liberal causes. He was a member of the League for Civic Unity and the American Civil Liberties Union. He participated, along with his wife, Dorothy Varian, in the activities of the Sierra Club. By the time of Varian Associates' founding, most members of the group belonged to a housing cooperative, Ladera, in the hills overlooking Stanford University.[6]

In addition to their unusual backgrounds and political outlook, most of these men belonged to different technical subcultures than Eitel, Litton, and McCullough. While the founders of Eitel-McCullough and Litton Industries had been introduced to electronics through amateur radio, three members of the Varian group had received advanced training in physics and electrical engineering at Stanford University in the 1920s and the 1930s. Russell Varian, an inventive and independent-minded man, had received bachelor's and master's degrees in physics from Stanford. While at the university, he had done research in spectroscopy and in x-ray physics. Stearns and Ginzton had also received a thorough education in electrical and electronic engineering. After obtaining undergraduate degrees in electrical engineering from the University of California at Berkeley and from the University of Idaho, Ginzton and Stearns had studied electronics under Frederick Terman at Stanford in the late 1930s. After receiving an engineering degree from Stanford in 1938, Stearns, who was remarkably energetic and entrepreneurial, found a position as a designer of television sets at Gilfillan Brothers in Los Angeles. Ginzton stayed at Stanford for further research in radio theory and received his doctorate the following year.[7]

The other members of the group had a strong mechanical orientation. Sigurd Varian, Russell's adventurous and fun-loving younger brother, was a pilot as well as a "mechanic." He had worked as a pilot for Pan American Airways. Because he repaired his airplanes on his own, Sigurd Varian had gained an excellent knowledge of the mechanical arts. He was also interested in precision manufacturing, or as he called it, the manufacture of the very small. Similarly, Snow and Salisbury were self-trained mechanics (they were also the only radio amateurs in the group). After working as a bank clerk, Salisbury had learned the machine trades and cultivated a taste for experimentation and innovation at Heintz and Kaufman. Snow was also a machinist by training and had worked as a mechanic at Federal Telegraph, before being trained in engineering in Heintz and Kaufman's shop. Snow and Salisbury had an excellent knowledge of metalworking techniques and were skilled glass blowers.[8]

These men had come together to develop and engineer klystrons and related microwave radar systems at Stanford University and the Sperry Gyroscope Company in the late 1930s and the early 1940s. Five members of the group—Russell and Sigurd Varian, Ginzton, Snow, and Salisbury—had participated in the early development of the klystron. The main impetus for this project was the Varian brothers' concern about the expansionist policies of the Third Reich. They were interested in developing a defensive weapon, the microwave radar, that would protect urban centers from bombing raids by the Luftwaffe. To develop the electron tube that would drive this microwave radar, Russell and Sigurd Varian partnered with William Hansen and the physics department at Stanford in 1937 (see previous chapter). With Hansen's theoretical backing, Russell Varian conceived the klystron in June 1937. His brother Sigurd reduced this idea to practice in the next few months.[9]

The klystron soon attracted the attention of Sperry Gyroscope. In 1938, Hansen and the Varian brothers secured a large contract from Sperry to further develop and patent their klystron invention. In return, Sperry got an exclusive license on Stanford's klystron patents. Sperry's patronage enabled Hansen and the Varians to substantially expand their klystron project. They hired doctoral students in the electrical engineering department, including Ginzton, as technicians. They also recruited

Snow and Salisbury from Heintz and Kaufman, the local maker of power-grid tubes and radio equipment, to build klystrons. Benefiting from this influx of talent, the Stanford group designed new klystrons for radar and blind landing systems. It also worked out the device's theory and devised novel microwave measurement techniques.[10]

At Sperry's urging, the Varian brothers, Hansen, Ginzton, Snow, and Salisbury moved to the East Coast in 1940 to strengthen the firm's microwave tube and radar operations. Recruited by Ginzton, Stearns joined the group at Sperry shortly thereafter. Although relatively little is known about their activities at Sperry, it is clear that these men directed many of the firm's tube and radar activities during World War II and in the immediate postwar period. While Hansen and Russell Varian were in charge of Sperry's "invention department," Ginzton rose to the leadership of three different laboratories devoted to klystrons, radar systems, and microwave measurements respectively. Similarly, Stearns managed large radar development programs during the war, before heading Sperry's microwave tube department in 1946.[11]

These men oriented Sperry Gyroscope toward the development and small-scale production of high-quality klystrons during World War II and in the immediate postwar period. This decision was motivated by the rapidly evolving competitive situation in the klystron and radar field in the early 1940s. In 1940, the National Defense Research Committee, which funded most research and engineering on microwave radar during the war, decided to concentrate its efforts on the development of pulse microwave radars. It also chose to use the magnetron to power these radar transmitters. The klystron, which could generate less power than the magnetron at that time, was used in the radar systems' receiving sets. These technological policies and the large-scale production of microwave radar systems during the war created a large market for receiving klystrons. RCA, Raytheon, GE, and Western Electric, which had substantial experience in the mass production of electron tubes, rapidly established themselves as leading suppliers of these klystron types during the war.[12]

Sperry Gyroscope, which could not compete with these firms in the mass production field but had strong engineering teams, adopted a specialty approach to the klystron business. The corporation developed and produced high-precision tubes in small quantities. The Varian group, for example, designed high-quality reflex klystrons. Reflex klystrons were single-cavity tubes (most klystrons had two cavities or more). In these reflex klystrons, the electron beam was reflected back through the cavity resonator by a repelling electrode with a negative voltage. These small and complex vacuum tubes were difficult to make. During the war, Sigurd Varian, Hansen, Snow, and Salisbury developed seven high-performance reflex klystrons. Some of these would be used in radar systems, others in microwave communication equipment. In addition, Hansen and Marvin Chodorow (a physicist who later would direct Stanford's klystron development efforts) developed a theory of the reflex klystron's operation at Sperry.[13]

The Varian group also designed high-power klystrons in conjunction with the development of novel radar transmitters during the war. They pioneered the development of Doppler continuous-wave radar. Unlike pulsed radar systems, Doppler radar sets relied on continuous waves. They could calculate the speed of a moving vehicle such as an airplane and distinguish targets from stationary background objects. In conjunction with this work on Doppler radar, the Varian group developed high-power klystrons. In particular, Salisbury and Ginzton worked on a three-cavity klystron in 1943. This klystron could generate 2 kilowatts of power. This was a very high output power for the time.[14]

When they were still at Sperry, the Varians, Ginzton, Stearns, Snow, and Salisbury made plans to establish their own company on the San Francisco Peninsula after the war. The Varian brothers initiated the project and carefully selected the firm's founding members. The group's motivations for setting up a new corporation were diverse. Ginzton, Stearns, Salisbury, Snow, and the Varians wanted to go back to California after the war. They longed for more congenial and familiar surroundings. This sentiment was not specific to the Varians and their friends. It was shared by many western engineers who had moved to the East Coast in the late 1930s and during World War II. Most of these men desired to go back to the West and were afflicted by what was often called at the time "Californiaitis"—the intense desire to relocate to California. In the postwar period, most western engineers who worked at Sperry, Raytheon, General Electric, Bell Telephone, and other electronics firms on the East Coast moved back to the West. Because the electronics industry on the San Francisco Peninsula was still embryonic and offered few jobs for people with their experience and training, the Varians, Ginzton, Snow, and Salisbury had to create their own. They decided to establish their own firm on the San Francisco Peninsula.[15]

The group also wanted to create a working environment to pursue their technical and scientific interests and be in control of their work. At Sperry, the Varians had chafed at the firm's tight rules and procedures and resented having to report to managers who had little knowledge or understanding of their work. Creating a new firm would enable them to be their own bosses and work on the science and technology of their choice. "The Varian brothers," one of their associates later recalled, "wanted to have independence with new research, new development, and earn enough money to pay for what they were doing, and have control over it."[16]

Finally, Ginzton and Russell Varian saw the formation of a new firm as an excellent opportunity to work out their ideas on how to organize and manage an engineering firm. The group wanted to build a small cooperative-like laboratory closely linked to university research. "The mode of operation of the new laboratory," Russell Varian wrote in 1946, "will be a partnership among five to ten members. The activities will be to develop new devices and to sell patents or manufacture the devices on a small scale, whichever appears to be the most advantageous."[17] The founders were particularly interested in transforming ideas coming from academic research into innovative and high-quality products. In particular, they wanted to exploit the results coming from Stanford's expanding research programs. "Many [academic] laboratories," Ginzton later recalled, "were being expanded and rebuilt, and these would uncover a number of practical applications. The unique characteristic of all of these projects at universities was that they were novel, interesting, but very difficult for a layperson to understand. Those of us who could hope to grasp the practical significance of novel discoveries in American universities were to be sure to be in close proximity to a gold mine."[18] To finance their research and engineering efforts, the group envisioned securing development contracts from the government and private corporations.

The founders, especially the Varian brothers, were also interested in building an engineering company different from Sperry and other traditional electronics corporations. "We did not want to have the hierarchy of a company owning facilities and employing employees," Ginzton later reminisced. "We wanted to be a cooperative. We wanted to create a cooperative organization."[19] The new firm would be owned and managed by engineers and scientists. It would also be an "association of equals" where "everyone would consider himself a true participation [*sic*]."[20] Hence its name: Varian Associates. Ginzton and the Varian brothers also wanted to build a social institution, which would fulfill the "human needs of its employees." In particular, they were interested in providing employment security and constructing "an environment conducive to satisfaction in one's daily work and pride in one's accomplishments."[21]

While these projects were shaped by their ideological commitments, Ginzton and the Varians believed that such a cooperative organization would be conducive to business success. Employee ownership and the creation of a good work environment would strongly contribute to employee morale and help bring forth each person's creative abilities. In other words, the firm's progressive policies would enhance its productivity and competitiveness. Russell Varian, Sigurd Varian, Ginzton, and Stearns were comforted in these beliefs by the example of General Radio, the Massachusetts-based electronic instrumentation corporation that had inspired Charles Litton to share half of his profits with his employees at Litton Engineering a few years earlier. General Radio was an employee-owned company. It belonged to its managers, engineers, and senior production supervisors. No outside investor could buy General Radio's shares, and employee-owners had to sell their stock back to General Radio when they left the company. The Varian brothers and most observers at the time believed that General Radio's unorthodox ownership structure was critical to the corporation's rise as a major manufacturer of electronic measurement instruments.[22]

Although the Varians, Ginzton, and Stearns shared a common institutional vision, the entrepreneurs had divergent views regarding the firm's future field of activity. In spite of their solid expertise in klystron design and small-scale production, Ginzton and Russell Varian felt that the future corporation should stay away from the microwave tube field. With its meager capital of $22,000, Varian Associates would not be able to compete with large tube firms such as Western Electric, Raytheon, and Sperry Gyroscope, which had large product lines, solid engineering teams, and efficient production facilities. "Since the laboratory would be quite small and have limited capital," Russell Varian later recalled, "I more or less eliminated klystrons from the proposed field of our activity. This was because I thought that in order to compete any company would have to have a considerable number of klystrons, and since I knew that they were quite expensive to develop I did not see any possibility at that time in entering the klystron business."[23] But Stearns and Sigurd Varian disagreed. They knew through their activities at Sperry that the government was increasingly funding klystron research and engineering. As many industrial laboratories had dropped their klystron development efforts immediately after the war, these men thought that there would be a demand for their tube-development skills. The group was also interested in exploiting other fields in high-frequency electronics. They envisioned working on microwave communication systems, electronics measuring instruments, and linear accelerators. Russell Varian was also interested in developing practical applications for nuclear magnetic resonance, which had been recently discovered at Stanford by William Hansen and Felix Bloch.[24]

Varian's founding group returned to California in two waves. In 1945, Russell Varian moved to the Peninsula to prepare for the establishment of the new firm. He was followed, in 1946, by Ginzton, who joined Stanford's physics faculty and assumed the associate directorship of the new Microwave Laboratory. The other founders, who had stayed at Sperry to bring the tubes they had developed during the war to production, moved to the Bay Area in the summer of 1948. At the same time, Varian's founders gathered a group of directors such as Hansen and Leonard Schiff, the new head of the physics department at Stanford. Richard Leonard, a lawyer, and Francis Farquhar, the head of an accounting firm, whom Russell Varian knew from the Sierra Club, also joined the board of directors, bringing much-needed legal and managerial expertise to the new company. (Terman, the dean of engineering at Stanford, and David Packard, of Hewlett-Packard, joined Varian's board in 1949.) Ginzton, Stearns, and the Varians also brought in consultants such as Marvin Chodorow, the Sperry physicist who had recently joined Hansen's and Ginzton's group in the physics department at Stanford.[25]

Contract Engineering

The entrepreneurs set up shop in San Carlos in July 1948. Their firm was located only a few blocks away from Litton's magnetron plant. Besides the establishment of suitable facilities and the acquisition of a security clearance for the new firm, their first task was to secure business contracts. The founders also needed to acquire additional capital. They soon discovered that the $22,000 they had invested in the venture was insufficient to establish a machine shop and vacuum tube laboratory. (This money came from their own savings and from the royalties the Varian brothers had received on the klystron patents.) The founders spent much of that sum solely to equip the shop with second hand machine tools and specialized equipment such as vacuum furnaces and a Litton glass lathe. They needed fresh funds to survive and put themselves on a more secure footing so that they would be able to undertake large projects. "We had to overcome some extremely difficult things in the early days," Stearns later recalled. "The first thing was just staying alive from a financial standpoint. It was really hand to mouth."[26]

To keep the firm going, the founders needed to raise additional capital. But since Russell Varian was adamant about building a firm controlled by its employees, they could not get funds from outside investors. (Some New York investors had shown interest in Varian Associates, but Russell Varian had vetoed their investing.) This led the entrepreneurs to raise funds from their friends. (In the fall of 1948, Hansen mortgaged his house in order to invest $17,000 in Varian Associates.) Varian's founders also decided to grant stock-ownership privileges to all current and future employees, in order to raise additional capital and, at the same time, realize their goal of building an employee-owned firm. Employees could buy these shares at the same price as the founders had bought them themselves.[27] The Varians, Stearns, and Ginzton also agreed that all the corporation's stock should be kept in the hands of employees, consultants, and directors. Shareholders who wanted to sell their stock had to offer it to other employee-owners, before selling it to the highest outside bidder. This ensured that no external investor would take control of the company and re-orient it in a more conventional direction. Relying on friends, employees, and local investors sympathetic to Varian's goals, the entrepreneurs raised $120,000 between October 1948 and June 1950. Snow and Salisbury decided not to participate in the management of the company. They were content to remain as technologists with no corporate responsibility. In essence, the company was run by only four of the original founders, Stearns, Ginzton, and the Varian brothers. Stearns ran Varian on a day-to-day basis and was the company's chief sales and marketing man. Sigurd Varian headed engineering, while Ginzton, who worked only part time at Varian advised Russell Varian, the company's head, regarding strategic issues.[28]

In parallel with the search for additional capital, the group looked for development contracts, especially in the military sector. The entrepreneurs sent announcements of the company's formation together with a brochure on its capabilities to government agencies. Stearns and Ginzton also visited military laboratories to solicit business for the new firm. "I went all up and down the East Coast, visiting laboratories, in the Air Force and Navy, and so on," Stearns later reminisced. "I visited as many of them as I could, just to touch base and see what was going on. Then I left Washington and I drove all the way across the country. I stopped at Wright Field [in Dayton] on the way out, and at the Air Force laboratories, where I knew a lot of people."[29]

Benefiting from the Department of Defense's rearmament efforts early in the Cold War, Varian Associates received two microwave tube contracts in September 1948. The firm, which had been recommended by Charles Litton, obtained a small production order from the Airborne Instruments Laboratory, an East Coast avionics firm. Under the contract, Varian was to produce six traveling-wave tubes (another type of velocity-modulated microwave tube), which had recently been designed by a research group in the electrical engineering department at Stanford.[30]

Varian also obtained a highly classified contract from the Diamond Ordnance Fuse Laboratory to develop an exotic reflex klystron, the R-1. The R-1 was the main component of the radar system that monitored an atomic bomb's distance from the target and triggered the explosion. "This radar," Chodorow, a Varian consultant, later recalled, "would transmit a radar signal. When it got close to the target, it could tell from the reflected signal how close it was, and when it was close enough, it would detonate the [bomb]."[31] Because of its unusual application, the tube had to be extremely reliable. It had also to withstand high shocks and vibrations. Because no existing reflex klystron met these requirements, the DOFL solicited bids from RCA, Sperry, and GE for such a tube in 1948. Varian Associates was not one of the firms that the Diamond Fuse Lab contacted for this project.[32]

It was because of Ginzton's consulting activities with DOFL that Varian bid on and secured the development contract for the fuse of atomic bombs. In addition to his assistant professorship at Stanford, Ginzton worked as a consultant for the Diamond Ordnance Fuse Laboratory. "I was asked as a consultant to come over to the Diamond Ordnance Laboratory to tell them whether the proposals by Sperry and RCA [for the tube] were sound," Ginzton later recalled. He continued:

Each one said they could do what was needed but it would cost a million dollars for the job. I looked at the proposals and decided that the problems were not nearly so severe as Sperry and RCA perceived them to be and I said: "let me think about it." I then went home, showed the requirements to Sigurd Varian and the rest of the group, and we all agreed that what needed to be done could be done very easily. In a week's time, our laboratory was able to build a model of a klystron which I took back to the East with me in my pocket. When I was once again asked if it was possible to develop the kind of tube they needed for $1 million, I said: "Yes of course, here's a sample. You can have it for free." That was dramatic staging, but that was how Varian got involved in its first major contract.[33]

Benefiting from Ginzton's showmanship, Varian Associates received a $21,400 cost-reimbursement contract from DOFL to establish the tube's feasibility. This contract was later renewed and enlarged several times. By the summer of 1950, Varian had received nearly half a million dollars to

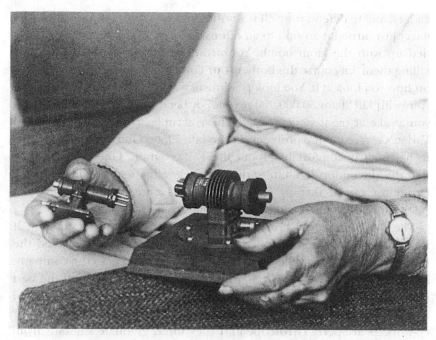

Figure 3.2
The R-1 reflex klystron (right) and its successor, the R-2 (left). Courtesy of Varian, Inc. and Stanford University Archives.

develop the R-1. This klystron became the company's mainstay and established Varian firmly in the microwave tube business.[34]

One might ask why the Varian brothers, Stearns, and Ginzton, who had progressive political views and had justified their earlier tube work on the grounds that klystrons were defensive weapons, accepted such a contract and became involved in the development of weapons of mass destruction. One can speculate that obtaining this contract was a matter of necessity. The firm had to secure and renew government contracts, whatever contracts, in order to survive. The DOFL contract also offered challenging technical problems, especially in fabrication techniques, which these men relished. In addition, Stearns, Ginzton, and the Varian brothers were patriots. They were not Marxist socialists, and they had little sympathy for the Soviet Union. There is substantial evidence, however, that some members of the group, especially Russell and Sigurd Varian, later sorely regretted their involvement in the development of fuses for atomic bombs. "In Russell's and my opinion," wrote Sigurd Varian with much anguish in 1958, "we thought that the klystron could be nothing but a defense gadget, and to defend yourself is a noble process, but it's not so. [The military] just turned it around to an offensive weapon and we helped. It was tied up with the atom bomb. You aren't saving anybody, you are just killing them. Of course this bothers our conscience quite a bit. It depends on how you look at it. You look pessimisticly [*sic*], as I do once in a while, you [will] kill 20, or 30,000,000 people, or been partly responsible. Keeps you awake at night sometimes."[35] These scruples contributed to Sigurd Varian's severe mental problems in the late 1950s. Ginzton himself seems to have felt uncomfortable about the R-1 program. One can speculate that this discomfort led him to direct the development of a linear accelerator for cancer therapy at Stanford and commercialize it at Varian Associates in the 1950s and the early 1960s.[36]

The R-1 contract was critical in terms of business and technology. It funded the acquisition of tube-making equipment. It also enabled the founders to recruit excellent microwave tube engineers to the company. For instance, they hired Cliff Gardner, a western engineer and the best microwave tube technologist at Raytheon. At Raytheon, Gardner headed all klystron development work. Varian's founders also recruited former co-workers at Sperry Gyroscope and they hired graduate students from Stanford—especially those who worked on their doctorates under Ginzton and Chodorow. These students had an excellent mastery of microwave theory and klystron design. By 1950, half of the Varian group was composed of Stanford faculty members and recent students. The rest was made up of experienced engineers and mechanics from Sperry and Raytheon.

The R-1 contract also led Varian Associates' founders and their recruits to make innovations in tube fabrication. The reflex klystron had unusual reliability and testing specifications. "The R-1 had to be able to survive the equivalent of having a half-ton weight fall on it, or having it drop from the top of a two-story building onto a hard surface," Chodorow, the tube designer, later reminisced. "That was the test. It was put in something like a shock machine."[37] Reflex klystrons built with traditional production techniques inherited from the receiving tube industry did not withstand such rough treatment: the tube envelope would bend, and cathodes and grids would fall off.[38]

Snow, Salisbury, and other Varian engineers experimented with new fabrication processes under the overall guidance of Sigurd Varian. In particular, they built the R-1 out of machined parts. This was an important shift in klystron practice. Until then, klystrons, like conventional receiving tubes, had been made out of stamped sheet metal parts. Machined

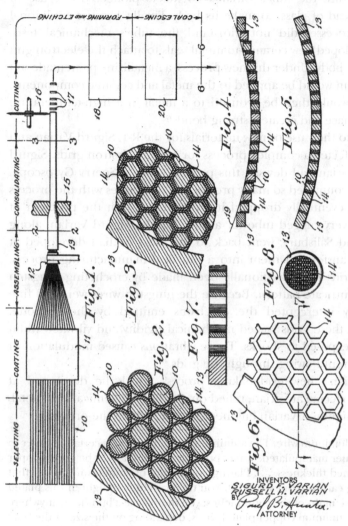

INVENTORS
SIGURD F. VARIAN
RUSSELL H. VARIAN
BY
ATTORNEY

Figure 3.3
Sigurd Varian's process for making honeycomb grids. Source: Sigurd Varian and
Russell Varian, "Method of Making a Grid Structure," U.S. patent 2,619,438, filed
April 16, 1945, granted November 25, 1952.

components enabled the construction of rigid tube structures capable of withstanding the shocks and vibrations to which the R-1 was subjected. To further ruggedize the tube, Varian's engineers pioneered the use of ceramics, instead of glass, to make its body. Finally, because traditional cementing processes did not withstand the tube's mechanical tests, Gardner developed new ceramic-to-metal seals to attach the electron gun to the ceramic body. Under this new process, a metallizing paint made out of molybdenum would be applied to the metal and ceramic components. This coating would then be reduced to a metal in a high-temperature hydrogen furnace and create a strong bond.[39]

In parallel to their use of new materials for the R-1, Sigurd Varian and his group perfected a complex process for making klystron grids. Sigurd and Russell Varian had devised this process while at Sperry Gyroscope. But Sperry encountered so many production difficulties with the process that the firm eventually dropped it. When faced with the problem of developing a very rugged tube for atomic bombs, Sigurd Varian, along with Snow and Salisbury, went back to the ideas he had developed at Sperry and transformed them into a practical manufacturing process. Until then, grids had traditionally been made by crocheting tungsten wire into an intricate pattern. Because the tungsten wires were not fine enough, they intercepted the electrons emitted by the cathode. Furthermore, these grids lacked mechanical rigidity and vibrated when the tube was exposed to shocks. These vibrations caused modulation of the signal current flowing through the grids.[40]

Sigurd Varian's new grid-making process, which he developed at Sperry and perfected at Varian, relied on a variety of chemical and metallurgical techniques. A Varian engineer later described the process:

You start with aluminum wire. The aluminum wire must be of a certain temper so that it can be later manipulated. You copper plate it and it must be good copper plating to defined thickness. . . . Plating copper out of aluminum was not that normal a thing early on, but we did it. You then carefully straighten this plated wire, cut it into lengths of six, eight inches; you then make a bundle of anywhere from 20 to 70 aluminum copper-plated wires, depending on the size of the final grid. You now compact that bundle, swage it, hammer it, and draw it through drawing dyes, to finally get it into a dense mass of copper and aluminum that is inside a copper tube. You have to saw it into disks. Then you etch out the aluminum in a chemical bath.[41]

This process enabled the making of a new type of grid, the "honeycomb grid" (so called because of its honeycomb-like structure). Because the new grid structure had webs as thin as 0.0001 inch, it was "transparent" to electrons. This grid could take the heat out as the spent electrons were collected in it. It was also much more rigid than traditional grids. As a result, it did not introduce noise into the electromagnetic signal. The honeycomb grid was an important departure in klystron design. It made small rugged reflex klystrons possible and accounted for the R-1's spectacular reliability.[42]

In January 1949, Varian Associates received a contract for the development of another high-end klystron from General Electric. GE contracted with Varian to develop the prototype of a high-power klystron for ultra-high-frequency (UHF) television transmitters. The Federal Communications Commission had recently approved UHF television broadcasting, and GE was eager to establish itself in this new market. To do so, GE had to solve an important bottleneck. No conventional power-grid tube could generate enough power at UHF frequencies for television broadcasting. It needed a klystron to do the job. But GE's mass production-oriented microwave tube division had little interest in developing such a tube. It viewed the demand for this tube as limited and estimated that manufacturing it in short production runs would not make the most efficient use of its facilities. The division was also in short supply of engineers, as a substantial share of its professional staff had been diverted to the development of color television. As a result, GE had to look for an outside supplier of tube engineering services.[43]

Varian Associates received the contract for the development of the UHF klystron because of Stearns' contacts at General Electric. Stearns had built good relations with GE's microwave tube division during the war. He was also well acquainted with the firm's upper management through his participation in the vacuum tube standard committee of the Electronics Industry Association. The tube's power (5 kilowatts) and its other unusual specifications led Snow and Salisbury to make further innovations in klystron design and processing. "This project was a challenge to us," a Varian employee reminisced, "the engineers ran into a lot of difficult technical problems; Sigurd Varian and Salisbury sweated out how to put their 200 pound, four foot tube together."[44] In particular, the tube's high temperatures led to the release of occluded gases from the metallic elements. These gases poisoned the oxide cathode and led to short tube life. To overcome this, Sigurd Varian, Snow and Salisbury designed a new cathode, the bombarded cathode. The cathode was made out of sheet of tantalum, an element, which absorbed, rather than released gases at high temperatures. (The use of tantalum in vacuum tubes had been pioneered at Heintz and Kaufman in the early 1930s.) By bombarding electrons on the backside of this sheet of tantalum, they heated up the cathode to an emitting temperature and created a powerful stream of electrons. In addition to this novel cathode design, they also engineered a water-cooled collector, which allowed for the tube's proper cooling. This tube represented an important breakthrough in klystron design and performance. It attracted considerable attention when the tube, along with GE's UHF transmitter, were shown at the annual meeting of the Institute of Radio Engineers in the winter of 1951.[45]

This development contract along with the contract from the Diamond Ordnance Fuse Laboratory helped the Varian group to further strengthen the design and fabrication expertise in reflex and high-power klystrons, which they had acquired at Sperry during the war. They gained a unique competence in the engineering and production of reliable and rugged tubes for demanding military applications. These contracts enabled the firm to hire expert engineers and technicians, acquire expensive processing and testing equipment, and develop new skills and techniques. But the R-1 and UHF television contracts also gained Varian Associates considerable visibility in the military and commercial sectors. The UHF-TV tube established Varian as an important player in the emerging field of high-power klystrons, and the R-1 project made it an important supplier to the Department of Defense.[46]

Advancing the Klystron Art

The Korean War led to rapid expansion of Varian Associates' custom engineering business and oriented the firm almost exclusively toward the military sector. In the early 1950s, the Department of Defense greatly enlarged its support for vacuum tube research and engineering. It needed advanced klystrons to drive novel radar systems. The radar sets developed and deployed during World War II had clear limitations. They could not detect large bombers more than 200 miles away. Neither could they detect aircraft flying at low altitude and distinguish these planes from stationary objects. These shortcomings left major gaps in the air defense system. To strengthen its air defense systems, the Department of Defense funded the development of a new technology, Doppler radar. This technology, which had been pioneered at Sperry Gyroscope (among other places) during World War II, relied on the Doppler effect, namely the fact that the steady motion of a target caused a shift in the frequency of a reflected radio signal. Doppler radars could distinguish moving vehicles from their stationary background and they could calculate their speed.[47]

The development of Doppler radars required entirely new types of transmitting tubes. Magnetrons, which had been used in radar transmitters during World War II, could not be easily employed in Doppler equipment. Their frequencies were not stable enough for Doppler radar. Only klystrons had the desired frequency stability. But the klystrons used in radar transmitters in the late 1940s could not generate more than a few kilowatts. Major engineering work was required to obtain higher power klystrons. At the same time, the Air Force and the Navy were interested in funding research and development projects on reflex klystrons. They needed new reflex klystrons for the radars used in aircraft and missiles. The reflex klystrons employed in airborne radars in the 1940s were highly unreliable. They did not withstand severe pressure and temperature differentials. Neither did they did survive the shocks and vibrations characteristic of jet aircraft. As a result, reflex klystrons failed regularly and incapacitated the planes' and missiles' avionics systems. The military had an urgent need for more rugged and reliable reflex klystrons.[48]

Because Varian Associates had a unique competency in klystron engineering, the Department of Defense pressed the firm to tackle these engineering problems and take on new R&D projects in the early 1950s. It also asked Varian to produce its tube designs, including the R-1, in small quantities at this time. Although these military demands went against the founders' original goal of slow and careful expansion, Varian's management complied with this request and enlarged the scale of their operation considerably in the early 1950s. In 1951, one of the firm's directors wrote the following:

Fortunately for the nation but unfortunately for our private plans, Varian Associates' basic technical abilities are of great value to the national defense effort and critically needed. Although we would have much preferred to continue on a moderate expansion program, reaching a modest size with interesting and profitable commercial business, I can state that the officers and directors of Varian Associates recognize fully their obligation to contribute to the maximum extent to the defense effort. Extreme pressure has been put upon us to expand at as rapid a rate as our management ability will permit. We are trying to comply with this urgent need for increased facilities by pushing our expansion program just as rapidly as we possibly can.[49]

To finance this rapid expansion, Stearns, Ginzton, and the Varians needed more capital. They again collected funds from employees and friends of the company. They also sold the sales and manufacturing rights to a line of microwave measurement instruments, which Varian's engineers had recently developed, to Hewlett-Packard for $20,000. Hewlett-

Packard was then a small electronic measurement and testing equipment corporation. The sale of Varian's family of microwave measurement instruments enabled Hewlett-Packard to enlarge its product line and substantially increase its revenues in the 1950s. Because the founders of Varian Associates were still short on funds, they resorted to public capital. In early 1951, they applied for a Victory loan to set up their own plant and purchase tube processing and testing equipment. V-loans had been established at the beginning of the Korean War to finance new factories that were deemed of crucial importance to the nation's defense efforts. The support of the Diamond Ordnance Fuse Laboratory (which needed Varian's reflex klystron, the R-1, for A-bomb fuses) enabled Varian to receive a loan of $1,520,000 from the Defense Production Administration in August 1951. One third of the loan was earmarked for the purchase of production equipment for the shop in San Carlos. The rest was allocated to the construction and fitting of an engineering building on the Stanford campus. Stearns, Ginzton, and the Varian brothers were interested in relocating the engineering staff to Stanford to reinforce the firm's close connection with the university's research programs.[50]

In tandem with the construction of a new facility on the Stanford campus, Varian Associates enlarged its engineering staff to develop the klystrons specified by the military. By the fall of 1952, Varian Associates had 125 physicists, engineers, and technicians in its engineering division. The firm's managers raided Sperry's engineering staff and hired Ginzton's and Chodorow's best doctoral students. They also recruited East Coast microwave tube engineers from Raytheon, Sylvania, and Western Electric. The technicians came from Raytheon, Eitel-McCullough, and the Microwave Laboratory at Stanford. Stearns, Ginzton, and the Varian brothers also recruited skilled contract officers from the Diamond Ordnance Fuse Laboratory to deal with the complexities of military contracting. These men were attracted by the caliber of the founding group, the corporation's ideology and progressive policies, and the fact that it was owned and managed by technologists. Some were also enticed by the prospect of living on the San Francisco Peninsula.[51]

To organize this much enlarged engineering workforce, Varian's management set up new tube development teams. These teams were patterned after the model established for the development of the R-1 and UHF tubes in the early days of the company. Varian's klystron-development teams differed sharply from those of Litton Industries, the other notable microwave tube firm on the San Francisco Peninsula at the time. The goal of Varian's engineering teams was to push the state of the art in klystrons and establish close relations with customers. In contrast, Litton's teams focused on the development of microwave tubes, which could be easily put into production and maintain a high vacuum over their lifetimes. These different goals led to divergent team composition and organization. At Litton Industries, engineering teams were made up of electrical engineers, mechanical engineers, and manufacturing specialists. In contrast, Varian's teams were composed of mechanics and PhD physicists. While the physicists on the team worked on the klystron's electrical design, the mechanics or technicians concentrated on its mechanical construction. Most engineering groups at Varian were directed by engineers and physicists. But expert mechanics such as Snow and Salisbury headed their own teams and directed the work of PhDs. Each group was responsible for a specific tube project. Their mandate was to produce the tube in small quantities and to establish close communication with their customers. These groups were largely autonomous—much more so than the ones at Litton. "We had the usual security classifications," a Varian engineer later remembered, "and then there was 'project private' where even the management did not know what was going on. Some of these engineering groups were like that, where the management did not know what was happening in them."[52] Varian's distinct team organization gave the firm an important competitive advantage over other corporations in the high-quality tube development business. It enabled the firm to engineer high-end tubes rapidly, to work closely with its customers, and to know what the customers needed.[53]

In the early 1950s, these teams made major innovations in klystron engineering and processing. It was a group directed by Snow that invented the external-cavity reflex klystron in 1951. Snow's goal was to further ruggedize reflex klystrons, possibly for the fuses of atomic bombs. Besides their tungsten grid and glass and sheet metal construction, the main weakness of conventional reflex klystrons (of the sort produced during World War II) was their tuning mechanism. These tubes were gap-tuned. In other words, their frequency changed by varying the space between the two grids, which formed part of the resonating cavity. The grid space was adjusted by flexing a thin metal diaphragm. The tuning mechanisms used to flex this diaphragm were microphonic or "noisy" when subjected to severe shock, vibration, and temperature variations. To solve this problem, Snow's group developed a new tube configuration. Under this new configuration, the resonating cavity was external rather than internal to the tube envelope. This enabled the tuning of the klystron by a screw inserted in the external cavity, rather than by flexing a fragile diaphragm. As a result, external-cavity reflex klystrons were remarkably free of "noise" under conditions of extreme shock and acceleration. This was a major innovation, which, along with the honeycomb grid and tube construction out of ceramic and machined parts, made possible the making of very reliable reflex klystrons. In the next few years, Varian's engineering teams developed more external-cavity reflex klystrons for airborne radar systems, especially those used in missiles. These tubes could operate at different frequencies and different powers than Snow's first external-cavity reflex klystron.[54]

Varian's engineering teams also advanced the art of the power klystron. They designed high-power klystrons for Doppler radar. They did so in close collaboration with Ginzton's and Chodorow's klystron research group in the Microwave Laboratory at Stanford. In the late 1940s, Ginzton and Chodorow developed a very powerful klystron to drive their linear accelerator. They did so in collaboration with Charles Litton. The klystron these men designed could deliver up to 14 megawatts in short pulses. At Ginzton's suggestion, engineers from Varian teamed up with the group at Stanford to develop a smaller version of this tube for pulse Doppler radar systems. This project was funded by the Office of Naval Research. To engineer such a tube, the Varian-Stanford group used new materials such as ceramics. Because this tube, unlike its accelerator counterpart, had to operate without vacuum pumps, the team also developed innovative procedures to bake out the gases occluded in the klystron's metallic parts. This project enabled engineers from Varian to gain a high level of competence in the design and processing of high-power klystrons for radar applications. In the next few years, these men, along with former students of Ginzton's and Chodorow's, designed a high-power tube for Doppler radar under a cost-reimbursement contract from Hughes Aircraft. This klystron was a six-cavity tube and could operate at 1 million watts (a megawatt). Engineering teams at Varian also developed a series of powerful klystrons for the Distant Early Warning (DEW) Line, a series of radar outposts in Northern Canada, which was to detect incoming bombers from the Soviet Union. They did so with funding from MIT's Lincoln Laboratory.[55]

In the early 1950s, Ginzton, Chodorow, and the Varian engineers shared their expertise in high-power klystrons with Eitel-McCullough. Bill Eitel and Jack McCullough, the manufacturers of tetrodes and other power-grid tubes, were eager to diversify into klystrons. They saw klystrons as a potential threat to their radio communication and broadcasting tetrodes and wanted to establish themselves in this new field. The

Varian group helped them to do so. In 1951 and 1952, Ginzton and Chodorow gave a series of lectures on klystron design and theory to Eitel-McCullough's engineering staff. Varian Associates' engineers also shared the special techniques they had developed for the fabrication of klystrons with Eitel-McCullough. Ginzton, Chodorow, Stearns, and the Varians decided to share their knowledge of high-power klystrons with Eitel-McCullough for several reasons. Ginzton and Chodorow were eager to dispel any appearances of impropriety or conflict of interest in regard to Stanford. "We felt that we had to be politically very careful [with high-power klystron technology]," Chodorow later recalled. He continued:

We had to have an industrial partner, and we had to be very careful not to have the exclusive partner be Varian. Varian agreed that since Ginzton and I had developed this tube with government money, we could not give [Varian] all of our know-how and exclude everybody else. We had to have a record that was completely fool-proof. We wanted to show that even though we personally had very close affiliations with Varian, what we were doing at Stanford was going to have an impact on the entire tube industry.[56]

Stearns, Ginzton, and the Varian brothers also thought that the potential market for high-power klystrons was so large that there would be enough business for several companies. One can also speculate that the technical policies of the Renegotiation Board (which claimed excess profits from military contractors) may have induced Varian to help Eitel-McCullough. The Renegotiation Board encouraged military contractors to share technical information with other military suppliers and took these collaborative activities into account when it set their profit level and decided how much money they had to give back to the federal government. The Varian-Stanford group enabled Eitel-McCullough to enter the klystron business. They later regretted it as Eitel-McCullough emerged as a major competitor to Varian in high-power klystrons in the mid and late 1950s.[57]

Varian Associates began manufacturing tubes during the Korean War. To meet the military demand for small tubes, the firm started producing the R-1 and other rugged reflex klystrons. To do so, Varian's founders set up a small production organization independent of engineering in 1951. They recruited Emmet Cameron, an engineering manager with manufacturing experience at Federal Telegraph, to head it. In turn, Cameron astutely relied on the group of men who had joined Varian in the late 1940s and who had participated in the development of the R-1 as well as GE's UHF tube to supervise the firm's new manufacturing operations. To man the plant, the firm recruited more than 200 technicians, operators, and skilled machinists between 1950 and 1953. In accordance with the founders' ideological commitments, the firm made a substantial effort to hire and train African-American operators and technicians. While a number of Varian's new hires had worked previously at Eitel-McCullough and Litton Industries, many had to be trained in the specialized processes characteristic of klystron manufacture.[58]

To build a manufacturing department congruent with their cooperative ideology, Cameron and Varian's founding group made managerial innovations. Unlike Eitel-McCullough and Litton Industries, which had hierarchical production organizations, these men constructed Varian's manufacturing operation around teams. These teams were composed of assemblers, technicians, and "leadpersons." They were in charge of a specific process or product. Varian's founders also established a new organization, the Management Advisory Board, in January 1953. The goal of the MAB was "to establish the best possible two-way communication between management and all groups within the company for information, criticism, and evaluation of major policies."[59] Nominated by fellow employees, the board's members came from all departments. The MAB had the authority to study any issue, suggestion, or idea pertaining to the management of the company. "The effort was really to get a better interface between employees and management and try to solve organization problems inside," a former member of the board later recalled. "The MAB accomplished quite a bit in getting management's attention to problems in the organization. It was also a sounding board for employees. The MAB was known as a place where a person could complain or praise or what have you."[60] The MAB addressed a variety of topics ranging from coffee service to injuries on the job as well as stock bonus plans. For example, it examined Varian's layoff policies and compared the firm's wage structure and employee benefits to those of other companies in the area. Varian's management adopted many of its recommendations.[61]

In conjunction with these management innovations, Varian's founding group and the firm's manufacturing engineers developed new facilities and work arrangements to manufacture the firm's exotic klystrons. In particular, Sigurd Varian, Gardner, Salisbury, and other engineers at Varian Associates built an innovative facility for the assembly and testing of rugged reflex klystrons. Their goal was to assemble klystrons, especially R-1s, in a clean environment. When an R-1 was assembled in the open air, loose particles were trapped in its envelope. These particles interfered with the tube's electron flow and ruined its operation. In 1951, with funding from the Diamond Ordnance Fuse Laboratory, these men con-

A 251

Figure 3.4
Production of power klystrons, San Carlos, 1954. Courtesy of Varian, Inc. and
Stanford University Archives.

structed one of the first clean rooms in the US electronics industry. (The
first one had been set up at the Bell Telephone Laboratories a few years
earlier for the production of repeaters for transatlantic telephone
cables.) "The whole thing was filtered air," a Varian engineer later
recalled. "In addition we had vented benches with filters that cleaned air,
[along with] air showers. The dust counts were amazingly good until the
people came in."[62] To control dust brought in by the operators, Varian's
engineers instituted a new kind of manufacturing discipline. Smocks and
hats were required, and operators were expected to clean their work-
benches each morning with a damp sponge to collect the small amount
of dust that would have accumulated overnight. Furthermore, new
employees (most of them women) were hired to collect the particles in
the clean room. "The particle picker," an engineer remembered, "had a
little vacuum pump with an hypodermic needle and under a microscope
she would find any speck of ceramic or whatever and suck it out of the

tube with her little vacuum."[63] With these and other techniques, Varian's management increased production from 250 klystrons per month in December 1950 to 1,000 per month in 1953.[64]

As a result of its entry into tube manufacture and its expanding research programs, Varian Associates grew rapidly during the Korean War. In the summer of 1950 the firm employed 75 engineers, technicians, and machinists; by 1953 its workforce had grown to 600. During the same period, its sales expanded from $461,000 to $5,197,000. This made it the largest manufacturer of microwave tubes on the San Francisco Peninsula. By 1953, Varian had also emerged as the second-largest klystron company in the United States. Only Raytheon had larger klystron sales. Sperry Gyroscope and Western Electric, long established in the klystron business, each had klystron revenues equal to about half of Varian's.[65]

Security Crisis

Varian Associates' new-found position in the klystron business was endangered in the summer of 1953 when Stanford's University's faculty members and administrators, many of whom were on Varian's board of directors, were investigated for possible communist affiliations. This investigation was part of a much larger effort fueled by the Federal Bureau of Investigation, the House Committee on Un-American Activities, and the Government Committee on Operations of the Senate. The House Committee on Un-American Activities, which had been set up to investigate threats to national security in 1938, looked aggressively for communist subversion in the late 1940s. It organized a much publicized inquiry of the motion picture industry and investigated a spy ring in the Department of State. This committee was later overshadowed by Joseph McCarthy, chairman of the Senate's Government Committee on Operations. In a series of speeches and Senate hearings, McCarthy charged that many US institutions, including the military services, had been infiltrated by communists and Soviet agents (Schrecker 1998).

In the spring of 1953, McCarthy and the House Committee on Un-American Activities started to examine universities. As a result, military investigators descended on Stanford and examined the backgrounds of its faculty members. They accused Edward Ginzton of being a communist sympathizer because he had befriended Frank Oppenheimer, J. Robert Oppenheimer's brother, a former communist, at Stanford in the late 1930s. Ginzton was also accused of associating with communist artists in San Francisco and with communist students at Stanford. Furthermore, he had lost secret Air Force documents entrusted to him. Similarly, Leonard Schiff, the head of the physics department, and Frederick Terman, the dean of engineering, were considered possible security risks. Because these men were on the board of Varian Associates, the company was itself under serious investigation. As a result, Varian lost a large grant from the Air Force to construct a klystron plant on the Stanford campus, and the Navy delayed the renewal of Varian's facility clearance and intimated that it was not going to be renewed. In view of its heavy reliance on classified military contracts, Varian's survival was at stake.[66]

To keep the firm's security clearance and reverse its sagging fortunes, Varian's management contacted the Secretary of Commerce through a board member who had connections in the Eisenhower administration. "We had to have a hearing at the highest levels of government," Myrl Stearns later recalled, "to find a way to explain our position and explain [that] we were not bad guys. But we had to have this hearing at a high enough level so that a decision could be made and we could show how important we were to the defense effort. The entry was Arthur Flemming, the former president of a Midwestern university which was heavily funded by one of our directors. He was our entry, and we carefully prepared for this, and [myself and one of the directors] were selected to go to this hearing. So we carefully prepared our case, presented it to the Secretary of Commerce, and he in turn led us to others in the Defense Department. He made several presentations within bureaucratic channels, and a decision came down, nothing formal, that 'if you take these people off your board of directors, you will probably be OK.'"[67]

In September 1953, Terman, Schiff, and Ginzton resigned from Varian's board. Their resignation and the firm's strong position in the klystron field ensured that the drive against Varian tapered off in the next few months. Ginzton, Schiff, and Terman lost their security clearance in the fall of 1953, but they were skillfully defended by one of San Francisco's best lawyers and were rapidly reinstated. Ginzton later rejoined Varian's board of directors.[68]

Dominance

After temporary setbacks due to the security crisis and a downturn in the military demand for klystrons at the end of the Korean War, Varian Associates experienced another period of rapid growth starting in 1955. This expansion was brought about by growing military spending, due largely to the intensifying arms race with the Soviet Union. Procurement

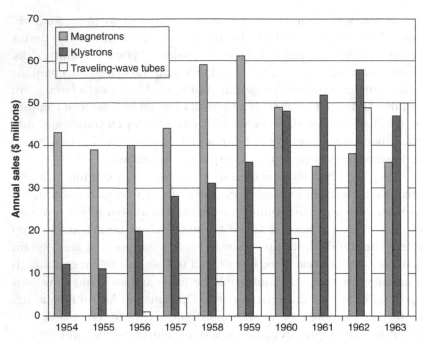

Figure 3.5
The market for microwave tubes, 1954–1963. Source: Edward Ginzton, speech at shareholders' meeting, February 20, 1964, folder: speeches 1960–1965, box 14-A, Edward Ginzton Papers, SC 330, Archives and Special Collections, Stanford University.

appropriations of the Department of Defense grew from $35.5 billion in 1955 to $48.3 billion in 1963. In particular, the purchase of electronic components and systems by the military more than doubled during this period. In 1955 the Department of Defense spent about $3.2 billion on electronic parts and equipment; in 1963 it spent about $7.6 billion.[69]

The microwave tube industry, which produced electronic components for many weapon systems, benefited handsomely from the surge in military spending. Military procurement of klystrons, magnetrons, and traveling-wave tubes grew from $58 million in 1954 to $146 million in 1962. And the composition of the military demand for microwave tubes changed radically during this period. In 1945, 88 percent of the microwave tubes bought by the Department of Defense had been magnetrons; in 1962, it was only 26 percent. During the same period, klystrons' market share increased from 22 percent to 38 percent. (Total sales of klystrons grew from $14 million in 1954 to $53 million in 1963.)

Traveling-wave tubes experienced the largest increase in sales, from less than 1 percent in 1954 to 36 percent by the early 1960s. The makeup of military procurements within each tube class also changed dramatically. During the Korean War, the Department of Defense had procured low-performance magnetrons and klystrons in large quantities; in the mid and late 1950s, it increasingly shifted its procurement dollars to the purchase of advanced tubes.[70] These dramatic shifts were due to a major transformation in military procurement in the mid 1950s. Whereas during the Korean War the Department of Defense had bought improved versions of World War II equipment in large quantities, it increasingly purchased advanced weapon systems in small quantities. In particular, the Department of Defense procured advanced aircraft and missile systems. Missile procurement increased from $604 million in 1954 to $3 billion in 1960. Similarly, building upon its research programs during the Korean War, the military shifted its procurement of radar systems from pulsed microwave radars in large volumes to sophisticated Doppler systems in shorter production runs. These new military systems required highly reliable high-performance microwave tubes in relatively small quantities. Missile radar systems, for example, relied on extremely rugged reflex klystrons. Doppler radar sets also required sophisticated klystrons.[71]

The rapidly growing demand for high-quality klystrons in the mid 1950s offered a major business opportunity to Varian Associates. It also led to heightened competition in the klystron field. RCA, Sperry Gyroscope, and GE enlarged their klystron operations, and Sylvania and Litton Industries began manufacturing klystrons. But most worrisome for Varian was that Bill Eitel and Jack McCullough sought to establish their firm as a major manufacturer of klystrons. To do so, they put a large share of their tetrode profits into the klystron business. They built a klystron laboratory and organized a klystron engineering group around Harold Sorg, the firm's patent lawyer and vice-president for research, and Don Preist, a British radio engineer whom Eitel and McCullough had met during World War II. To build this group, Sorg and Preist hired college-trained engineers who were practically oriented and had a background in amateur radio. Employing college-trained engineers was a major departure from past practice at Eitel-McCullough, whose technical staff had consisted almost exclusively of radio amateurs (Preist 1992).

Taking advantage of the technology transfers from Stanford and Varian, Eitel-McCullough engineers were soon competing with Varian

Associates. They got second-source contracts to produce Varian's reflex klystrons. They also entered the manufacture of high-power continuous-wave klystrons for UHF television, a field which Varian had pioneered in the late 1940s. To develop these tubes, they took a different approach from the one at Varian. Whereas Varian's engineers used internal cavities for their UHF klystrons, the Eitel-McCullough group developed power klystrons with external cavities—that is, cavities that were outside of the tube envelope. They chose this approach for several reasons. Because of their background in amateur radio and the firm's prior orientation toward power-grid tubes, Eitel-McCullough's engineers had been conditioned to view electronic circuits (the cavities in the case of klystrons) as external to the vacuum tubes. But external-cavity klystrons were also well suited for applications which required modest power at the lower frequencies. They also had the advantage of being cheaper and easier to make than internal-cavity tubes. In 1953, Eitel-McCullough introduced a family of three UHF power klystrons to the market. This led the Varian group to cease all collaborative activities with Eitel-McCullough.[72]

When it became clear that television manufacturers did not want to move into UHF transmission and that the TV market for UHF klystrons would be limited, Sorg and Preist created a new market for their tubes in tropospheric scatter communication. In the mid 1950s, the Department of Defense became interested in building a new global communication network that would use the radio phenomenon of tropospheric bounce. (The troposphere, the lowest portion of the atmosphere, reflects VHF signals.) Exploiting the military's interest in tropospheric communication for commercial advantage, Eitel-McCullough secured a contract from the Naval Research Laboratory to design a tropospheric communication transmitter in mid 1954. The firm asked one of its best tetrode customers, Radio Engineering Laboratories (REL), to build this transmitter, with the tacit understanding that REL would grant klystron subcontracts to Eitel-McCullough when it received large production contracts for tropospheric communication transmitters from the military. This shrewd move opened up a vast market for Eitel-McCullough. In the second half of the 1950s, the military gave REL a series of sole-source production contracts for tropospheric communication transmitters. REL granted Eitel-McCullough sole-source subcontracts for klystrons. As a result, Eitel-McCullough saw its sales of continuous-wave klystrons grow substantially. By the mid 1950s, Eitel-McCullough had become a significant manufacturer of microwave tubes.[73]

   To compete with Eitel-McCullough and at the same time take advantage of the growing military demand for high-quality tubes, Stearns, Ginzton, and the Varians set out to develop a wide range of advanced klystrons and manufacture them in quantity. To do so, they looked aggressively for military development contracts. Beginning in the late 1950s, Stearns hired retired generals to help him identify funding opportunities in the various military services. He also organized yearly trips to Washington for Varian engineers and managers so that they could present Varian's latest products and learn more about military requirements. As a result of this intelligence and the firm's growing reputation in the military services, Varian Associates secured a growing number of military R&D contracts. In 1954 the firm had received $2,283,000 in development contracts from the military services and from military system contractors; in 1962 it received $6,100,000. Lavish military patronage enabled Varian to enlarge its engineering division and to build the largest microwave tube engineering and research group in the United States. In 1955 Varian had employed about 70 engineers and physicists in its tube engineering division; by 1961 it employed 200. Many of these engineers were among the most creative and talented microwave tube technologists in the country.[74]

   Varian's technologists designed and fabricated more than 200 klystron designs from 1954 to 1962. Their goal, as Emmet Cameron, the new general manager for microwave tubes, put it in 1958, was to "supersede [the firm's current] products with improved, cheaper ones when competition on first product got tough."[75] A significant number of these tubes were classified; but other klystrons became standard items, listed in Varian's product catalog. Of the 110 tubes in the catalog in 1960, nearly half were rugged reflex klystrons, and the rest were high-power klystrons. (Varian also designed medium-power klystrons for use in airborne radars.) By 1958, practically every US missile used Varian's reflex klystrons.[76]

   Varian's engineering teams developed klystrons with unheard-of frequencies and power outputs. In addition, they significantly enlarged the bandwidth of these tubes. Because high-power klystrons were difficult to exhaust, Varian's engineers developed a new vacuum pump, the VacIon pump, which produced a much higher and cleaner vacuum than the oil pumps then on the market. The VacIon pump (invented in 1956 by Lewis Hall, a Varian engineer) sucked gases electronically and then entrapped the gas molecules through the formation of chemically stable compounds. It could also work as a vacuum gauge. It enabled Varian's engineers and mechanics to obtain very high vacuums and at the same time avoid contamination of the cathodes with oil compounds, which often occurred with oil pumps. In addition, they attached VacIon pumps permanently to large tubes. When the klystrons were heated and running, these pumps would remove gases released by their metallic elements. As a result, VacIon pumps significantly extended the life of big power tubes.[77]

These advances in tube design and vacuum pumps enabled the Varian group to take the lead in the development of superpower tubes for the Ballistic Missile Early Warning System (BMEWS). In the mid 1950s, the Department of Defense became interested in procuring powerful radar systems capable of detecting ballistic missiles coming from the Soviet Union. Because radar range was directly related to power output, these systems required klystrons with unprecedented power characteristics. (The rule of thumb was that doubling the range of radar systems required 16 times the initial power.) Eitel-McCullough, now Varian's arch-competitor, was the first firm to engineer klystrons for ballistic missile detection. In 1955, it received a contract from the Air Force to study the feasibility of very powerful UHF klystrons for BMEWS radars. This feasibility study enabled Eitel-McCullough to obtain a very large development contract for this tube. To meet the tube's specifications, Eitel-McCullough engineers scaled up their external-cavity design. But, they soon ran into significant technical difficulties. It became clear that the external-cavity approach was not suitable for the generation of high power. Only internal cavities would do the job, but Eitel-McCullough's engineers, who were practically oriented, did not have the theoretical skills required to design these complex tubes.[78]

Eitel-McCullough's technical difficulties created an opportunity for Varian. Because Eitel-McCullough was 18 months late in its engineering development, the Air Force became nervous and gave another development contract for the same tube to Varian Associates. The Air Force had an urgent need for the BMEWS klystron. The Soviet Union had recently launched Sputnik, the first artificial satellite. This technological feat made it clear that the Soviets had the capability of launching missiles that could carry nuclear weapons. It was imperative that the Air Force be able to detect these missiles before they would reach the United States. At the request of the military, Varian started a crash program to engineer a klystron for BMEWS in 1958. Within 6 months, Varian's engineering team had a working prototype. This tube was the world's largest klystron. It was 11 feet long and had four internal cavities. It could deliver 1.5 million watts of power in short pulses. Unlike Eitel-McCullough's klystron, it met the performance and reliability specifications of the Air Force. The BMEWS tube consolidated Varian's dominance in high-power klystrons for radar applications and enabled Varian to get development contracts for superpower klystrons in the early 1960s.[79]

Because of its unique strength in klystron engineering, Varian Associates captured a large share of all klystron production contracts in the late 1950s and the early 1960s. The Department of Defense and military system contractors asked Varian Associates to produce many of the tubes that Varian had engineered for them. Most of these contracts were awarded on a sole-source basis. The military services and their system contractors needed these tubes in a hurry and did not have the time to open these production contracts for bids and create other sources of supply. Myrl Stearns and his managers also made skillful use of the fact that these tubes had been developed at Varian to obtain these sole-source manufacturing contracts. They influenced the writing of the military specifications for the production versions of these klystrons. As a result, when the military asked other manufacturers to bid on these contracts, Varian almost invariably had the best proposal and the best price. Varian had defined the specs in such a way that it would be the best bidder for the tube. Obtaining sole-source contracts was critical. Because they involved no competition, sole-source contracts tended to be highly profitable. In the late 1950s, Varian Associates made profits of up to 60 percent on sole-source contracts for reflex and high-power klystrons. These profits were so high that in the early 1960s the General Accounting Office, the investigative arm of Congress, scrutinized Varian's pricing practices and examined its accounting books.[80]

To meet demand, Stearns, Ginzton, and the Varian brothers decided to scale up production. This required an infusion of capital. Varian could not finance the large sums required for processing and manufacturing equipment solely out of profits. Nor could it resort to employees and consultants, as it had done in the past. In other words, the company was growing too fast for its ownership structure. At the same time, Varian's ownership structure came under attack from its employees. Starting in 1954, some employee-owners became increasingly restive. They had invested their savings in Varian, but the company, which reinvested all its profits in the business, did not distribute dividends to its stockholders. This led a few junior executives to resign and sell their shares back to the company, thereby further worsening its cash crisis. Other stockholders wanted to sell their Varian shares on the open market. Around this time, Wall Street's banks and brokerage firms discovered the growth potential of the electronics industry—which led to a surge in the valuation of electronic securities. Many employee-owners were eager to benefit from this speculation and sell some of their Varian shares to outside investors.[81]

Because of these internal tensions and the company's dire need for fresh funds, Stearns, Ginzton, Cameron, and the Varian brothers, in a major policy shift, decided to open the firm to outside investors. In June 1955 they put the company's stock on the over-the-counter market. A few years later, they decided to list Varian Associates on the New York Stock Exchange. (Eitel-McCullough and Hewlett-Packard went public around the same time.) These moves enabled Varian's managers to raise the funds for their expansion program. In 1956 alone, Varian secured more than $3 million from the sale of stock and convertible debentures. It has been estimated that the firm raised more than $10 million in this way by the early 1960s. At the same time, because the price of Varian's shares went from $2 in 1955 to $68 in 1960, Varian's entrepreneurs and employee-owners made significant capital gains. By the time of his death in the summer of 1959, Russell Varian had accumulated a fortune of $1.8 million. When Sigurd Varian died, in 1961, his estate was valued at $800,000. Because of the Varians' unusual stock policies, the firm's consultants, engineers, and technicians, who owned a very large share of the company's stock, grew rich in the second half of the 1950s. Some technicians and mechanics who had invested in Varian's stock since the early 1950s became wealthy and retired.[82]

The decision to go public altered Varian's character. Employees had owned 80 percent of the corporation in June 1955; by 1958, they owned only 40 percent. In 1956, to maintain the firm's cooperative orientation, the founders established an "employee stock purchase plan" that subsidized the purchase of stock by employees. This measure helped keep the percentage of employees who were shareholders around 80. Management also resorted to more traditional ways of achieving worker participation. In 1957, for example, a profit-sharing program was instituted, and substantial portions of the company's profits were distributed to employees. Health and retirement benefits were increased. By the late 1950s, Varian Associates was more traditional in its outlook and in its ownership structure, and it had resorted to participatory forms of management reminiscent of Litton Engineering, Eitel-McCullough, and Hewlett-Packard.[83]

Stearns, Ginzton, Russell Varian, and Cameron used the funds raised on the financial markets to scale up production. The firm hired and trained more than 800 machinists, operators, and technicians between

1955 and 1962. It also enlarged its inventory of tools. In 1956 alone, Varian bought more machine tools and specialized tube processing machinery than during its previous 8 years of operation. The company also built a new state-of-the-art factory on the Stanford campus in 1957. By this time, Varian had two different production facilities. In the job shop, klystrons were custom-made by teams of highly skilled and versatile technicians. In the factory, klystrons were manufactured in larger volumes, ranging from twenty to several thousand. (Unlike Litton Industries and most East Coast firms, Varian never produced more than a few thousand units of the same tube type.) Varian also subcontracted a significant share of its machining jobs to a local machine shop. This production setup improved Varian's flexibility and enabled it to fill the diverse requirements of military customers.[84]

With its efficient and flexible manufacturing organization and its unique competence in klystron engineering, Varian Associates gained the lion's share of the expanding klystron business and became the largest microwave tube firm in the United States in the late 1950s. Its sales of klystrons and other velocity-modulated tubes surged from $5.5 million in 1954 to $36.8 million in 1962. Its sales of power klystrons alone grew from $3.3 million in 1959 to $14.7 million in 1962. During the same period, employment in Varian's tube group in Palo Alto grew from 545 to 1,700. By the early 1960s, Varian controlled roughly half of the klystron market in the United States. Along with Bomac, a Massachusetts-based manufacturer of klystrons and magnetrons which it acquired in 1959, Varian had about 30 percent of the market for microwave tubes. Its microwave tube business also made significant profits. By 1960, net profits were close to 10 percent of sales.[85]

Varian's rapid growth relegated long-established firms in the microwave tube business, including RCA, Sperry Gyroscope, GE, and Raytheon, to secondary status. Raytheon, which had been the largest manufacturer of klystrons and magnetrons in the early 1950s, saw its sales plummet from $38.8 million in 1953 to $19.6 million in 1958. Even the other microwave tube firms on the San Francisco Peninsula did not do as well as Varian. For example, Eitel-McCullough established itself as the industry leader for television and tropospheric communication klystrons and grew its microwave tube sales to $13.6 million in 1960, but it lost money in microwave tubes every year in the second half of the 1950s and the early 1960s. (For much of this period, only its power-grid tube business kept Eitel-McCullough profitable.) Varian Associates not only emerged as the leading microwave tube manufacturer in the United States; it also expanded

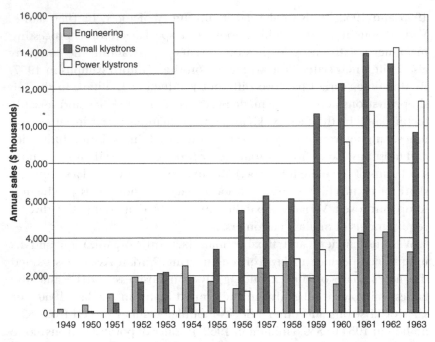

Figure 3.6
Varian's klystron sales, 1948–1963. Source: "Final Performance Analysis of Fiscal Year 1963," folder: Varian studies, box 5, Varian Associates Records, MSS 75/65c, Bancroft Library, University of California, Berkeley.

into foreign markets. In 1955 it established a subsidiary in Canada, and in 1962 it gained a strong presence in Europe through a joint venture with the French electronics manufacturer Thomson-Houston.[86]

Conclusion

"We are different because the tubes we make are different," reflected one of Varian's directors in the early 1950s. "They are newly developed types, and they are very difficult to manufacture. Varian tubes do things few other tubes can do and things which are needed in advanced electronic systems."[87] Because of the imperatives of the specialized tube business and the founders' unusual ideological commitments, Varian evolved into an unusual institution characterized by employee ownership, a flexible organizational structure, and a culture that fostered innovation and experimentation. Two reasons for Varian's swift rise in the microwave tube business were its focus on advanced klystrons and its organization around flexible development and manufacturing teams. This organizational structure enabled the firm to respond rapidly to the changing needs of military customers and to produce advanced tubes in small production runs. As a result, Varian was well positioned to take advantage of the Department of Defense's growing procurement of high-quality klystrons in small volumes in the second half of the 1950s.

The firm's management also made important social innovations, notably the ways in which it related to employees. Varian Associates was the only employee-owned company in the microwave tube industry. It also developed innovative profit-sharing programs and offered excellent benefits to employees. This helped bring forth the creative abilities of its engineering and manufacturing workforce and improve its productivity. "Everybody [at Varian]," an engineer recalled, "was a stockholder, or practically everybody. Engineering people had enough stock so that having the company make money was to their personal profit as well as a matter of pride."[88] These traits helped Varian to expand rapidly and to partially displace East Coast microwave tube manufacturers during the Cold War. Thus, in less than a decade a company that had started as a cooperative engineering laboratory became the largest manufacturer of microwave tubes.

Varian's growth and the rapid expansion of Eitel-McCullough, Litton Industries, Watkins-Johnson, and other microwave tube operations (including those started by Sylvania and GE) made the San Francisco Peninsula the center of the American microwave tube industry, with 6,600 engineers, technicians, and operators employed there by 1961. The rapidly expanding microwave tube business contributed to the building of a solid industrial infrastructure on the Peninsula. Vacuum tube firms trained thousands of operators and technicians in glass blowing, vacuum processing, and chemical handling. They also generated a significant number of precision machine shops and other specialized industrial establishments that were independent of Varian, Litton, and Eitel-McCullough. Many of these machine shops were established by former Varian and Eitel-McCullough employees. Others were started by precision machinists who had left the East Coast to take advantage of the new business opportunities in Northern California. In addition, Varian, Litton, and Eitel-McCullough spun off ceramics shops and small firms that made vacuum pumps and vacuum components. The microwave tube firms also attracted vendors of specialty chemicals, which set up facilities in the area to serve the needs of tube manufacturers. Perhaps more important, the microwave tube firms offered models of how to build electronic companies in the West. Varian, Litton, and Eitel-McCullough showed other electronics startups how to compete with large East Coast firms—mostly through innovative manufacturing and through the development of high-quality products. Varian Associates also showed the importance of treating one's workforce well. Varian's example persuaded other electronic corporations in the area that giving "a piece of the action" to employees, especially engineers, was a source of competitive advantage. Stock ownership aligned the interests of the engineers with those of the founders and gave the engineers a strong incentive to innovate and to introduce new products rapidly to the market.[89]

Thus, in the late 1950s and the early 1960s the microwave tube corporations facilitated the growth of another industry: semiconductor manufacturing. Semiconductor firms established in the second half of the 1950s, such as Shockley, Fairchild, and Rheem, made use of the area's precision machining facilities and its glass and ceramic shops. They also took advantage of the district's vacuum equipment businesses. "It was vacuum heaven around here," Julius Blank, a Fairchild Semiconductor founder, later recalled. "So there was an infrastructure that built heavy duty vacuum fittings and flanges, and things like that. We used to buy these and build our equipment with them."[90] Semiconductor firms also hired skilled technicians from the tube industry. These workers brought with them a knowledge of chemical handling, glass working, vacuum techniques, and the fabrication of processing equipment. Though few microwave tube design engineers converted to semiconductor technology, the tube firms provided process engineers to the new industry. For example, much to Eitel's and McCullough's chagrin, a large group of chemical engineers left Eitel-McCullough to join Rheem Semiconductor in 1960. In addition, some of the semiconductor startups, especially Fairchild, consulted with Varian Associates and other makers of microwave tubes to learn how to manage and mobilize a highly skilled workforce. In short, the vacuum tube industry and the ancillary businesses that grew from it helped Fairchild, Rheem, and other semiconductor firms reduce their startup costs and their business risks. They also helped these semiconductor corporations bring new products to market at record speed, which in turn gave these firms a substantial advantage over competitors based in less favorable locations.[91]

4

Revolution in Silicon

The semiconductor industry in Santa Clara Valley grew around Fairchild Semiconductor. Fairchild Semiconductor was not the first corporation to make semiconductor components in the area. Hewlett-Packard fabricated silicon diodes and Shockley Semiconductor worked on transistors before Fairchild. Nor was it the firm that did the initial work in silicon electronics. (The main design techniques and fabrication processes were pioneered at the Bell Telephone Laboratories, and silicon devices were first commercialized by Texas Instruments and Hughes Semiconductor.) But Fairchild was the source of most semiconductor corporations on the San Francisco Peninsula. It helped define the new industry in the late 1950s and the very early 1960s. Fairchild delineated the semiconductor industry's main silicon products and processes. It developed innovative sales and marketing practices as well as new organizational structures that were later widely adopted by other semiconductor firms. Finally, Fairchild pioneered the quantity production of advanced silicon components. How did Fairchild Semiconductor revolutionize silicon electronics? What role did military patronage and procurement play in the shaping of silicon technology?

Most accounts of Fairchild's rise present it as the creation of the "visionary" technologist-turned-businessman Robert Noyce. These narratives portray Noyce, the co-inventor of the integrated circuit and the firm's second general manager, as the main force behind the establishment of Fairchild. Noyce, rather than Fairchild's founding group, is also often seen as responsible for the firm's spectacular success.[1] The role of the Department of Defense in the shaping of semiconductor technology and industry has been much debated. Some scholars have argued that the military had little influence on the evolution of silicon and germanium technology in the 1950s and the early 1960s (DeGrasse 1984; Asher and Strom 1977; Seidenberg 1997; Ceruzzi 1998). The development of the planar process at Fairchild happened without government support, as did other innovations. Instead, these authors have viewed the evolution of semiconductor design and processing as driven mostly by internal forces. As a result, the military appears in their accounts as a selecting mechanism for industrial success rather than as a primary driver of technological change. Others have claimed that the Department of Defense shaped the process of technical change in semiconductors through its procurement policies and the financing of manufacturing facilities. They argued that military demand for reliable and compact electronics equipment led to major semiconductor innovations such as the silicon transistor and the integrated circuit and that large-scale government procurement accelerated the development of these innovations (Golding 1971; Levin 1982; Misa 1985; Holbrook 1995, 1999).

Newly available historical materials concerning Fairchild Semiconductor show that the shaping of silicon technology and the related industry on the San Francisco Peninsula was a group effort rather than the creation of "heroic" individuals such as Noyce. It was also closely coupled with military procurement and the establishment of reliability and performance standards by the Department of Defense. A group of highly creative scientists and technologists (including Noyce) established Fairchild Semiconductor in alliance with a New York investment bank and a medium-size military contractor on the East Coast. Fairchild's founders reshaped the semiconductor industry's products and manufacturing methods by adopting the diffusion process recently developed at the Bell Telephone Laboratories. As a result, Fairchild was the first commercial firm to introduce high-frequency silicon transistors to the market. These transistors met a rapidly growing demand for fast components in the emerging field of digital-based guidance and control systems for jet aircraft and missiles.

However, at first Fairchild's transistors did not meet the reliability criteria of the Minuteman program. Adapting their technologies to meet these reliability standards, Fairchild's founding engineers made a second round of major innovations. One was the planar process and another the integrated circuit. Fairchild's engineers also perfected their production systems and tightened their control of the manufacturing process with the help and under the scrutiny of the Minuteman program. This enabled Fairchild to gain a large share of the military market for high-quality silicon components. The firm emerged as the leading manufacturer of advanced silicon components and the second-largest maker of silicon devices after Texas Instruments in the late 1950s and the very early 1960s.

Rebellion

Fairchild Semiconductor was established in Palo Alto in October 1957 by a group of physicists and engineers from Shockley Semiconductor. Sheldon Roberts, Eugene Kleiner, Jean Hoerni, Gordon Moore, Jay Last, Victor Grinich, Julius Blank, and Robert Noyce were creative and independent-minded scientists with impeccable professional credentials. Five had PhDs in the physical sciences and had done noted research work in spectroscopy, metallurgy, and solid-state physics. Noyce, a solid-state physicist with a doctorate from MIT, was the only one with a strong semiconductor background. He had worked on semiconductors at Philco, where he had developed jet etching, a new process for making high-frequency germanium transistors. Last, trained as an optical engineer, had done graduate work in solid-state physics at MIT and had produced a dissertation on ferroelectric materials. Like Noyce and Last, Roberts, a metallurgist, was an MIT PhD. Moore, a physical chemist from Caltech, had worked on the spectroscopy of flames for the missile program at the Applied Physics Laboratory at Johns Hopkins. Hoerni was a theoretical physicist with two PhDs, one from Cambridge and one from the University of Geneva.[2]

The group also included three engineers. Kleiner, an Austrian émigré, had studied mechanical and industrial engineering before designing cigar-making machinery at the American Shoe Foundry and making relays at Western Electric. Blank, a mechanical engineer by training, had held a variety of technical and engineering positions at Babcock & Wilcox and at Western Electric, where he produced power machinery and telephone exchanges. Grinich, the group's only electrical engineer, had received a doctorate in circuit theory from Stanford University. At the Stanford Research Institute, he designed transistor circuits for color television and for ERMA (the first computing machine for banking applications).[3]

This unusual group had joined Shockley Semiconductor in the spring of 1956. Although small and removed from major scientific and technological centers, Shockley Semiconductor represented an interesting opportunity for young and ambitious scientists with an interest in the promising semiconductor field. William Shockley, the firm's founder, had considerable scientific stature. In 1948 he had developed the junction transistor, an invention for which he later received the Nobel Prize in physics. Shockley also had directed much of the Bell Telephone Laboratories' extensive solid-state physics activities and had helped make Bell Labs preeminent in US semiconductor research. Besides the

Figure 4.1
Fairchild Semiconductor's founders in the transistor plant in Mountain View,
circa 1960. Standing, left to right: Victor Grinich, Gordon Moore, Julius Blank,
Eugene Kleiner. Center group: Jean Hoerni (standing), Jay Last (sitting on the
left), Sheldon Roberts (sitting on the right). Addressing the group: Robert Noyce.
Courtesy of Fairchild Semiconductor.

opportunity of associating themselves with a man of Shockley's bril-
liance, Shockley Semiconductor offered work in the intellectually
promising field of solid-state electronics.[4]

The eight Shockley recruits were also attracted to the Bay Area by its
beauty and its proximity to the Sierra Nevada Mountains. Shockley had
established his firm in Mountain View for a number of reasons. Frederick
Terman, Stanford's provost, who was interested in attracting new
electronics-based businesses to the area, had encouraged Shockley to
locate his new firm near the university. Shockley was a native of Palo Alto,
and his mother still lived in the area. Though Arnold Beckman of
Beckman Instruments, who financed Shockley's venture, would have preferred to locate it in Southern California, he did not strongly object to
Shockley's decision. Beckman had a subsidiary, Spinco, that designed and
produced ultra-centrifuges in the recently created Stanford Industrial
Park. Spinco would provide Shockley with management expertise.[5]

The new firm was to develop and manufacture advanced transistors
and other components based on two technologies in which Shockley saw

much industrial potential: silicon and solid-state diffusion. Silicon, although less used than germanium, was considerably cheaper and allowed the fabrication of more durable and temperature-resistant devices. Solid-state diffusion, recently developed at the Bell Labs, promised to revolutionize the manufacture of silicon components. Up until that time, silicon devices had been fabricated by two different techniques. In the grown junction process, dopant atoms were introduced into the silicon crystal at the time of its growth to form transistor junctions. In the alloy junction process, a pellet of indium was alloyed to a thin slice of silicon crystal. Although these two processes allowed the fabrication of operational transistors and diodes, they did not allow much precision in the formation of the devices' junctions. As a result, these processes did not enable the manufacture of high-frequency components. Solid-state diffusion offered a much more precise and controllable way of forming transistor and diode junctions. In the diffusion process, the dopants, which controlled the semiconductors' electrical characteristics, were introduced onto the surface of the semiconductor. As the crystal was heated, the dopant atoms diffused into the silicon slice in a manner that could be tightly controlled as a function of time and temperature. In other words, the solid-state diffusion process allowed the formation of much thinner base regions and therefore made possible the manufacture of faster and higher-frequency devices.[6]

Shockley set out to develop double diffused silicon transistors. "It was pretty much research on developing the basic technology," Moore later recalled, "trying to repeat some of the things that had been done [at the Bell Labs], understand some of the problems that we still had because neither the processing nor the physics of [silicon] was well understood. We were just exploring the technology and figuring out what could be done and we had a lot of things to make work before we could try to build something. We were far from developing a commercial device."[7]

Besides fundamental studies on currents in silicon crystals, much of the group's work was on diffusion phenomena. They calculated diffusion curves for dopants such as gallium, boron, and phosphorus. They also tested these curves experimentally, using different concentrations, temperatures, and times to see whether the theoretical formulas were met empirically. Diffusion curves, plotting dopant concentration against junction depth under certain time and temperature conditions, helped them calculate the diffusion rate and thereby adjust the heating time and furnace temperature for obtaining the desired junction depths. The group also experimented with mesa transistor structures, which had recently been developed at Bell Labs. In the mesa process, two layers of dopants were diffused beneath the surface of a silicon slice or wafer. A patch of wax was then applied to the top of the wafer. The final step was to etch away some of the upper surface, except under the patch. This left a small platform—a "mesa"—over a wider bottom layer. Adopting the Bell Labs process, the Shockley recruits built and tested experimental mesa structures.[8]

Working with Shockley did not prove easy. In less than a year, seven of his recruits, Roberts, Last, Hoerni, Moore, Grinich, Kleiner, and Blank rebelled against him. (Noyce was the exception.) They asked Beckman, Shockley Semiconductor's backer, to remove Shockley from day-to-day management. They wanted to replace him with "a professional manager to whom [they] could report."[9] The group's rebellion was motivated by Shockley's heavy-handed management actions as well as by sharp disagreements about the firm's direction. While the rebels admired his physical intuition and command of solid-state physics, they were increasingly unhappy with his mercurial temperament and his poor treatment of employees. Shockley not only routinely threatened employees with immediate dismissal; he also staged public firings and called for lie-detector tests to be given for trifling matters.[10]

Moreover, Shockley supervised their projects closely, which left them with little creative liberty and no real sense of pride in their work. The insurgents were further disconcerted by the apparent lack of focus. Wavering between engineering commercial devices and doing advanced research in solid-state physics, Shockley regularly shifted his staff back and forth from research projects to product development. Moreover, Shockley constantly promoted new ideas and directed his staff to abandon the projects they were working on to turn to new ones.[11]

The group also objected strongly to Shockley's decision to shift the firm's focus from transistors to PNPN diodes, specialized devices recently invented by Shockley for telephone switching applications. Because of the tight tolerances PNPN diodes required, they could not be manufactured reliably and reproducibly. As a result, they offered rather poor prospects as a first product. Instead, the rebels urged that the lab return to its original goal of manufacturing double diffused silicon transistors. No corporation, they argued, had yet been able to produce and commercialize these devices. Double diffused transistors, the insurgents further claimed, would outperform Texas Instruments' most advanced silicon transistor, a combination of a grown junction for the base and a diffusion for the emitter. As a result, the double-diffused transistors would meet the high speed requirements of digital circuits, an application which members of the group had been sensitized to through their research projects at MIT, Philco, and the Stanford Research Institute.[12]

When the dissidents failed to convince Beckman to remove Shockley and redirect the lab toward the production of diffused silicon transistors, they found themselves in an uncomfortable position. Unwilling to stay at Shockley Semiconductor, they could not easily find other jobs in the area. The local tube and instrumentation firms had little demand for people with their skills. Intent upon staying in the area, keeping the group together, and pursuing their work on diffused silicon transistors, the rebels contacted Hayden Stone & Company, a small investment bank in New York with which Kleiner's father had an account. In a bold move, the rebels asked the bank to help them find a corporation interested in hiring them collectively and in setting up a silicon operation on the San Francisco Peninsula.[13] "Because of seemingly insuperable problems with the present management," the Shockley insurgents wrote to Hayden Stone in June 1957,

this group wishes to find a corporation interested in getting into the advanced semiconductor device business. If such suitable backing can be obtained the present group can reasonably expect to take with them other senior people and an excellent supporting staff totaling about thirty people. Thus a backer has the opportunity to obtain at one time a well trained technical group by supplying enlightened administration and support. It is the aim of the group to negotiate with a company which can supply good management. We believe that we could get a company into the semiconductor business within three months which would represent a considerable saving in cost and time. The initial product will be a line of silicon diffused transistors of unusual design applicable to the production of both high-frequency and high-power devices. It should be pointed out that the complicated techniques necessary for producing these semiconductors have already been worked out in detail by this group of people, and are not restricted by any obligation to the present organization. We [also] have an excellent supporting staff. Because of this experienced staff, and also because of the group's own attachment to the lower San Francisco Peninsula area, we would want to establish the operation here south of San Francisco. It is estimated that the establishment of this new enterprise and its efficient operation in the first year will require an expenditure in the neighborhood of $750,000.[14]

Sent to the clerk in charge of the elder Kleiner's account, the letter soon attracted the attention of a young security analyst, Arthur Rock, and a managing partner, Alfred Coyle, who had a keen interest in science-based industries. While Hayden Stone had been traditionally oriented toward established industrial sectors, the bank, under Coyle's leadership, had recently expanded into electronics. In particular, Rock and Coyle had arranged the public financing of General Transistor, a manufacturer of germanium transistors, and seen a small investment in the firm's securities multiply over a short period of time. These men had also followed the emergence of the venture capital industry and in particular firms such as J. H. Whitney & Co. and American Research and Development, which had provided financing to new chemical and instrumentation firms since the late 1940s. Inspired by these examples, Rock and Coyle were interested in developing new types of financial services for such firms and, in particular, in helping new electronics firms secure early rounds of financing from established corporations. While riskier than the underwriting of securities, these activities might bring in large financial returns.[15]

Rock and Coyle flew to California in July 1957. They were impressed by the defectors' intellectual abilities and their capacity to work as a group and were aware of the potential of the semiconductor business. The Hayden Stone representatives made them an unusual proposition. At a time when few scientists and engineers started new business enterprises, Rock and Coyle suggested that the group establish its own corporation rather than look for collective employment. Furthermore, they offered to secure capital for the new venture among corporate backers. In return, they asked for a small interest in the new company. Their proposal was a startling one for the group. Socialized in academic science rather than entrepreneurship, they had never thought of establishing their own firm and assumed that they would always work in corporations. The group, however, found Rock's proposal appealing for personal and professional reasons. They saw the formation of a new firm as a way of staying in California, controlling their own technical work, and, as Last put it, "being their own boss."[16]

Acting as the group's agent, Rock approached more than thirty potential corporate backers, including many East Coast electronics firms and large aerospace corporations such as IT&T, Sperry Rand, and North American Aviation. He also contacted Litton Industries and Eitel-McCullough. Rock encountered considerable difficulties in raising the needed capital. Some firms approached by Rock were taken aback by the risks involved in supporting a group with no product and little management experience. Others had already started their own research programs in silicon and did not feel a need for the defectors' expertise. Most corporations deemed the project impractical if not disruptive. Because they had never financed a company outside of their own business, many managers did not see how they could structure such a deal. They also worried about the effect on their own employees. They feared in particular that it would lead them to set up their own businesses and compete with their former employer.[17]

Only Fairchild Camera and Instrument, a medium-size military contractor based on Long Island, expressed an active interest in the idea. Lacking internal silicon expertise, it was willing to consider unorthodox proposals. Established in the early 1920s by Sherman Fairchild (the wealthy son of an IBM executive, and then IBM's largest stockholder) to manufacture aerial cameras, the firm had expanded in the immediate postwar period into other military businesses such as high-precision potentiometers for analog computers and military avionics equipment. After undergoing rapid growth during the Korean War, Fairchild Camera had experienced substantial setbacks in the mid 1950s. It had suffered from severe cutbacks in the military procurement of aerial cameras and incurred heavy losses on military development contracts. As a result, Fairchild Camera had seen its sales decline from $42 million in 1954 to $36 million in 1956. Its earnings dwindled from $1.6 million to $260,000 during the same period.[18]

To reverse the sagging fortunes of Fairchild Camera and Instrument, Sherman Fairchild redirected the company toward electronics and especially toward technologies for gathering, transmitting, and storing data. He hired a group of young managers such as John Carter from Corning Glass. He also hired Richard Hodgson, a Stanford electrical engineer who had headed a small West Coast company that made television tubes. The Fairchild firm then acquired a number of system businesses, including the teletypesetter division of AT&T and a manufacturer of magnetic tape transports for digital computing. These men were also looking for opportunities in the transistor business, but were taken aback, as they later recalled, by "[the scarcity of] qualified personnel, the very large capital investment [required], and the prospect of years of research before production might be feasible."[19] Hence their interest in the Shockley rebels. The defectors offered what Fairchild Camera was looking for. They were acquainted with the latest silicon techniques, they had nearly mastered the manufacturing processes, and they were reasonably close to production. They provided a fast and relatively cheap entry into the semiconductor industry.[20]

Exploiting Fairchild Camera's keen interest in the rebels' proposal, Hayden Stone negotiated, on their behalf, one of the first venture capital agreements on the West Coast. Fairchild Camera financed the establishment of a new firm, Fairchild Semiconductor Corporation, with a loan of $1.38 million for its first year and a half of operation. The new firm was jointly owned by Hayden Stone, the seven rebels, and Noyce, the shrewd and charismatic assistant director of research at Shockley, who joined the group as its technical leader.[21] Hayden Stone owned roughly one-fifth of Fairchild Semiconductor; the remaining shares were distributed equally among the eight entrepreneurs. Fairchild Camera controlled the firm's board of directors and, in conjunction with Hayden Stone and the California group, had the right to choose its general manager.[22] The contract further specified that, in case the new firm became successful and met certain profitability requirements, Fairchild Camera had the option of acquiring it for $3 million after 2 years or $5 million after 8 years. The group signed the agreement with Fairchild Camera in September 1957.[23]

Around the same time, the entrepreneurs approached James Gibbons, one of their co-workers at Shockley Semiconductor, and asked him to join the group as the company's ninth founder. Gibbons, a junior faculty in the electrical engineering department at Stanford, worked at Shockley Semiconductor on a part-time basis. Frederick Terman, Stanford's provost, and John Linvill, the head of the Solid-State Laboratory, had recently apprenticed Gibbons to William Shockley. They had asked Gibbons to learn the techniques required for the fabrication of silicon devices from Shockley and then transfer these techniques back to the university. This was not the first time that Terman had sought to appropriate process technologies from local firms. Twenty years earlier, he had asked Karl Spangenberg, a young instructor in the electrical engineering department, to learn all he could from Charles Litton about making vacuum tubes—a move which enabled Stanford to emerge as an important player in microwave tubes after World War II. At Shockley Semiconductor, Gibbons learned silicon processing from Noyce, Moore, Kleiner, and their technicians. He also developed applications for Shockley's PNPN diodes and made calls on customers. When Fairchild's founders approached Gibbons regarding joining their startup, Gibbons, who was interested in an academic career, declined their offer. Over the next few years, Gibbons reproduced Shockley's laboratory on campus. As a result, Stanford was probably the first university to have the capability of making silicon diodes and transistors. This processing expertise enabled Gibbons and Linvill to build a large teaching and research program in solid-state electronics at Stanford in the late 1950s and the first half of the 1960s.[24]

Fairchild Semiconductor, the eight founders and their backers agreed, would make high-performance silicon transistors for the military sector. Although the group had done no formal market research prior to the firm's establishment, they sensed that the military sector offered by far the most promising market. Interested in the high-temperature operational characteristics of silicon components, the Department of Defense was the largest user of silicon transistors and diodes by the mid 1950s and had expressed a strong interest in the application of solid-state diffusion to the manufacture of silicon transistors. The Signal Corps, for instance, had given large engineering development contracts to firms such as Hughes and Pacific Semiconductors to develop and produce diffused silicon transistors in 1956. Finally, only the military and large weapon system contractors, Fairchild's founders reasoned, would have the necessary financial resources to buy the complex and expensive products.[25]

Besides its orientation toward the military market, Fairchild Semiconductor, the entrepreneurs decided, would be a manufacturing organization. Unlike Litton Industries and Varian Associates, which had started as contract engineering firms, Fairchild was to concentrate on the profitable high-volume production of silicon transistors. Indeed, the founders had no interest in military research contracts—for a number of reasons. Benefiting from generous private financing, Fairchild Semiconductor did not need military patronage to finance research and product development. Furthermore, the entrepreneurs deemed military research contracts detrimental because they would give the Department of Defense control of the firm's research program and product line, leading it in directions of direct interest to the military but of little industrial potential. Finally, because of their one- to three-year duration, military research contracts would restrict the firm's ability to adjust rapidly to new technical and market opportunities in a fast-evolving industry.[26]

Bringing a New Product to Market

The eight entrepreneurs set up shop in Palo Alto in October 1957. Besides the establishment of suitable facilities and the building of specialized equipment such as crystal growers, vacuum evaporators, and diffusion furnaces, their first task was to build a strong technical and management team. To do so, the group hired former co-workers from Shockley's lab such as David Allison, a solid-state diffusion expert. The founders also recruited local electronics technicians. Those who had worked in the Peninsula's tube industries brought with them knowledge of chemical handling, glass working, and vacuum techniques. Under Hodgson's guidance, the founders also recruited managers from component firms in Southern California. They appointed Thomas Bay, a bright and easy-talking salesman who had worked as marketing manager at Fairchild Camera's potentiometer division in Los Angeles, as the head of sales and marketing. Edward Baldwin, a Hughes executive, was recruited as the firm's general manager in February 1958 by giving him a share of the new company. Baldwin was an important recruit. A physicist by training, he had directed product engineering at Hughes Semiconductor, one of the largest silicon diode manufacturers in the United States. Baldwin brought with him a large contingent of manufacturing and instrumentation engineers from Hughes. Bay also hired some of Hughes Semiconductor's best and most aggressive salesmen.[27]

The new team proceeded to identify potential users for the firm's products before developing their first transistors. To determine the component requirements of military system manufacturers, Noyce and Bay, benefiting from Sherman Fairchild's close contacts with IBM, visited its Federal Systems division. This division specialized in the design and manufacture of advanced military computers. Noyce and Bay also toured large military contractors on the East Coast and Southern California such as Hughes, TRW, Arma, and Sperry Gyroscope. While conferring with design engineers and project managers at these firms, they discovered that there was an emerging market for double diffused silicon transistors in military avionics, especially in new digital-based guidance and flight-control systems.[28]

Starting in 1956, the Air Force championed the digitalization of avionics equipment to bolster the reliability and capabilities of its weapon systems. Until that time, aircraft and missiles had been controlled by analog techniques, especially analog computers. These computers solved numerical problems by setting up equivalent electric circuits or mechanical systems—to calculate the position of airplanes, control their mechanical subsystems, and determine their best flight trajectory for weapon delivery. Analog-based avionics systems, however, depended on failure-prone vacuum tubes and a multitude of moving parts that were sensitive to vibration and wear and tear. As a result, analog autopilots, bombsights, and navigation instruments failed on average every 70 hours, which severely impaired the military's operational readiness and led to enormous repair and maintenance costs for the Department of Defense. Furthermore, analog computers offered approximate and often inaccurate results and were therefore unfit for use in guidance systems of intercontinental ballistic missiles, which required the utmost precision. Analog computers had the further disadvantage of lacking flexibility, as they had to be torn apart and reconfigured to solve new problems.[29]

To improve the reliability and capability of jet airplanes and ballistic missiles, the Air Force encouraged avionics firms to digitalize their flight-control and navigation systems and to incorporate digital computers into their equipment. Digital computers, the Air Force reasoned, had a number of advantages. They were more accurate and could calculate the trajectory of ballistic missiles more precisely. Digital computers could operate at speeds far outstripping those of mechanical and electro-mechanical machines and could therefore control complex weapon systems in real time. They were general purpose machines and could be used for a variety of functions: the calculation of the aircraft's position, the determination of the best flight trajectory for bomb delivery, and the monitoring of a plane's or a missile's various subsystems. Furthermore they promised to be more dependable than analog machines.[30]

As Sperry, Arma, Hughes Aircraft, and other manufacturers of navigation and flight-control systems gradually shifted from analog to digital techniques, they started building high-speed digital computers. As a result, a market for high-performance digital components emerged in the military avionics industry. Going further, the Air Force insisted that avionics firms employ silicon transistors as much as possible. Vacuum tubes and germanium components did not meet the reliability and miniaturization requirements of the military. Vacuum tubes were fragile and had short lifetimes. Germanium transistors failed when exposed to high temperatures and, as a result, required heavy and bulky air conditioning equipment. Only silicon components withstood the high temperatures of airborne systems. Because they required no air conditioning, silicon devices also lent themselves to the design of miniaturized avionics equipment—an important characteristic for aerospace systems where space and weight were at a premium.[31]

Within this emerging market for high-frequency silicon components in military avionics, Bay and Fairchild's founders identified a pressing need for switching transistors in computer memories. The engineers at IBM's Federal System Division were designing a prototype navigational computer for the B-70 aircraft and urgently needed a transistor that could drive the machine's core memory. No germanium transistor could fulfill the application's temperature specifications. Similarly, none of the silicon transistors then on the market met the designers' performance requirements. The devices had to operate at high switching speeds and drive large currents in order to store data in the core memory. IBM needed a device faster and more powerful than any silicon transistor then on the market. Such a device, Fairchild's founders realized, would fill "a vacant area in transistors"[32] and find a ready market in military airborne computers.[33]

The Fairchild group decided to develop "core driver" transistors that would meet IBM's engineering specifications and negotiated the terms of a purchase order with IBM's Federal System Division. Because the engineers who designed the navigational computer had concerns about the firm's production capability and financial soundness, the Fairchild group enlisted the help of Richard Hodgson and Sherman Fairchild. These men went to see Thomas Watson Jr., IBM's chief executive officer, to promote Fairchild Semiconductor's offer and to "convince him that [buying transistors from the new firm] was a safe thing to do."[34] In February 1958, as a result of this timely visit, Fairchild received a purchase order for 100 mesa transistors at the hefty price of $150 each. Although the device's exact configuration (NPN or PNP) was left for the group to decide, IBM's engineers carefully specified the transistor's electrical parameters. Furthermore, they impressed upon the Fairchild group the importance of supplying reliable components. They asked that Fairchild "cut out all random catastrophic failures" in its production lots, use high-quality packages for its transistors, and apply special procedures to test them.[35]

To engineer these complex transistors, the Fairchild eight organized themselves in a loosely integrated fashion reminiscent of Varian's team approach to klystron engineering. They could draw on their varied skills to be responsible for their part of development of the first transistor. Moore (assisted by Allison) worked on diffusion, metallization, and assembly techniques. Roberts was responsible for crystal growing. Last and Noyce developed photolithographic processes, which had been recently developed at the Bell Labs and the Diamond Ordnance Fuse Laboratory to control the lateral dimensions of diffused regions. Last also developed the wax deposition techniques to delineate the etched mesa dimensions, worked on crystal surface preparation and on assembly problems. Hoerni worked on solid-state diffusion. Grinich, with a greater appreciation of the operating requirements needed for transistors than other members of the group, provided needed input for device design. He also developed transistor test equipment. (The founders had asked Hewlett-Packard to design and make a transistor tester for them, but the quote was so high that they decided to develop their tester in house.) Kleiner designed the equipment needed for device development and eventual production and outsourced its construction to local machine shops. Blank, in addition to working on equipment design, installed the facilities, such as gas and electrical distribution, and clean rooms, to allow the development work to proceed rapidly. After the diffusion furnaces were installed and in operation, two transistor configurations were developed in parallel. The main effort, headed by Moore, worked on the NPN version of the IBM core driver. Hoerni engineered its PNP counterpart.[36]

Developing these radically new transistor products presented substantial difficulties. The group could rely on sophisticated physical theories developed at Bell Labs for the design of mesa transistors, but the production techniques required to make them were fraught with difficulties. To develop a process applicable on a production scale, the Fairchild founders made risky and innovative process choices, which contrasted sharply with those of Western Electric, Bell's manufacturing arm. While Western Electric employed proven techniques such as metal masking and gallium diffusion, the Fairchild group chose to exploit advanced processes that had been developed at Bell Labs. Specifically the group decided to use oxide masking and photolithography. Although these techniques were more complex and difficult to use than metal masks, they promised more precise control of the lateral dimensions of transistor structures. It was expected, as a result, that they would yield transistors with better electrical characteristics. In addition to photolithography, the Fairchild group settled on boron and phosphorus diffusion as well as aluminum deposition, all techniques that later became standard in the silicon industry.[37]

In less than 5 months, the team transformed these and other laboratory techniques into reproducible and economical fabrication processes. Experimenting by trial and error, and relying on their rich and varied scientific and technological skills, these men made numerous process innovations ranging from improvements in crystal growing techniques to the development of a novel procedure for attaching gold wires to the transistor chips. They also developed a controllable photolithography process and its attendant equipment. In particular, Noyce and Last designed, along with Kleiner, a step and repeat camera to make masks and devised an innovative method for aligning them (Moore 1998).

In conjunction with their work on masking, Last and Noyce improved upon existing photoresists, the photographic emulsions used to selectively etch the wafers. Originally developed by Eastman Kodak for the manufacture of printed circuit boards, photoresists did not meet the exacting requirements of silicon processing. They did not stick to the wafers and as

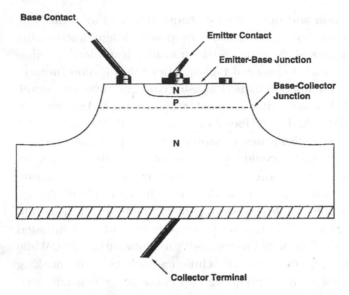

Figure 4.2
Fairchild's first NPN transistor. Source: Moore 1998. Copyright 1998 IEEE.

a result could not be used to etch the oxide layer properly. They also introduced impurities into the silicon crystal. To solve these problems, Last and Noyce, working in collaboration with Eastman Kodak, transformed the photoresists' chemical composition and purified them so as to eliminate contaminants and make them adhere to the silicon wafer.[38]

The group sought to better understand the boron and phosphorus diffusion processes and to engineer controllable and economic diffusion techniques. Hoerni, for instance, developed a new boron process using a gaseous source rather than the boron oxide powder then currently employed. "The diffusion of boron," Hoerni recalled, "was very difficult because people used it as a source of boron oxide. They would put it in a boat, put the boat at high temperature [in a diffusion furnace], and the vapor of boron would diffuse. The only problem was that it would create a lot of boron oxide in the tube, and after a couple of runs the tube was completely soggy with all that oxide and we could not push the boat to put the wafers in because it was sticking."[39] To avoid constant replacing of the tube (which was expensive), Hoerni developed a complex gaseous process. Under the new process, pressurized gases of nitrogen, oxygen, hydrogen, and boron trichloride were introduced into the diffusion tube to form boron oxide. By carefully controlling this explosive mixture, they could create boron oxide in large enough quantities to dope the silicon wafer but not enough to clog the tube. The new process had the additional advantage of being more reproducible and controllable than the boron oxide powder process.[40]

Finally, the Fairchild group developed a novel fabrication technique to deposit contacts on top of the transistor dice, or chips, in order to connect them to the package's wires. While Western Electric used two different metals, aluminum and silver, for its contacts, Moore developed an all aluminum process. It had the advantage of diminishing the number of manufacturing steps and therefore reducing the potential for errors in the production process. Roberts recalls heated discussion between Hoerni and Moore regarding which metal to use. Hoerni wanted to use silver but Moore had decided upon aluminum. Moore's argument prevailed. Using aluminum for both contacts was a counterintuitive move: aluminum, a P element, could be reasonably expected to make a good contact to a P-type material but create a junction when alloyed to a N-type region. To avoid making the contact with the N-type material a rectifying one, Moore developed a complex process in which alloying an aluminum film at 600°C eliminated the unwanted PN junction. He also heavily doped the emitter layer as a way of avoiding the formation of an undesired P-type layer. This was an important innovation as aluminum later became the metal of choice for making contacts in the semiconductor industry.[41]

In addition to the development of reproducible manufacturing processes, Fairchild's engineers faced the problem of packaging double diffused transistors. The transistors needed to be shielded from outside contaminants and had to withstand the high temperatures and vibrations that are characteristic of military aircraft. At IBM's request, Fairchild's engineers used the highest-quality packages then on the market: hermetic metal cans with advanced seals originally developed for the manufacture of power-grid tubes. Fairchild engineers also devised new ways of attaching the transistor chip or die to its package. Texas Instruments had attached the silicon chips to their cans through dangling wires, which made its transistor products sensitive to shock and vibrations and earned them a reputation for unreliability. Instead, Fairchild engineers directly soldered the chips to their containers. As a result of these innovations and the use of high-quality packages, Fairchild's transistor products were much more rugged than those of its competitors.[42]

Because the NPN version of the core driver was more advanced in its design than its PNP counterpart, Fairchild first put it into production in

Figure 4.3
Diffusion ovens at Fairchild Semiconductor, circa 1958. Courtesy of Fairchild
Semiconductor Corporation.

May 1958.[43] To produce this revolutionary transistor, Kleiner and the
engineers from Hughes (notably Frank Grady) borrowed practices and
management techniques from the silicon industry, including the use of
process manuals, the employment of relatively unskilled female workers,
and the establishment of a pre-production engineering group whose sole
responsibility was to scale up the process. Nevertheless, these men made
important innovations in production. They, in effect, pioneered the
development of a new form of silicon manufacturing.

First, they built an unusual production line with both a batch component and a continuous flow element. In the first part of the manufacturing process, the devices were batch produced. Groups of wafers
containing hundreds of transistors were diffused and oxidized at the
same time. The second part of the process was an assembly affair, reminiscent of the production lines in the mechanical industries. The wafers
were sliced and the individual transistor dice or chips were hand soldered
to the base of the transistor cases. Finally, gold wires were attached to the
transistor contacts and each unit was hermetically sealed.[44]

Second, Fairchild's manufacturing engineers gave considerable attention to cleanliness. They dust proofed and air conditioned the firm's pro-

cessing and assembly areas. They also enforced strict no-smoking rules to limit environmental contamination to a minimum. As a result, Fairchild's production line was substantially cleaner than those of other silicon corporations. It had more stringent cleanliness standards than the manufacturing areas of other electronics component firms on the San Francisco Peninsula such as microwave tube firms which had pioneered the building of clean rooms in the late 1940s and the early 1950s.[45]

Third, Kleiner and Grady tightly controlled a complex and unforgiving process by imposing an unprecedented level of industrial discipline on the workforce. They developed process manuals that were much more detailed than the specification books used by other silicon corporations, carefully specifying the complex procedures operators had to perform in the factory. "The specification books [were produced] in several editions," recalled Eugene Kleiner, the first head of manufacturing at Fairchild. "We wrote them and then followed them in the shop and then we saw where we did not describe the process as well as we should. So very often the manuals had to be revised several times, before they became clear enough to be followed by a relatively unskilled person."[46] These process manuals described in excruciating detail the production operations from crystal growing, to diffusion, photolithography, metallization, assembly, and testing.[47] To ensure that operators followed the manuals exactly and did not introduce variations in the process, Fairchild's engineers imposed highly regimented work tasks. "We had to have a great deal of discipline on the production line," Kleiner later reminisced, "so all we tried to do was make sure that, whatever the process manual said, it was sacred, that the operators did not try to change it."[48] As operators resisted and added to the process, Fairchild's physicists and engineers closely supervised them and dismissed those who did not carefully follow the procedures. The production supervisors were very strict and controlled every aspect of the line. Any infractions by workers were subject to firing.[49]

To ensure that their transistors met the performance and reliability specifications of IBM and other military contractors, Fairchild's manufacturing engineers emphasized quality control. They closely monitored production by checking devices at critical junctures in the process. As variations at the beginning of the process had a magnifying effect on the latter steps, the quality-control department inspected wafers before the first diffusion. It also measured the distribution of the electrical parameters over the wafer surface after each diffusion step. Finally, quality-control technicians examined the wafers after masking and metallization.

In addition to in-process inspections, Fairchild's quality-control department electrically tested all transistors before shipment. In accordance with military specifications, it also applied shock and vibration tests to the firm's products.[50]

Fairchild Semiconductor made its first product delivery to IBM in the summer of 1958. It introduced its NPN transistor to the market at Wescon (the Western electronics trade show) in August. At Wescon, the Fairchild group discovered with relish that they were the first ones to commercialize a double diffused silicon transistor. "We scooped the industry," Noyce reported gleefully at a Fairchild policy meeting a few days after the trade show. "Nobody [is] ready to put something like this on the market." There was, he added, "no prospect of anybody getting in our way in the immediate future."[51] Only Western Electric had succeeded in manufacturing double diffused transistors, but it did not pose any competitive threat as its devices were used exclusively by the Bell system. In effect, Fairchild Semiconductor had become the only supplier of double diffused silicon transistors available on the open market. Because other silicon firms encountered major difficulties in working with the new technology, Fairchild Semiconductor kept its monopoly on these devices for nearly a year and a half.[52]

Engineering Ultra-Reliability

The standard sales practice in the electronics component industries at the time was to approach the procurement officers of large military contractors. Reasoning that these officers were interested mostly in supplier dependability and viewed small, untried firms such as Fairchild with suspicion, Thomas Bay and his salesmen circumvented them. They approached, instead, the design engineers in charge of developing airborne computers and avionics systems. These engineers, Bay surmised, were concerned mostly with performance and reliability and were looking for the most advanced components. Because few design engineers then knew about transistors, let alone diffused silicon ones, Fairchild made another marketing innovation. Unlike other manufacturers of silicon components, it gave strong technical support to its customers. Fairchild's applications engineering laboratory, under Grinich, wrote notes describing the unusual electrical characteristics of the firm's device and explained how it could be used and tested. These notes even proposed circuit arrangements around the firm's transistor, which customers could copy in their own designs.[53]

Fairchild rapidly built a substantial business for its transistors by using these sales and marketing tactics. The firm also benefited from the fact that the NPN transistor, originally designed for IBM, fortuitously met the requirements of a wide range of avionics applications. While Fairchild had $65,000 in sales in August and September 1958, its revenues jumped to $440,000 in the fall of the same year, reaching $2.8 million in the first 8 months of 1959. An overwhelming share of these sales was in the avionics sector, as Fairchild secured large purchase orders from Hughes, Sperry, Arma, and other companies developing digital computers and avionics systems for jet aircraft and missiles.[54]

Fairchild received an especially important procurement contract from Autonetics, a division of North American Aviation, which was developing the guidance and control system of the Minuteman missile. This system had the unusual performance specifications of guiding a missile to a target in the Soviet Union with a precision of a few hundred meters. It also had to be extremely reliable: the Air Force wanted the Minuteman to be fired without long warm-up periods, be easy and inexpensive to maintain, and operate without failures for more than a year at a time. This implied a mean time between failures of 7,000 hours, a hundredfold improvement over the average reliability of avionics systems at the time. To meet these extraordinary reliability and performance specifications, Autonetics selected an all digital design. It also chose to use the fastest and most reliable components: solid-state devices and, wherever possible, silicon diffused semiconductors. Autonetics granted a contract to Fairchild, the only supplier of double diffused silicon transistors, with the understanding that the company would improve the reliability of its devices by several orders of magnitude.[55]

At this stage, Fairchild Semiconductor ran into a serious crisis. A customer, possibly Autonetics, discovered that Fairchild transistors had a major reliability limitation. Merely tapping the transistor cans with a pencil would produce unstable voltage characteristics and make the transistors unfit for operation. After months of unsuccessful inquiries, a Fairchild technician found out that these failures were caused by particles shaken loose in the transistor can. Attracted by the high electric fields at the transistor junctions, the particles would short the junctions and cause their premature breakdown. Although the occurrence of transistor failures was reduced by applying tapping tests at the end of the production line, Fairchild's engineers could not eliminate the problem entirely: it was an intrinsic failure mechanism linked to the transistor's very structure. A major issue for a startup depending on a single product, the tap failure problem was made even more acute by the fact that Fairchild's customers, especially Autonetics, demanded highly reliable components. Solving the tapping problem was imperative. The survival of the company was at stake.[56]

To eliminate tap failures and meet Autonetics' reliability criteria, Fairchild's engineers, in a burst of technological creativity, developed revolutionary products and processes. Indeed, solving the tapping and contamination problems of mesa transistors led Hoerni to make the most important innovation in the history of the semiconductor industry. He developed a new process, the planar process, which made possible the manufacture of ultra-reliable transistors and diodes, as well as the development of a new component, the integrated circuit. The Fairchild group learned at a meeting of the Electrochemical Society in May 1958 that a research group at Bell Labs had discovered that a thermally grown oxide passivated or electrically stabilized the silicon surface. Independently, Hoerni, along with Noyce and Moore, had experimented with these oxides at Shockley Semiconductor. They had found that a clean oxide layer on the junction between the base and emitter regions of the transistor could reduce leakage or, in other words, diminish the undesired reverse flow of electrons at the junction. Following up on this work, these men developed at Fairchild a transistor in which the oxide layer had been left on top of the emitter junction after processing. They discovered that this new device had much improved electrical parameters.[57]

Building on these results, Hoerni developed, on a bootlegged basis and without much input from the firm's other founders, a new manufacturing process that used the masking and passivating properties of silicon oxide. Hoerni, a proud and temperamental man, was motivated, in part, by the rejection of his PNP transistor by Moore a few months earlier. Under the mesa process the oxide layer was deposited after the making of the base in order to mask the emitter diffusion and was later removed. In contrast, Hoerni grew an oxide layer on top of the wafer at the beginning of the process and used it to mask both the base and the emitter. In an innovative move, Hoerni left the oxide layer on top of the wafer after transistor processing. This went against all accepted knowledge in the silicon community: it was widely believed at the time among practitioners of the silicon art that the oxide layer used to mask dopants was dirty and, as a result, had to be etched away. Instead, Hoerni, a self-professed contrarian, left the oxide layer on top of the wafer. He then made the startling discovery that far from contaminating the wafer, it passivated the crystal's surface and protected the transistor junctions from outside contaminants.[58]

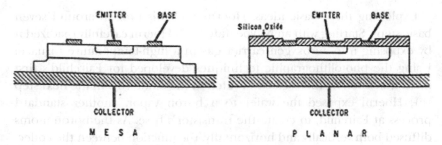

Fig. 1.   Transistor Structures

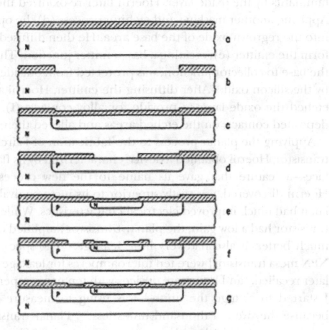

Fig. 2.   Steps in Planar Transistor Fabrication

Figure 4.4
The planar process. Source: Hoerni 1960. Courtesy of Fairchild Semiconductor and Stanford University Archives.

Exploiting these basic ideas, Hoerni built his process around seven basic steps. Starting with an N-type wafer (a), Hoerni carefully oxidized it by exposing it to an oxygen carrier gas in a high-temperature furnace. Using the photolithographic techniques developed for Fairchild's first mesa transistor, he then selectively etched the oxide (b). In the next step (c), Hoerni exposed the wafer to a boron vapor, another standard process at Fairchild, to create the transistor's base. As the boron atoms diffused both vertically and horizontally, the junction between the collector and the base moved laterally and was protected from outside contaminants by the oxide layer. Hoerni later re-oxidized the exposed area. Applying another masking and etching process (d), he opened a window into the regrown oxide of the base area. He then diffused phosphorus to form the emitter (e), creating a base-emitter junction. This junction, like the base to collector junction, was protected from outside contamination by the silicon oxide. After diffusing the emitter, Hoerni again selectively etched the oxide layer to provide for alloy contacts (f). Finally (g), he deposited contacts on the etched areas and alloyed them in.[59]

Applying the planar process to the fabrication of Fairchild's first NPN transistor, Hoerni obtained a device characterized by its flat or planar surface—a feature that gave its name to the new process. This device, Hoerni discovered, was vastly superior to its mesa equivalent. In particular, it had much-improved electrical characteristics. While Fairchild's first transistor had a low gain, the planar transistor amplified electrical signals much better. It also had very little leakage. "The specs [for Fairchild's NPN mesa transistor] were ten microamperes for leakage count," Hoerni later recalled, "and here were devices with ten nanoamperes. I remember I started to change the units I was using to measure leakage counts because they were a thousand times less."[60] Planar transistors also were much more reliable than their mesa counterparts. "The most interesting thing [with these devices]," Hoerni reminisced, "was once they were sealed then you could tap forever, nothing would happen."[61]

Noyce and Moore, who had emerged as the company's main managers, decided to proceed with the development of the planar process and bring it to production. Although this decision might appear obvious with hindsight, it was not evident at the time and involved much technological risk. In particular, planar transistors were extremely difficult to make. "The yield," Hoerni later recalled, "was terrible."[62] At an early stage in their development, the yield of Fairchild's first NPN mesa transistors had been in the 25–30 percent range. The yield of planar devices did not exceed 5 percent, a fact that cast doubts about their ultimate manufacturability.

Indeed, the Bell Telephone Laboratories, which devised a process similar to Hoerni's about the same time, decided to forego its further development because of its seeming lack of manufacturability. Fairchild's management chose otherwise however. It saw the planar process as the solution to the company's severe tap failure problem and as a way to manufacture the highly reliable transistors specified by Autonetics. "The reason why [Noyce and Moore] eventually went to the planar," Hoerni later reminisced, "was because they had to solve this damn tap test problem. Marketing was after them."[63] Unlike Bell's component groups, which served the needs of a telephone monopoly, Fairchild was oriented mainly toward avionics applications, which had very high reliability requirements and put enormous pressure on the company to eliminate its products' catastrophic failures.[64]

Two other factors further encouraged the decision to bring the planar process to production. First, Autonetics—impressed by the reliability potential of the planar process—pushed Fairchild to perfect the new process and bring it to production. "Autonetics was a big force behind the development of the planar," Moore later recalled. "They wanted us to go with planar technology."[65] Crucial in the decision was Autonetics' willingness to buy Fairchild's future planar transistors. A large market for the new components would help Fairchild recoup its heavy investments in process and product engineering. Second, Fairchild faced new competition in the mesa transistor field. Motorola and Texas Instruments had active engineering programs in mesa transistors and introduced copies of Fairchild's first NPN mesa transistor to the market in late 1959. The main competitive threat, however, came from Fairchild's own engineers. Enticed by the large profits in the silicon business, Baldwin (the firm's general manager) and the engineers he had brought in from Hughes left Fairchild to start Rheem Semiconductor in March 1959. Bringing with them Fairchild's process manuals, these men set out to produce the firm's mesa transistors at lower cost than Fairchild. To compete with Rheem, Fairchild had to introduce new and better devices. Bringing the planar transistor to market would give the company a major competitive edge.[66]

Because of these pressures, Noyce and Moore dedicated substantial engineering and financial resources to the further development of Hoerni's fragile process and the design of planar products. Although little is known about Fairchild's effort to transform Hoerni's lab techniques into a stable and reproducible manufacturing process, it is clear that the firm's research laboratory devoted substantial efforts to understanding this complex process. They especially needed to isolate and analyze its faulty steps. A group headed by Hoerni discovered that the poor transistor yields were partly due to imperfections in the oxide layer. Tiny holes in the oxide let phosphorus atoms diffuse into the crystal and create highly doped regions in the collector, putting in effect an emitter on the collector side. This finding led to the development of new oxidation techniques which enabled the growth of cleaner and more uniform oxide films on top of the silicon wafers.[67]

In conjunction with the development of a stable manufacturing process, Hoerni and other engineers at Fairchild designed a series of ultra-reliable planar products. These men planarized Fairchild's transistor line. They designed planar variants of the firm's first transistors as well as versions of fast-switching, gold-doped mesa devices the company had introduced for logic applications in mid 1959. (Hoerni had developed the new technique of gold doping to improve the switching speeds of silicon transistors. Going against the conventional wisdom that saw gold as a contaminant reducing transistor gain, Hoerni diffused gold on the back of the silicon wafer. Much to the surprise of the other Fairchild founders, gold-doped transistors turned off faster (2 microseconds, versus 1 nanosecond). David Allison and some other Fairchild engineers soon applied this technique to new switching transistor products used in computing. They later designed planar versions of these transistors.) Fairchild's planar transistors had exceptional electrical parameters, such as high gain and low leakage currents. Furthermore, while the electrical characteristics of mesa transistors drifted during operation, those of planar devices were remarkably stable. Finally, planar transistors were highly resistant to shock and impervious to outside contaminants. As a result, they were extremely reliable. Planar devices represented a major breakthrough in transistor technology.[68]

Hoerni knew that mesa silicon diodes manufactured by Hughes and Pacific Semiconductors had the same tap failure problem as Fairchild's first transistors. He then went one step further in the commercial exploitation of his process in the spring of 1959 by designing a planar diode for computing applications, which was faster and more reliable than any mesa product on the market. Seeing a ready market for such a diode at Autonetics and in the avionics industry, Hoerni and Frank Grady, the new head of operations, urged Fairchild's management to enter the diode business. As a result of their efforts, Fairchild's engineering groups designed a large family of ultra-reliable planar diodes. The company also set up a new plant dedicated exclusively to diode manufacture in Marin County, north of San Francisco, in October 1959.[69]

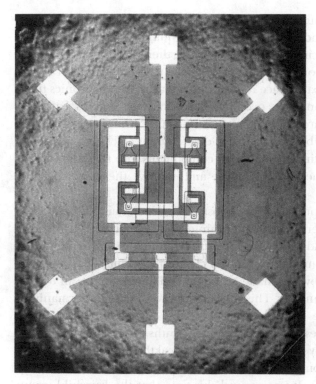

Figure 4.5
The first planar integrated circuit (1960), designed and built by Lionel Kattner
and Isy Haas under the direction of Jay Last. Courtesy of Lionel Kattner.

Fairchild's engineers also developed a radically new planar component,
the integrated circuit. Unlike discrete devices, the integrated circuit incorporated an entire electronic circuit, which included transistors, diodes,
capacitors, and resistors into the silicon crystal. Avionics reliability was
once again the main driver of this development. By integrating a whole
circuit into the silicon slice, these men sought further improvement in
the reliability of avionics and other electronics systems. Many system failures were due either to faulty connections between the silicon die and
the electric wires of the component packages or to errors in the assembly
of electronics components to the systems' printed circuit boards. By integrating a whole circuit into a silicon die and by depositing an aluminum
film to interconnect its various components, the potential for such failures was substantially reduced.[70]
Fairchild's engineers also saw integrated circuits as a way to further
miniaturize electronics systems. The use of discrete silicon devices had

helped avionics manufacturers miniaturize their systems by dispensing with the bulky air conditioning equipment required by germanium components. Integrated circuits represented a further step in that direction. By cramming more components into the same package, Fairchild's engineers reduced the board space required for a particular electronics function and helped diminish the overall size of electronics systems. Fairchild also saw integrated circuits as a way of "automating" transistor and diode assembly. The assembly of transistor dice to their packages was labor intensive. By integrating a number of components into the same silicon die, they would reduce the workforce and improve the productivity of assembly stations.[71]

But competitive pressures also fostered and sustained Fairchild's research and engineering program on integrated circuits. In 1958, Texas Instruments, Fairchild's archrival, constituted a new group around Jack Kilby to work on the miniaturization of electronic circuits. By September of the same year, Kilby had made an integrated linear circuit, an oscillator, using mesa techniques. This mesa circuit was made of germanium. To interconnect the various devices on the germanium chip, Kilby employed flying wires. In the next few months, Kilby and his group informed the military services about their breakthrough and received a research contract from the Air Force to further develop the concept of the integrated circuit. It was around this time that the Fairchild group first heard of Kilby's circuit. Texas Instruments soon publicized its integrated circuit program. In March 1959, the company organized a news conference to promote its mesa integrated circuits. At this conference, Patrick Haggerty, TI's chief executive officer, claimed that the microcircuit was a major invention and that integrated circuits would soon be used in weapon systems and consumer electronics products. Kilby's work and especially the publicity that surrounded it put considerable pressure on Fairchild's managers to start their own integrated circuit project and obtain fast results for it. They had to show that, like Texas Instruments, Fairchild Semiconductor was working on the technical frontier.[72]

Noyce, who was well aware of these market and competitive pressures, conceived of the planar integrated circuit in January 1959. By then, Hoerni had just developed the planar process and Noyce was the first one to learn about it (it was only a few weeks later that Hoerni revealed his process to the whole founding group). Exploiting his advance knowledge of the planar process, Noyce thought of using it to fabricate a complete circuit rather than individual components. This process, Noyce reflected, had two unusual features: it permitted the manufacture of hundreds of devices on the same slice of silicon. It also left a layer of oxide on top of the wafer. This oxide layer, in addition to its masking and passivating properties, could also act as an electrical insulator. As a result, Noyce reasoned, one could electrically connect the different components on the same wafer by evaporating and properly etching an aluminum film on top of the silicon oxide. To obtain a functional electronic circuit, Noyce also devised ways of making other standard components such as resistors and capacitors with diffusion and photolithographic techniques. Finally, he conceived of using diodes mounted back to back to electrically isolate the different devices on the same die.[73]

Noyce filed a patent for his planar integrated circuit idea in July 1959. This idea was put into silicon and productized in the next 2 years by a group directed by Jay Last. Interestingly, Noyce played no role in this effort. It was Last and his engineers in the micrologic section of the research laboratory who made the revolutionary step of engineering and fabricating planar integrated circuits. Building a functional integrated circuit, Last and his group discovered, was extremely difficult. To obtain the close tolerances that were required, they had to substantially improve Fairchild's diffusion and photolithographic processes. They also had to develop new techniques to evaporate aluminum so as to properly interconnect the different components on the same wafer. The main problem they encountered, however, was how to electrically isolate the transistors, diodes, resistors, and capacitors on the same die or silicon chip.[74]

To solve the isolation problem, Last and his engineers explored two different approaches. In the fall of 1959, Last thought of electrically insulating transistors on the same die with plastic fillings to make a flip-flop, a digital circuit used for counting purposes. "It was a rather implausible device," Last later recalled. "You would diffuse in all the transistors [by using the oxide layer as a mask]. You would then turn the device over and you would etch all the way down to the silicon until you got to the oxide. So the device was supported only by the oxide."[75] In the closing step of the process, the openings, which had been etched between the transistors on the back of the wafer, were filled with an epoxy resin. This insulated the devices electrically and gave mechanical strength to the silicon die. In the spring and summer of 1960, Last's engineering team further developed this technique. Lionel Kattner, whom Last had recently recruited from Texas Instruments' manufacturing organization, experimented with different materials, glass powders and epoxies, for the filling of the grooves in the silicon crystal. The material had to be non-conductive, contaminant-free, and at the same time have the same coefficient of expansion as silicon (this was particularly important since the oxide layer, which supported the whole device, was very fragile). Other engineers in Last's group developed an infrared jig to properly align the etching of the grooves on the back of the wafer with the interstitial space between the transistors on the top of the wafer. Last and his group made hundreds of physically isolated devices. But these circuits had poor yields and continued to be plagued by breakage of the oxide layer.[76]

In September 1960, Kattner and Isy Haas (a circuit engineer at Fairchild) began work on an alternative solution to the isolation problem. The physically isolated microcircuits could not be manufactured in volume. Kattner and Haas decided to use diffusion regions to isolate the different devices on the same wafer. They did so with Last's approval and support and with the help of Allison, Fairchild's foremost diffusion expert. Critical to this approach was their use of a new boron diffusion technique developed in the pre-production group and refined by one of their co-workers in the micrologic section. Under this process, a carrier gas passing through liquid borate was mixed with oxygen and then burned at the end of the diffusion furnace. The combustion produced a cloud of boron gas, which was carried over the wafer. This technique permitted the diffusion of P-type dopants for more than 24 hours without destroying the silicon oxide layer.[77]

Taking advantage of this new diffusion technique, Kattner and Haas developed a complex fabrication process to make a planar flip-flop. At first, they etched thin bands or windows in the oxide layer on top of an N-type wafer. Kattner and Haas then diffused boron atoms, a P-type element, both through these openings and from the entire back of the wafer for more than 20 hours. The atoms diffused simultaneously from both sides of the silicon wafer. They met in the middle, creating good isolation regions. Having made these isolation regions, Kattner and Haas then processed planar transistors, diodes, and resistors in the pockets of N-type material left by the boron diffusion. They later interconnected these devices by depositing an aluminum film on top of the wafer. Kattner and Haas made the first operating flip-flop in the last week of September. In the next few months, Kattner, Haas, and other engineers in Last's group cleaned up and troubleshot the process. By the end of the year, it was clear that they had a solid method for making planar integrated circuits.[78]

Using this new process, Kattner, Haas, and other engineers at Fairchild developed a family of digital integrated circuits in late 1960 and the first half of 1961. In addition to the flip-flop, they designed a gate, a buffer, a half-adder, and a half-shift register circuit. Noyce and Moore announced this micrologic family at a news conference in New York in March 1961. By this time, Fairchild was the only firm with a family of planar integrated circuits on the market. (Texas Instruments introduced its own family of planar microcircuits in October 1961.) In 1961, Fairchild developed a small market for its microcircuits. Because of their high price and low performance, these circuits could only be used in a small number of applications. However, they constituted an important breakthrough in silicon technology and promised major improvements in system reliability, performance, and miniaturization.[79]

Manufacturing Ultra-Reliability

In tandem with designing reliability into its products at the component and circuit level, Fairchild's engineers perfected their production systems to meet the tight reliability specifications of military avionics. Autonetics forced Fairchild to improve its manufacturing operations by setting high reliability criteria. It also financed and closely supervised Fairchild's efforts to enhance its production systems by instituting a comprehensive "reliability improvement program." The goal of this program, which applied to all Autonetics' subcontractors, was to drastically improve the reliability of solid-state components by transforming the ways in which they were manufactured. Autonetics sought to reinforce its suppliers' manufacturing disciplines, tighten their process controls, and augment their testing procedures.[80]

Autonetics' program was patterned after similar efforts at the Bell Telephone Laboratories, which developed reliable repeater tubes for undersea telephone cables in the 1940s and the early 1950s. Adopting many of Bell's methods and applying them to semiconductor manufacturing, Autonetics required all its suppliers to implement a comprehensive "reliability improvement program" that touched all aspects of their manufacturing operations. Suppliers of solid-state devices were asked to document carefully their manufacturing processes. They also had to build "high reliability lines" using such techniques as "assembly in dust-free environments, carefully spelled out operator instructions, and close monitoring by Quality Control personnel." (Smith 1963; Scheffler 1981)[81]

Modeling its program after Bell's, Autonetics requested its vendors to apply serial numbers to their components in order to follow the history of each part so they could trace those that had failed to the materials and processes with which they had been produced. This required an enormous record keeping effort as Autonetics' suppliers made hundreds of thousands of components for the Minuteman program. And Autonetics asked its subcontractors to isolate their products' failure modes and identify their causes, demanding that they apply statistical control techniques to their manufacturing processes in order to monitor process variations and achieve narrower distributions of parameter characteristics (Scheffler 1981).

Autonetics further required its subcontractors to establish strict testing procedures to screen out defective products. They demanded that the subcontractors apply the life tests pioneered by Bell for vacuum tubes to solid-state components to evaluate their reliability and determine the best temperature and voltage conditions under which these devices should be operated.[82]

To ensure that its suppliers carefully implemented this extensive reliability improvement program, Autonetics closely supervised their activities. Resident inspectors were stationed at its subcontractors' plants. These inspectors were "free to walk into the line and look at every step" and they had the authority to make changes in the manufacturing process.[83] Autonetics also established an in-house component engineering group. This group closely followed the subcontractors' engineering efforts and evaluated their progress toward the Minuteman reliability goals.[84]

Autonetics had a major impact on manufacturing at Fairchild. To meet Autonetics' specifications and fulfill the requirement of its reliability program, Fairchild made substantial advances in manufacturing that were lavishly funded by Autonetics. Over the course of the contract, Autonetics spent more than $4.5 million to improve the reliability of Fairchild's transistors. "The Minuteman contract," Eugene Kleiner recalled, "was a very good contract, not only from the income point of view but to advance the state of the art of manufacturing. It certainly advanced our knowledge and increased our experience. It increased our quality control procedures, and testing procedures, which we had to some degree. But the quality requirements of this Minuteman program were much higher than [those] you have in commercial use."[85]

To meet Autonetics' requirements for a cleaner, more disciplined manufacturing operation, Fairchild set up a separate area in the factory with precise air control. This area produced the NPN and PNP mesa transistors procured by Autonetics. The firm also tightened its manufacturing discipline. Fairchild's management strengthened the plant's supervisory apparatus. It also introduced the wearing of smocks and hairnets.[86] In addition to these traditional forms of regimentation, Fairchild, like other Minuteman suppliers, established a "training and motivation program" for its operators, foremen, and manufacturing engineers. "The techniques, the attitude, and the desire for high reliability," a Fairchild reliability brochure later reported, "was spread throughout the entire organization. [The motivation program] included 52 courses and conferences involving 11,765 man-hours of training as well as newspaper articles, posters, movies, lectures, and seminars."[87]

The Minuteman contract also helped Fairchild's engineers to acquire a solid competence in testing and enabled them to gather vast amounts of data on the reliability of its products. While the firm had relied in its early days on a rather limited set of testing instruments, Fairchild's engineers designed a wide variety of electrical and mechanical testers for the Minuteman program. Fairchild set up a Reliability Evaluation Division, the sole objective of which was to appraise the dependability of its transistors. This division conducted extensive life tests on hundreds of thousands of mesa and, later, planar transistors. By 1961, it had accumulated more than 150 million transistor hours, which "gave Fairchild experience orders of magnitude higher than those results from previous evaluations in the industry."[88]

As a result of these systematic efforts, the reliability of Fairchild's mesa transistors improved enormously between 1959 and 1961. In 1959, the failure rate of Fairchild's first NPN transistor had been 0.1 percent per thousand hours. In early 1961, it reached 0.004 percent per thousand hours. By the time the Minuteman missile was fully operational, the failure rate had dropped to 0.00009 percent. This was an average of less than one failure in 10,000 years. Because of these extraordinary improvements in reliability, Fairchild's devices became, by far, the most dependable transistors used in Autonetics' guidance and control system. Autonetics' program also had another important effect: it helped Fairchild tighten process control and, thereby, improve its yields and reduce its manufacturing costs. These advances were quickly applied to the firm's other mesa transistors and, later, to its planar products as well.[89]

Growth

The Minuteman program sanctioned Fairchild as a manufacturer of high-quality products. This gave it considerable visibility in the military sector in the late 1950s and the early 1960s. "The Minuteman program," Thomas Bay later recollected, "took a company that was a little nothing and put us on the map if you will—because we were compared to TI and Motorola. We were a contender in the semiconductor business. It was not a big contract in terms of lots of devices and lots of dollars. But it was a contract that proved the reliability of our device. The value of the [Minuteman contract] to the company was much more in the reputation it gave us than it was in the dollars that it generated by itself, although by the time the program was going and the Minuteman was being produced, it was substantial volume."[90]

As a result, Fairchild was well positioned to take advantage of the rapid growth of the military market for silicon components in the late 1950s. This market for silicon transistors almost tripled from $32 million in 1958 to $90 million in 1960. Avionics contractors, which had used silicon transistors in their digital prototypes in the mid 1950s, required them in large volumes when they moved these new systems to production at the end of the decade. At the same time, a new military market emerged for silicon transistors in ground-based equipment. Intent upon improving the reliability of radio communication, telemetry, and other ground-based systems, the military required its suppliers to employ silicon rather than germanium transistors.[91]

Aggressively exploiting these rapidly expanding markets, Fairchild saw its sales grow from $500,000 in 1958 to $21 million in 1960. To meet this demand, Fairchild greatly enlarged its manufacturing facilities. In addition to the diode plant in San Rafael, Fairchild opened a new factory dedicated to transistor production in Mountain View in the summer of 1959. The firm rapidly expanded its workforce by relying on the local pool of electronics operators and technicians and by recruiting engineers from other silicon corporations such as Texas Instruments. In February 1959 Fairchild Semiconductor had 180 employees; a year later it had more than 1,400. By that time it was one of the biggest electronics component manufacturers on the San Francisco Peninsula.[92]

To handle this rapid growth and construct a cohesive organization, Fairchild's managers adopted many of the corporatist management techniques developed at Varian Associates and other Peninsula firms. These older firms emphasized collaboration and mutual obligations between managers and employees. When they adopted these techniques, Robert Noyce and Jack Yelverton (Fairchild's head of human resources) did not share the Varians' deep interest in abolishing the distinction between capital and labor. They were primarily motivated by self-interest and the exigencies of the semiconductor business. Like vacuum tube corporations, Fairchild Semiconductor was dependent on a highly skilled work force to control complex manufacturing processes and design advanced products. This work force, the company's main asset, had to be mobilized and stretched to its utmost capacity. Noyce, Yelverton, and other managers at Fairchild were also interested in keeping unions out of the plant. Even more than their counterparts in the tube industries, they saw unionization as a deadly threat to their business. Fairchild's managers believed that unions would prevent the imposition on the work force of the extremely tight discipline that they thought crucial. Only through highly regimented work tasks could Fairchild's manufacturing engineers control a highly unstable manufacturing process. Fairchild managers also opposed unionization because semiconductor technology was changing so rapidly that the work force had to be flexible and able to shift tasks quickly, a practice that Fairchild's managers deemed impossible in a unionized plant.[93]

To motivate the employees and build a cohesive and classless firm, Noyce sought, as Yelverton later put it, to "encourage communication and the feeling that there is a great sense of participation." To learn how to accomplish that, Noyce and Yelverton consulted with their counterparts at Varian Associates, Watkins-Johnson, and Hewlett-Packard. These firms' practices inspired many of Fairchild's employee relations programs. Encouraging communication and developing a feeling of participation required breaking the social barriers among managers, engineers, and operators. To do so, Yelverton organized company sports teams and outings. Fairchild's managers carefully avoided all outward symbols of power and cultivated an egalitarian style. At the same time, Noyce and Yelverton encouraged communication between managers and operators. Whereas Varian Associates had established the Management Advisory Board for that purpose, Fairchild's managers adopted a more informal approach. They let it be known that any engineer or operator could approach them regarding grievances and problems on the job without the knowledge of their supervisors. They also had regular lunches with a randomly selected group of employees to keep in touch with what was happening in the laboratories, the fabricating area, and the assembly plant. In addition, they assigned human resource managers to walk through the production lines, getting to know operators individually. In that way, they could detect problems early, providing Yelverton with advance warnings of unionization efforts.[94]

In conjunction with the development of smooth communication processes, Fairchild's management sought to ensure that employees would have a sense of participation in the firm. Noyce and other managers delegated significant authority to their subordinates and pushed decision making as far down in the organization as possible. Like their counterparts at Varian, engineers were given considerable autonomy and had a say in the course of their projects. They often worked in teams. In 1960, Fairchild Semiconductor took the innovative step of giving stock options to some of its engineers. These options, previously confined to upper managers at East Coast firms, offered the right to buy stock at a predetermined price at a future date. They enabled the optionees to make substantial capital gains if the stock increased in value. Noyce and Grady, the head of operations at Fairchild, sought to give stock options to the firm's 40 most important engineers. They wanted them to participate in the company's financial success and have a strong financial incentive to stay at Fairchild. But their stock-option proposal encountered considerable opposition from Fairchild Camera, where it was seen as a socialist plot. As a result, the number of options given to each engineer was substantially reduced, thereby substantially diminishing their monetary value and ability to lock the best engineers into the firm. Although Noyce and Grady did not go as far as they had wanted, the distribution of stock options to a large group of engineers was an important social innovation. In the second half of the 1960s and the early 1970s, many semiconductor firms in the area followed Fairchild Semiconductor's example. They gave stock options to their employees in order to involve them in the entrepreneurial process and its financial rewards and at the same time align their interests with those of the entrepreneurs and their financial backers.[95]

Conclusion

Fairchild's founders and the engineers and managers they hired defined the semiconductor industry's new silicon products, processes, manufacturing systems, and management-employee relations in the late 1950s and the very early 1960s. The key to the reshaping of the industry was their adoption of the solid-state diffusion process recently developed at Bell Laboratories. To bring this complex process to quantity production, the Fairchild group drew upon a wide variety of industrial practices and bodies of scientific and technical knowledge. In addition to solid-state physics, they relied on their expertise in optics, metallurgy, chemistry, and electrical and mechanical engineering. These men also drew upon techniques coming out of the metalworking, printed circuit, and vacuum tube industries. Out of this vast array of skills and practices, the Fairchild group built radically new products and designed innovative fabrication methods and manufacturing systems. They also developed creative sales and marketing techniques, crafted new ways to handle relations with employees, and closely coupled product development with market demands.

Because of their focus on military avionics, they were constantly under enormous pressure to meet the reliability standards of avionics system manufacturers and, more indirectly, of the Air Force. Because of these demands, competition from Texas Instruments, and the group's internal dynamics, Fairchild's entrepreneurs made major manufacturing process and design innovations in the late 1950s and the early 1960s. In close succession, they developed the planar process and the integrated circuit, both of which made the engineering of highly reliable electronics systems possible. Finally, under the guidance and scrutiny of Autonetics, Fairchild's manufacturing engineers perfected their production systems and tightened their control of the manufacturing process in order to produce highly reliable diodes and transistors.

In a few years, Fairchild profoundly transformed semiconductor manufacturing. It revolutionized the industry's products by introducing high-performance, high-reliability silicon devices that other firms later copied. As an example, by the summer of 1960, versions of the firm's first transistor were produced by Rheem, Motorola, Texas Instruments, Pacific Semiconductors, and Hoffman Electronics, among others. Fairchild also developed radically new device designs such as the integrated circuit, which became the industry's mainstay in the 1960s. In conjunction with these revolutionary product changes, Fairchild radically transformed manufacturing methods. Whereas Texas Instruments, Hughes, and other firms had used the grown junction and alloy junction techniques in the mid 1950s, Fairchild's founders re-oriented the industry toward solid-state diffusion. The corporation also developed the planar process, which revolutionized the manufacture of diffused components and rapidly became the standard process for making silicon components. Corporations that did not readily adopt Fairchild's innovations went out of business in the early 1960s.[96]

In addition to its influence on the evolution of the semiconductor industry, Fairchild Semiconductor had a significant impact on electronics manufacturing on the San Francisco Peninsula in the early 1960s. It played an essential role in the formation of the region's semiconductor equipment industry. In the early 1960s, Fairchild's founders urged their technicians to establish their own semiconductor-manufacturing equipment firms. For example, in 1960 Eugene Kleiner encouraged Arthur Lasch, a technician in Hoerni's group, to start his own business, asking him to make glass capillaries (used in the making of gold bonds) for Fairchild. In July 1960, Lasch incorporated Electroglas to produce these capillaries. He later diversified into the making of bonding machines and soon afterwards into the manufacture of diffusion furnaces. In 1963, Electroglas employed 50 people. It had $550,000 in sales, roughly 80 percent of which were in glass capillaries and bonding equipment and the rest in diffusion furnaces. Other entrepreneurs rapidly joined the fray. For example, Frank Christensen, a Stanford graduate who came from a family engaged in the manufacture of diamond drills for the petroleum industry, started a new business, the Tempress Research Company, in 1962. Tempress specialized in bonding equipment and scribing machines, which were used to cut silicon wafers into chips. Because of the high precision and quality of its machines, Tempress Research grew rapidly. By 1965, it had 80 employees and more than $2 million in revenues.[97]

In business developments of a different type, Fairchild Semiconductor brought venture capital and venture capitalists to the San Francisco Peninsula. Financiers and engineers involved in the establishment of Fairchild Semiconductor set up some of the first venture capital partnerships in the area in the 1960s and the early 1970s. In 1961, Arthur Rock, who had been instrumental in Fairchild's formation, left Hayden Stone and moved to Northern California. He allied himself with Thomas Davis, a vice-president at the Kern County Land Company who had invested in the microwave tube maker Watkins-Johnson, to establish a new venture capital partnership that would invest in electronics startups on the West Coast. To muster the required capital, they raised monies among East Coast investors and Bay Area industrialists. Fairchild's founders, each of whom had received $250,000 after the sale of Fairchild Semiconductor to Fairchild Camera in 1959, were among the first potential investors whom Davis and Rock approached. Four Fairchild entrepreneurs (Hoerni, Last, Kleiner, and Roberts) as well as a founding employee of Varian Associates invested in Davis and Rock's fund.[98] In the next few years, this partnership invested in electrical and electronic component startups in the Bay Area. Among these were the General Capacitor Company and the Microwave Electronics Corporation, a maker of microwave tubes. Davis and Rock also took minority positions in electronic system firms in Southern California. For example, they bought into a variety of microwave radar and instrumentation companies. They also invested in Teledyne, an avionics corporation, and Scientific Data Systems, a maker of minicomputers.[99]

Because of Teledyne's and Scientific Data Systems' rapid expansion in the 1960s, Davis,and Rock's fund had spectacular returns. By the time of the partnership's dissolution in 1968, the $3 million Davis and Rock had invested in electronics firms had grown to $90 million. Twenty percent of the returns went to the venture capitalists. The rest was distributed among their investors. These enormous capital gains encouraged others to enter the venture capital business. Among these was Kleiner, a limited partner in Davis and Rock's fund. In the second half of the 1960s, Kleiner invested in new science-based companies in collaboration with Roberts. He later formalized these activities by establishing a venture capital partnership of his own, Kleiner Perkins, in 1972. The growth of the venture capital business and the emergence of an independent semiconductor equipment industry on the Peninsula facilitated the establishment of numerous semiconductor firms by former Fairchild employees in the 1960s and the early 1970s.[100]

This second wave of entrepreneurship was made possible, like the first, by the area's competency in semiconductor manufacturing. Fairchild transformed the San Francisco Peninsula into a major center of expertise for the processing and design of advanced silicon components. The firm gained and retained a significant advance over its competitors in the control and understanding of manufacturing processes. In addition, it trained hundreds of engineers and technicians in these new techniques. At the same time, the formation of Fairchild Semiconductor introduced new attitudes toward entrepreneurship on the San Francisco Peninsula. After the rebellion of the eight Fairchild founders, it became increasingly acceptable (and thinkable) for engineers to start their own corporations. The wealth of the group that established Fairchild Semiconductor encouraged others to try their luck and start their own semiconductor businesses in the area. These new attitudes combined with the local processing expertise and the rise of venture capital led to the rapid expansion of the semiconductor industry on the Peninsula in the second half of the 1960s and the early 1970s.

5

Opening Up New Markets

The environment for electronics entrepreneurship had changed significantly in a few years, observed Thomas Davis of Davis and Rock in March 1964. "The government," he noted, "does not award cost-plus-fixed-fee contracts as liberally as it once did and does not take the tolerant attitude toward cost over runs as it did in the past. The 'population explosion' in electronics companies gives the government and prime contractors a broader spectrum of suppliers. Sole-source awards are fewer. Competition is intense." (Davis 1964) The world of military contracting changed dramatically in the early 1960s. The Department of Defense instituted a severe cost-reduction program and restructured its procurement of weapon systems. As a result, the military markets for advanced electronic devices became highly unstable and they contracted in the first half of the 1960s. These cutbacks and the reform of the military procurement process had a major impact on Bay Area manufacturing. This halted the rapid growth of electronics firms in the area. Fairchild, Varian, Litton, and Eitel-McCullough, which had expanded and prospered with military contracts during the 1950s, saw their sales shrink. Many firms became unprofitable.

The electronics entrepreneurs on the San Francisco Peninsula developed various strategies to weather this crisis. In order to reduce financial losses, they downsized and laid off engineers, technicians, and operators by the thousands. They also sought to consolidate the electronics industry by acquiring or merging with other firms. This was especially the case in vacuum tube manufacturing, which was particularly hard hit by the recession. Many entrepreneurs sought to reduce their reliance on military patronage and procurement by entering civilian markets. They began to capitalize on the new business opportunities that had been brought about by the rise of consumer society and growing federal outlays for science, health care, and space research. In order to exploit these opportunities and develop commercial markets for their products, local firms reorganized themselves. They acquired new bodies of technical knowledge and developed new sales and marketing techniques. What they discovered was that serving commercial customers was indeed different from dealing with the Department of Defense. In order to establish themselves as civilian producers, electronics firms on the Peninsula had to learn about commercial markets. They had to restructure their manufacturing, sales, and marketing organizations to meet the requirements of commercial customers.

To explore these profound changes in the electronic component industries in the first half of the 1960s, I will focus on Varian Associates, Eitel-McCullough, and Fairchild Semiconductor. Varian, Fairchild, and Eitel-McCullough were the largest makers of electronic components on the San Francisco Peninsula. In the first half of the 1960s, they consolidated the local electronics industry and led its partial reconversion toward commercial markets. And they were emblematic of the divergent paths of the two main component industries on the Peninsula: microwave tubes and semiconductors. During this period, semiconductor firms developed civilian markets for silicon devices, while microwave tube corporations expanded into commercial electronic systems and began supplying manufacturing equipment to semiconductor corporations.

Varian Associates and Eitel-McCullough were not immune to the sharp recession that occurred in the first half of the 1960s. Facing an abrupt decline in the military demand for klystrons, Edward Ginzton, Varian's chairman, sought to consolidate much of the microwave tube industry on the Peninsula. In 1965 he merged Varian with Eitel-McCullough. He also intensely focused on reducing Varian's dependence on the military. In order to expand into commercial markets, Ginzton acquired instrumentation businesses. He also supported the development and commercialization of NMR spectrometers and clinical linear accelerators. In addition, Ginzton built a large vacuum equipment business, supplying vacuum pumps and systems to the semiconductor industry and the space program. In contrast, Robert Noyce and Gordon Moore at Fairchild developed new markets for their transistors in the computer and consumer electronics industries. To meet the price and volume requirements of commercial users, they transformed Fairchild into a mass-production organization. The techniques of mass production were adapted from the electrical and automotive industries. To achieve additional cost savings, Fairchild's management set up plants in Hong Kong and in other regions where labor costs were low. To accelerate the development of commercial markets, they also created new marketing techniques. As an example, the firm's applications laboratory developed electronic systems and gave these designs away to its customers at no cost, thereby seeding a market for its products. These moves enabled Fairchild to expand its commercial sales significantly during this recessionary period.

The "McNamara Depression"

The procurement reforms and financial cutbacks of the Department of Defense were initiated by the Eisenhower administration and greatly accelerated by Robert McNamara, the Secretary of Defense in the Kennedy administration. In the first half of the 1960s, the Department of Defense canceled weapon systems already under development and halted the deployment of others. For example, McNamara eliminated Skybolt and other missile programs. In conjunction with these cutbacks at the system level, the Department of Defense limited its purchase of electronic parts and subsystems. It also eliminated excess inventories. Starting in 1960, the military services reduced their stocks of electronic components and other parts. McNamara reinforced this policy by introducing improved inventory control methods. As a result, the Department of Defense discovered that the military services were overstocked with many types of vacuum tubes and semiconductor components. This led to a further reduction in military orders for electronic components as these inventories were being worked off.[1]

In addition to cutbacks in spending, the Department of Defense reformed the procurement process to effect price reductions. In the 1950s, the Department of Defense had heavily used sole-source contracts. In the following decade, it increasingly fostered competition among military contractors. The military opened a growing number of contracts to competitive bidding. It also developed alternative sources for certain components and subsystems. To do so, the military claimed rights to its contractors' intellectual property and manufacturing drawings. These tactics enabled the military to hand drawings to competing firms and obtain competitive bids from these corporations. The Department of Defense also pressed more forcefully for second sourcing and price redetermination clauses in its procurement contracts. In addition, the Department of Defense increasingly used fixed-price contracts instead of cost-plus fixed-fee contracts. In cost-plus contracts, the military reimbursed the contractor for its expenses. It also gave the contractor a modest fee. In contrast, fixed-price contracts required that the contractor complete a certain task at a predetermined price. Because cost-plus contracts gave military suppliers little incentive to cut costs, McNamara shifted a growing share of military procurement toward fixed-price contracts. As a result, the percentage of cost plus contracts in total military procurement fell from 35 percent in 1960 to 15 percent in 1965.[2]

In addition to fostering competition, the Department of Defense diminished the bargaining powers of its suppliers in price negotiations in order to reduce the price of military components and systems. To do this, the Department of Defense increasingly required prime contractors to disclose their manufacturing costs for fixed-price contracts. It further demanded that the system contractors obtain cost information from their subcontractors and negotiate prices on the basis of these costs. Finally, the Department of Defense required access to its suppliers' accounting books and audited their cost estimates as well as their actual costs as a way of reducing their profits. The Department of Defense required these cost breakdowns increasingly aggressively in 1959. When McNamara became Secretary of Defense in 1961, he required even greater cost disclosure from military contractors. These efforts culminated with Congress passing a new military procurement law in December 1962. This law stipulated that defense suppliers give cost breakdowns to military auditors on fixed-price contracts and negotiate these contracts on the basis of their costs. This law and the previous regulations of the Department of Defense succeeded in substantially diminishing the bargaining powers of military contractors. They also significantly reduced their profits (Klass 1962).[3]

The reform of military procurement and severe cutbacks in procurement spending had severe repercussions for the US electronics industry. Among the firms that were the most affected by these changes were those that produced electronic components. The market for electronic components and related measuring instruments shrunk in the first half of the 1960s. It also became unstable. For example, the military demand for silicon transistors, which had grown from $1.8 million in 1955 to $99 million in 1960, declined to $96 million in 1961 and increased at a slow pace thereafter. Similarly, the military market for microwave tubes became highly unstable. It had doubled in size from $51 million in 1955 to $113 million in 1959. In 1960, it plateaued around $115 million. The military market grew to $146 million in 1962 before falling to its 1960 level in 1964. Superimposed on this general trend was a switch in the military demand for the different types of microwave tubes. In the 1960s, the

Department of Defense and military system contractors procured fewer and fewer magnetrons. The military demand for klystrons grew rapidly in the early 1960s but fell off precipitously in 1963 and 1964. In contrast, the military market for traveling-wave tubes grew substantially in the first half of the 1960s.[4]

As a result of these fluctuations in military procurement, the electronic component industries experienced a severe crisis. Supply exceeded demand. Intense competition and the new procurement regulations led to falling prices and dwindling profits. To obtain military contracts, many tube and semiconductor firms had to cut their prices below cost. As a result, the electronic component business became highly unprofitable. A severe shakeout ensued. The component divisions of the large system firms on the East Coast either retrenched or closed down. Philco abandoned the transistor business. Bendix and Sylvania went out of microwave tubes. RCA, Raytheon, GE, and Westinghouse de-emphasized their component activities, focusing instead on their more profitable system businesses. Transitron, a merchant producer of diodes and transistors based in Massachusetts, also experienced heavy financial losses.[5]

The semiconductor and vacuum tube firms on the San Francisco Peninsula were similarly affected by the downturn. Fairchild Semiconductor experienced a substantial slowdown in the early 1960s. Weak military sales led to severe inventory overruns. As a result, the firm had to lay off 25 percent of its workforce in the spring of 1960. The Department of Defense's attempts to weaken the bargaining position and reduce the profits of its suppliers also adversely impacted Fairchild. Taking advantage of the emergence of new suppliers of advanced silicon transistors (such as Motorola and Texas Instruments), the Department of Defense and prime military contractors put growing pressure on Fairchild to submit pricing data and grant access to its books. While these demands were difficult for most military suppliers to accept, they were particularly unacceptable to Fairchild and its parent company Fairchild Camera and Instrument. Since its founding, Fairchild Semiconductor had tenaciously kept its independence from military oversight. In particular, it had consistently refused military research contracts because these gave the military rights to its technology as well as control of its research and engineering programs. Similarly, Fairchild Camera had traditionally refused recosting and price redetermination for off-the-shelf products. Although Fairchild Semiconductor strongly resisted military demands, the firm increasingly had to "cave in." As a result, it lost some of its autonomy. "The competition became so great," recalled Nelson Stone, Fairchild Camera's general counsel, "that we had to back off from our policy. We no longer had as many unique products. And more and more companies were agreeing to this sort of thing."[6]

The vacuum tube corporations on the Peninsula also suffered from the reforms and cutbacks of the 1960s. The cancellation of military system programs severely impacted the smaller microwave tube shops such as Huggins Laboratories and Stewart Engineering. The vacuum tube division of Litton Industries also suffered from the declining military demand for magnetrons and klystrons. It laid off 150 employees in May 1960 before cutting nearly half of its remaining workforce in 1963 and 1964 (685 employees out of 1,250). Despite these cuts, the division still lost money. Eitel-McCullough, the manufacturer of power-grid tubes and klystrons, was also hit hard by military cutbacks. Eitel-McCullough's power-grid tube and microwave tube businesses lost $1.5 million (out of sales of $28.3 million) in 1960. To turn the company around, William Eitel and Jack McCullough, the firm's founders, laid off a third of their workforce. They demoted their top managers and brought in Richard Orth, an experienced executive who had worked at RCA and Westinghouse, to run the firm on a daily basis. Orth, Eitel, and McCullough introduced tighter financial controls. They split the company's operations into two divisions specializing in power-grid tubes and microwave tubes respectively. These measures enabled Eitel-McCullough to return its power-grid tube business to profitability. Orth, Eitel, and McCullough obtained decidedly mixed results with their microwave tube division. This division saw its sales grow from $8.4 million in 1961 to $10.4 million in 1964. But it also lost $3 million during this period. As a result, the corporation as a whole was barely profitable in the first half of the 1960s.[7]

The firm hardest hit by the "McNamara depression" was Varian Associates. In the early 1960s, Varian ran into a severe conflict with the Department of Defense. It refused to give cost breakdowns to the Department of Defense and grant military auditors access to its accounting books. Emmet Cameron, the head of the vacuum tube group at Varian, was strongly opposed to the DoD's new procurement regulations. Giving cost breakdowns to the DoD would severely constrain the company's freedom of action and greatly reduce its profits. Myrl Stearns, Varian's president, and Edward Ginzton, who had recently succeeded Russell Varian as the company's chairman, concurred. In December 1959, the board of directors adopted a resolution whereby Varian would refuse to disclose cost information on fixed-price contracts to the DoD. (Hewlett-Packard followed a similar course of action at the time.[8])

Varian's unbending position put it in an increasingly difficult situation. Irritated by its defiant attitude, the Air Force punished the company severely. The Air Force first did this by creating second sources for Varian's high-power klystrons. It transferred Varian's engineers and tube designs to Litton Industries and Eitel-McCullough. In 1960, the Air Materiel Command persuaded a team of Varian engineers to leave the firm and move to Eitel-McCullough. The Air Force then awarded this team research and engineering contracts to develop radar klystrons that would compete with Varian's products. In addition, the Air Force gave Varian's BMEWS klystron to the vacuum tube division of Litton Industries. At the military's request, Litton's technologists reverse engineered the Varian tube. They later underbid Varian for the second wave of production contracts for the BMEWS klystron and obtained a substantial share of these contracts. As a result, Varian lost much of its high-power klystron business to Eitel-McCullough and Litton Industries in 1963 and 1964.[9]

The Air Force also asked the General Accounting Office to examine Varian's accounting books and production records. Varian obliged and opened its doors to GAO auditors in 1961. Hewlett-Packard took a different course of action. When faced with the same request, it refused to open its books to the GAO on the grounds that allowing its auditors to see production cost records would automatically make this information public and thus available to competitors. In 1963, the GAO sued Hewlett-Packard to force the company to open its books. The lawsuit went to the Supreme Court, where Hewlett-Packard lost its case in 1968.[10] At Varian, the GAO's auditors found substantial evidence of high and (in their view) excessive profits. In their report, they documented that Varian had made millions of dollars in "unreasonable" profits for ten different klystrons from 1958 to 1960. They also recommended that "the Secretary of Defense have the military departments take all possible action to recover the excess amounts paid Varian for klystron products." These findings led the Pentagon to blacklist Varian in the fall of 1962. The Department of Defense forbade military system contractors to do business with Varian Associates. The DoD also instructed its procurement offices to avoid giving contracts to the company.[11]

This blacklisting and the decline in the overall military demand for klystrons devastated Varian. In the last months of 1962 and the first half of 1963, the company saw its order rate for microwave tubes drop 75 percent. Its sales of klystrons and other microwave tubes plummeted from $36.8 million in 1962 to $18 million in 1964. This led to extremely painful readjustments. The company, which had a policy of no layoffs, had to let go nearly half of its tube workforce, including many skilled engineers and technicians. Between 1962 and 1964, its microwave tube labor force declined from 1,550 to 800. The division was forced to close down some of its manufacturing facilities and sell off surplus production equipment. The collapse of the microwave tube division undermined the company as a whole. Until then, the tube business had been Varian's main source of revenues and profits. Once the division was floundering, the company as a whole experienced a deep crisis. Its total sales declined from $71 million in 1962 to $53 million in 1964. Varian Associates also lost $3.5 million from June 1963 to May 1964. Only a good cash position at the beginning of the downturn and Ginzton's decision in the spring of 1963 to grant military auditors access to the company's accounting books saved Varian Associates from bankruptcy. When Varian caved in, the military removed the firm from its blacklist. The microwave tube orders came back trickling in the second half of 1964 and in early 1965.[12]

Varian Associates survived the "McNamara depression." But the crisis decimated its managerial team. Emmet Cameron, who had advocated Varian's no-disclosure policy, lost his job as head of the vacuum tube group. In early January 1964, he became vice-president of engineering, a staff function with limited responsibilities. At the same time, the board of directors granted Myrl Stearns, Varian's president and co-founder, a "leave of absence for three months for health reasons." An airplane accident in mid January forced Stearns to resign the presidency and retire. Edward Ginzton, chairman since 1959, barely survived an ouster attempt in 1964. Frank Walker, an investment banker who joined Varian's board in 1959, sought to remove Ginzton from the chairmanship. He was angered by the company's financial losses and by the steep decline in the value of its stock, which had tumbled from $72 a share in 1961 to $11 a share in the spring of 1964. Only by threatening to take the best engineers and managers with him and start Varian anew did Ginzton manage to keep his job. By 1965, he was the only one of the founders still in a position of executive authority at Varian.[13]

Consolidation

The recession that Varian's engineers dubbed the "McNamara depression" was a turning point in the history of the electronics manufacturing district on the San Francisco Peninsula. It brought about deep restructuring of the electronic component industries. Now that they had been instructed of the dangers of military procurement, the managers of vacuum tube and semiconductor corporations sought to reduce their firms' dependence on the Department of Defense and to insulate them from future fluctuations in the military market. To do so, they diversified their corporations by entering civilian markets. The downturn also led to a consolidation movement in power-grid and microwave tube manufacturing. The main impetus behind this wave of mergers and acquisitions was sheer economic survival. The microwave tube entrepreneurs had to lower their operating costs in order to stay in business. Acquiring other firms enabled them to eliminate overlapping research and engineering programs, close unneeded manufacturing facilities, and thereby realize substantial savings. These acquisitions also allowed the Peninsula industrialists to use their marketing and sales organizations more effectively. More important, mergers offered an excellent way to reduce competition in the specialty tube business. Mergers took competitors off the market. Reduced competition enabled the entrepreneurs to raise prices and return their firms to profitability. Consolidation was the Peninsula's best antidote to McNamara's reform of the military procurement process.

Between 1963 and 1965, the stronger microwave tube firms absorbed the weaker ones. In 1963, Watkins-Johnson acquired Stewart Engineering, a maker of specialized microwave tubes based in Santa Cruz. Eitel-McCullough purchased the Los Angeles division of EMI, a British manufacturer of vacuum tubes. MEC and Litton Industries bought Sylvania's microwave tube businesses. General Electric's microwave tube research laboratory in the Stanford Industrial Park was another acquisition target. The lab, which had opened with great fanfare in 1954, was doing poorly. Starting in 1960, its research contracts and production orders declined and it also became unprofitable. At the same time, GE's corporate management lost interest in microwave tubes. They decided to sell the Palo Alto laboratory. This led Chet Lob, the laboratory's manager, to approach Ginzton at Varian Associates. Lob felt that Varian would be a more congenial employer than other microwave tube firms on the Peninsula. In 1965, Varian bought the GE microwave tube lab.[14]

Varian Associates, Litton Industries, and Watkins-Johnson also sought to acquire Eitel-McCullough, the oldest vacuum tube corporation on the Peninsula. In the early 1960s, William Eitel and Jack McCullough let it be known that they would consider merging their firm with another electronics firm. After more than 30 years in the electronics business, they were ready to retire, but they had no attractive candidate to run Eitel-McCullough. A merger would solve this problem. It would bring in the top executive they needed. At the same time, Eitel and McCullough thought that a merger would enable them to solve the financial problems of their microwave tube division. This division consistently lost money and had no prospect of becoming profitable in the near future. The market was too competitive and the pricing structure was too depressed to make a profit in the klystron business. The only solution was to merge Eitel-McCullough with another microwave tube firm. Otherwise Eitel-McCullough would have to close its microwave tube division.[15]

Eitel's and McCullough's feelers raised considerable interest among the larger microwave tube companies on the Peninsula, especially Varian Associates, Watkins-Johnson, and Litton Industries. For more than a year, Eitel and McCullough negotiated with these corporations and with a few East Coast firms. In the spring of 1965 they settled on Varian. In their view, Varian was a good merger candidate. Eitel and McCullough knew Edward Ginzton well. They had been closely acquainted with him since he had lectured on klystrons at Eitel-McCullough in the early 1950s. They trusted his business judgment and managerial abilities. Eitel and McCullough also liked the fact that Varian was in their words a "decent corporation," a corporation which like Eitel-McCullough cared about the welfare of its employees. Ginzton, who was keen to merge Varian with Eitel-McCullough, made Eitel and McCullough an offer that was too good to refuse. He proposed that each share of capital stock of Eitel-McCullough be converted into half a share of capital stock of Varian. This would make Eitel and McCullough Varian's largest stockholders. With 17 percent of the company's stock, they would sit on the executive committee of the board of directors and have a major say in the direction of the company. Ginzton also promised that some of Eitel-McCullough's managers would run the combined tube operations.[16]

Satisfied with this agreement, Eitel and McCullough approved the merger with Varian in April 1965. Varian's board of directors immediately ratified the agreement. In early July, the shareholders of Varian Associates and Eitel-McCullough voted overwhelmingly in favor of the deal. But the San Francisco office of the Antitrust Division of the US Department of Justice learned about the merger and started to investigate the agreement for violations of antitrust laws. In June, the division asked the corporations to postpone the merger in order to give the Department of Justice enough time to complete its investigation. The DoJ's San Francisco office was concerned that the merger would reduce competition in microwave tubes, especially klystrons. In 1965, Varian and Eitel-McCullough produced about one-fourth of the microwave tubes and nearly one-half of the klystrons made in the United States. A merger would create a klystron monopoly.[17]

To overcome the opposition of the San Francisco office, in July 1965 Ginzton and Eitel visited the head of the antitrust division in Washington and sought to convince him of the innocuousness of the merger. They argued (quite disingenuously) that the merger would not diminish competition in microwave tubes and would not lead to higher tube prices. They also argued that a merger between Varian and Eitel-McCullough would be good for the defense posture of the United States. To bolster their case with the Department of Justice, the executives of Varian and Eitel-McCullough enlisted the support of the Department of Defense. A few weeks after their visit to the Department of Justice, Ginzton, Eitel, and McCullough made a trip to the Pentagon, where they met with Deputy Secretary of Defense Cyrus Vance. They presented their case in front of the Deputy Assistant Secretary for Procurement and the Assistant Secretaries of the Army and the Navy. Ginzton, Eitel, and McCullough also met at the Pentagon with generals and colonels in charge of procuring electronic equipment.[18] In an internal memorandum, Ginzton summarized these meetings:

In each case, [Eitel and McCullough] presented their views on the merger, emphasizing the gravity of Eitel-McCullough's situation, and the fact that a change in the status of Eitel-McCullough's management was inevitable. I explained the general reasons for Varian's interest in the merger and tried to speak, with the aid of various charts, to the point that competition between Varian and Eitel-McCullough was minimal despite the Department of Justice assertion that the companies would have 50 percent of the klystron market. I then spoke of the impact of business conditions in the klystron market upon the two companies, and of the ability of the two companies to conduct research and development work independently. I summarized this by claiming that the companies, after merger, would be stronger in the face of competition against the giants of the industry.[19]

In addition, Ginzton, Eitel, and McCullough made veiled threats to the Department of Defense. In a supporting memorandum, which they gave to their interlocutors at the DoD, Ginzton and Eitel suggested that Eitel-McCullough might leave the microwave tube business if the merger did not go through. The memorandum also claimed that, should the Department of Justice block the merger, Varian would have to "channel new capital investments, company R&D funds and its skilled technical personnel from microwave tubes into more promising lines of endeavor."[20] In other words, Varian and Eitel-McCullough would move out of the microwave tube business if the military services did not support the merger.

These visits enabled Ginzton, Eitel, and McCullough to muster enough support for the merger, and the Department of Justice approved it in August 1965. The combination of Varian and Eitel-McCullough created a vacuum tube giant. The combined firm, which kept the name of Varian Associates, employed 5,400 people in the summer of 1965. It was the largest maker of specialty vacuum tubes in the world. The merger also restored the financial health of Varian Associates' and Eitel-McCullough's tube-making activities. The company achieved substantial savings by eliminating overlapping research and engineering programs in microwave tubes. It cut sales expenses by eliminating Eitel-McCullough's network of manufacturer representatives. (Eitel-McCullough's direct salesmen joined the Varian sales organization.) The merger greatly reduced competition in the klystron business, enabling Varian to raise its prices on klystrons and to make substantial profits in the mid 1960s. In 1966, the vacuum tube group had its most profitable year ever.[21]

Ultra-High Vacuum

The "McNamara depression" led to the consolidation of the microwave tube industry on the Peninsula; it also re-oriented the area's electronic component firms toward the civilian markets. (System-oriented corporations such as Hewlett-Packard also significantly expanded their commercial businesses.) Starting in the early 1960s, corporations making microwave tubes and semiconductors reduced their dependence on the Department of Defense by diversifying into commercial markets. They did so in different ways. Fairchild and other semiconductor firms created commercial uses for their silicon transistors and diodes. On the other hand, Varian, Litton, and Watkins-Johnson, which found little new commercial demand for their microwave tube products, diversified into the production of electronic systems for commercial users. These were businesses where the tube firms could use their competence in high-precision manufacturing and vacuum processing as well as their knowledge of electronic components and circuits. For example, in 1963 Litton Industries' microwave tube division branched out into microwave ovens on the basis of its expertise in low-cost magnetrons. Watkins-Johnson also manufactured furnaces for the semiconductor industry.[22]

But it was Varian Associates that had the most ambitious diversification program in the first half of the 1960s. The firm expanded into vacuum equipment and medical and scientific instrumentation. Varian did so through acquisitions and internal research and development. Edward Ginzton, Varian's chairman, drove these diversification efforts. He was determined to wean the firm away from military sales and re-orient it toward the civilian sector. In Ginzton's view, vacuum equipment and scientific instrumentation were particularly promising commercial businesses for Varian. Beginning in the mid 1950s, the firm had acquired significant technical expertise in these areas. Ginzton also anticipated that large commercial markets would soon emerge for vacuum products and scientific instruments. After the launch of Sputnik, the Eisenhower and Kennedy administrations invested heavily in scientific research and in the training of engineers and scientists. They also considerably enlarged the US space program. In May 1961, President Kennedy committed the United States to landing a man on the moon. This decision led to the Apollo Program and the growth of NASA. In Ginzton's view, these various federal programs would stimulate the demand for scientific instruments. They would also enlarge the demand for ultra-high-vacuum components and systems that would simulate conditions in outer space. In the late 1950s and the early 1960s, the federal government also increased its investments in medical research and the provision of health care, which would open up new markets for medical instrumentation.[23]

Varian's diversification efforts did not go without conflicts. Ginzton's plans were strongly resisted by Myrl Stearns, Sigurd Varian, and other managers at Varian. Stearns and Sigurd Varian argued that Varian was a component company. Its strength was in the design and manufacture of microwave tubes. Expanding into other fields, especially scientific and medical instruments, would be costly and would divert financial and engineering resources away from the tube business. Stearns and Varian impeded Ginzton's expansion plans as much as they could. Mid-level managers also opposed Ginzton's plan of making medical instruments on the grounds that the market for them was too small. The tenacious Ginzton overcame such opposition and forced the firm to diversify into commercial electronic systems.[24]

To expand into the civilian market, Ginzton invested $16.8 million in the development of new system products at Varian from 1960 to 1965. This represented a large share of Varian's overall R&D spending during this period. A significant portion of these monies was invested into building the vacuum pump and system business. The basis of this new business was an important innovation made in the microwave tube division at Varian. In 1956, Lewis Hall, a research engineer at Varian, invented a new vacuum pump, the VacIon pump (also known as the sputter-ion pump), to evacuate klystrons. Unlike other pumps, the VacIon pump depended on electronic rather than mechanical means to create a vacuum. It operated by entrapping gas molecules and atoms through the formation of chemically stable compounds. The pump consisted of a magnet and of an enclosure containing an anode grid placed between two cathode plates. The process by which the pump entrapped gases was complex. At first, a voltage was introduced between the anode grid and the cathode plates. Electrons tending to flow toward the anode were forced into a spiral path by the presence of a strong magnetic field. The increased length of the electron path caused a high rate of collision between free electrons and gas molecules. These collisions produced gas ions. The positively charged gas ions bombarded the titanium cathode plates, which knocked the titanium atoms from the plate. These knocked or "sputtered" atoms were deposited on the anode grid, forming chemically stable compounds with gas atoms such as oxygen and nitrogen. Finally, chemically inert gases were removed by ion burial in the cathode and by entrapment on the anode. In other words, these gases remained within the pump enclosure. Because each collision produced an increasing number of electrons with long effective path lengths, the VacIon pump could evacuate an enclosure to very low pressures.[25]

The VacIon pump had several advantages over mechanical oil pumps. It created a much higher vacuum than any other device on the market. The VacIon pump also produced a very clean vacuum. This was a major improvement, as oil pumps tended to contaminate the inner surfaces of the enclosures they evacuated with hydrocarbons. In addition, the new pump did not require cooling water, which made it easily portable. The VacIon pump had no moving parts and was vibration free. These characteristics made the VacIon pump useful for space chambers. Space chambers simulated conditions in outer space and enabled the testing of the components of satellites and other space systems. VacIon pumps could also be used in vacuum metallurgy, high-energy accelerators for physics research, and the manufacture of semiconductor components. All these applications required a very high and very clean vacuum.[26]

Varian Associates rapidly exploited these market opportunities. Central to the company's aggressive exploitation of sputter-ion pump technology was Louis Malter, a physicist who had recently joined Varian Associates. Malter was a veteran of the electronics industry. He had headed a magnetron design team at RCA during World War II before becoming the chief engineer of the corporation's semiconductor division. Varian recruited Malter to direct its recently formed central research laboratory. But Malter rapidly became more intrigued by the VacIon pump and its commercial potential than by the challenge of building an industrial research laboratory. He thought that the pump offered the opportunity to build an entirely new vacuum products and systems business for Varian. He pressed hard for Varian's entry into the vacuum business and recommended that Varian commercialize the VacIon pump and related components and systems. (Until then, the VacIon pump was only used to evacuate Varian's klystrons.) Ginzton supported Malter's strategy and judged that Malter was the most qualified employee to carry it out. In January 1959, he asked him to direct the VacIon group. Six months later, Ginzton established a new division for vacuum products. Malter left the central research laboratory and became the new division's general manager. Ginzton gave Malter the mandate of building a large vacuum business and making it profitable within 3 years. These decisions angered Lewis Hall, the VacIon pump's inventor. Hall had hoped to head Varian's vacuum business. When his hopes were dashed, he left the company and started his own firm, the Ultek Corporation, in 1959. Ultek made and marketed VacIon pumps in direct competition with Varian.[27]

To establish itself in the vacuum market and to compete with Ultek, Varian Associates developed a broad product line. Much of the design work was done by engineers who had worked in klystron design and production and moved to the new division in the late 1950s. Under Malter's direction, these engineers designed numerous VacIon pumps of different sizes and pumping speeds. The original VacIon pump operated at 10 liters per second. In the late 1950s and the first half of the 1960s, engineers at Varian developed pumps with evacuation speeds ranging from 0.2 liter to 10,000 liters per second. Each pump was engineered with a different application in mind. In conjunction with this broad family of VacIon pumps, engineers at Varian also developed a new line of vacuum

Figure 5.1
Louis Malter (right) and William Lloyd with a large VacIon pump, early 1960s.
Courtesy of Varian, Inc. and Stanford University Archives.

components. Vacuum components offered by other manufacturers were not good enough for the ultra-high vacuum that the VacIon pumps could generate. As a result, Varian engineers designed new valves, flanges, fittings, feed-throughs, and compression ports. In 1962, the engineering group also developed a vacuum leak detector and an ion gauge system. In conjunction with the development of a line of vacuum components, Varian's engineers assembled their products into whole systems. At first, they made custom systems that met the specifications of particular customers, such as Republic Aviation or the Atomic Energy Commission. But in 1961 Varian also introduced standard vacuum systems. These systems were suitable for environmental testing and vacuum evaporation applications, especially in the semiconductor industry.[28]

Varian's vacuum product line did well in the marketplace. In the late 1950s and the 1960s, three markets emerged for ultra-high-vacuum devices and systems. The United States' growing investments in space technology, especially the Apollo Program, created a large demand for systems that could simulate conditions found in deep space. Physics and chemistry laboratories at universities and other research institutions constituted the second market for Varian's vacuum products. Many research groups embraced the VacIon pumps and coupled them with sensitive analytical instruments such as mass spectrometers, electron microscopes, and x-ray diffraction systems. High-energy physics laboratories in the United States and Europe also became large users of VacIon pumps.[29]

The VacIon pump was seeing more use in semiconductor manufacturing. Fairchild Semiconductor and other makers of silicon devices were buying Varian's pumps for their R&D laboratories. They were also purchasing the firm's vacuum evaporators in greater numbers than before. These evaporators were used for the deposition of aluminum contacts on silicon transistors, a technique that had been partially developed at Fairchild Semiconductor. Evaporators could also be used to make hybrid or thin film circuits. Hybrid circuits are miniaturized electronic circuits made of thin film devices and active semiconductor components. With these evaporators, manufacturers of hybrid circuits could deposit thin films on the circuits' ceramic substrates under a very high vacuum. These thin films formed passive elements such as resistors and capacitors. Transistors and diodes were then mounted on the substrate.[30]

Fierce competition with Ultek, Varian's spinoff, further stimulated the growth of the market for ultra-high-vacuum products. Starting in 1959, Ultek developed its own line of sputter-ion pumps. It also designed and fabricated environmental chambers simulating deep space environments.

To gain market share, Ultek introduced its products at a much lower price than Varian Associates. As a result, prices for sputter-ion pumps tumbled. As an example, Varian's five-liter VacIon pump, which sold for $795 in 1958, could be purchased for $195 in 1960. This price war substantially cut into Varian's and Ultek's profits. (Varian's vacuum products division lost money until 1961.) But falling prices significantly stimulated the demand for ultra-high-vacuum pumps and systems. As a result, Varian's sales of vacuum products increased from $211,000 in 1958 to $11,287,000 in 1966. By mid 1965, the vacuum products division employed 315 engineers, technicians, and operators. In 1965 and 1966, that division brought in $1.7 million in profits. By that time, Varian was the largest producer of vacuum equipment in the United States. Ultek carved out a significant market for its vacuum pumps and systems as well. Its sales reached $3.5 million dollars in 1965 and more than $4 million in 1966. In the mid 1960s, Varian and Ultek monopolized the high-vacuum market.[31]

Instruments

In tandem with their successful foray into vacuum products, Varian's managers built a large commercial instrumentation business in the late 1950s and the second half of the 1960s. Again their goal was to reduce the firm's dependence on military contracts. To build these businesses, they took advantage of innovative research programs in Stanford University's physics department. For example, Ginzton built an electron linear accelerator business on the basis of his work in the Microwave Laboratory at Stanford. In the late 1940s and the 1950s, Ginzton had directed the construction of ever larger and more powerful linear accelerators for high-energy physics at Stanford. These efforts culminated in the late 1950s with the planning of a two-mile-long linear accelerator to be built on the Stanford campus. Ginzton and his group also constructed a linear accelerator for cancer therapy. In the mid 1950s, Ginzton urged Varian Associates to enter the linear accelerator business. In spite of considerable resistance from Myrl Stearns and Sigurd Varian, Varian's management established a new division, the radiation division, to design and construct linear accelerators. This division made research accelerators for DuPont, the Danish Atomic Energy Commission, and the Frascati Laboratory in Italy. They also produced accelerators used in the treatment of deep-seated cancers. But these machines failed to find a large market in US hospitals. The medical accelerators were expensive. Small medical centers could not afford them, and few radiotherapists and radiation technicians had the skills needed to use them. Because of the commercial flop of the medical accelerator and the mixed success of its research machines, the radiation division was barely profitable, with sales of only $4.7 million in the mid 1960s (Ginzton and Cottrell 1995; Galison et al. 1992).[32]

Varian's chemical instrumentation business grew out of the research project of two Stanford physicists, William Hansen and Felix Bloch, on nuclear induction, also known as nuclear magnetic resonance (NMR). In 1946, Bloch, Hansen, and their graduate student Martin Packard first observed the phenomenon of nuclear magnetic resonance—an observation for which Bloch received the Nobel Prize in physics in 1952. In their experiment at Stanford, Bloch, Hansen, and Packard found that nuclei act like tiny magnets and that a strong magnetic field exerts a force that causes them to rotate (precess). When the natural frequency of the precessing nuclear magnets corresponds to the frequency of a weak external radio signal striking the material, energy is absorbed by the radio wave. This selective absorption (also called resonance) was produced by tuning the natural frequency of the weak radio wave to that of the nuclear magnets. Bloch's and Hansen's innovative experiment piqued the curiosity of Russell Varian, who had recently returned from the East Coast and worked as an unpaid research associate in the physics department. After discussions with Hansen and a number of chemists, Russell Varian came to believe that the phenomenon of nuclear magnetic resonance could be used to analyze chemical compounds. He urged Bloch and Hansen to file a patent on the "Method and Means for Chemical Analysis by Nuclear Induction." He also offered to prepare the patent application in return for an exclusive license to be transferred to Varian Associates once the company was established. Hansen and Bloch accepted the offer and filed the patent application in December 1946.[33]

Russell Varian's optimism bore fruit in the next few years. In the early 1950s, doctoral students in the Bloch lab discovered that the value of the magnetic field at the nucleus depended to some extent on the chemical environment, the so-called chemical shift. They also found that different organic compounds had different magnetic fingerprints (in the form of different NMR spectra). This was an important finding. Magnetic fingerprints could help analyze new organic compounds and determine the structure of these molecules. Spurred by these results in the Bloch lab, in 1951 Russell Varian, Sigurd Varian, and Edward Ginzton decided to start a new business around NMR. The company would build an entirely new chemical instrument, the NMR spectrometer. To develop this chemical instrument, Varian's managers relied heavily on the apparatus designed by Bloch, Hansen, and Packard for their Stanford experiments. They also recruited some of Bloch's most talented students, including Martin Packard and Emery Rogers. They also hired James Shoolery, an entrepreneurial physical chemist who had recently received his doctorate from Caltech. In the first half of the 1950s, the new recruits engineered a series of NMR spectrometers. To do so, they drew on the firm's expertise in radar circuits. (The circuits used in NMR spectrometers are similar to those used in radar.) They also incorporated the latest innovations in NMR technology: enhanced magnet technology and advances in frequency stabilization. Some of these innovations were developed internally at Varian. Others came from Stanford and the industrial laboratories that used the first NMR spectrometers. But these machines were difficult to operate. They were devices for research chemists with considerable knowledge of physics. Moreover, the spectrometers were expensive. As a result, Varian's chemical instrumentation business remained modest in the first half of the 1950s. In 1956, the firm sold only $410,000 worth of NMR spectrometers.[34]

To expand the market for NMR spectrometers, in the late 1950s and the early 1960s Varian Associates (at the instigation of Ginzton and Russell Varian) invested heavily in the development of new instruments. Central to these efforts was the design of the A-60 spectrometer. In 1957, Shoolery and Rogers made a proposal for a four-year program to construct an entirely new NMR spectrometer. The goal of this effort was to create an instrument simple enough for any organic chemist or graduate student to operate with the aid of a manual. It would be affordable, reliable, and conveniently sized and shaped. The company funded the project and formed a team of chemists, physicists, and electronic engineers to design the new machine. To tackle this complex task, John Moran, an electronic engineer who headed the project, adopted new management techniques that had been pioneered in the aerospace industry. He used critical path analysis and Program Evaluation Review Techniques charts to coordinate the design and production of all the spectrometer's features and components and at the same time keep the project on schedule and within budget. The end result, the A-60 (A for analytical and 60 for the frequency), easily fit into an ordinary chemistry laboratory. Varian introduced it to the market in 1961. The machine was reliable, generated reproducible results, and was user-friendly. In other words, it could be used in most organic chemistry laboratories. Its price, $23,750, also made it affordable for university chemists.[35]

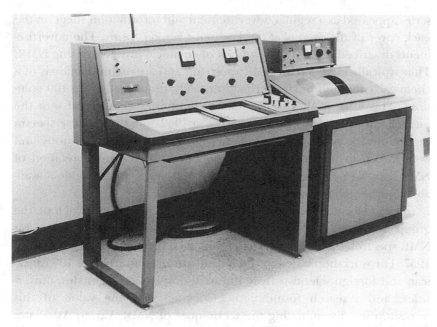

Figure 5.2
The A-60 NMR spectrometer. Courtesy of Varian, Inc. and Stanford University Archives.

In conjunction with the design of a user-friendly instrument, the Varian group developed innovative marketing techniques in order to create a demand for this machine. To convince chemists of the power and usefulness of NMR spectroscopy, Shoolery and other chemists at Varian opened several Applications Laboratories. The first Applications Laboratory was located at the firm's headquarters in Palo Alto. Shoolery and his group later established similar laboratories in Zurich and at the Pittsburgh International Airport. These laboratories had several functions. They helped outside chemists solve their chemical problems. These chemists would submit their samples to Varian Associates, which would then use its NMR spectrometers to analyze them. In addition to this service function, the Applications Laboratories devised new approaches for eliciting information from nuclear radiospectra and communicated this information to the chemical community. In the second half of the 1950s, Varian's scientists published 65 articles on NMR spectrometric methods and their applications to chemistry. To further publicize their work and promote NMR spectroscopy, Shoolery and his colleagues introduced the "NMR at work" series. Starting in 1957, this series appeared as a regular advertisement and information sheet on the back cover of the *Journal of the American Chemical Society*. The advertisements described exemplary solutions of chemical problems using NMR. They typically included an NMR spectrum and a structural analysis of a chemical compound. The series eventually numbered over 100 solutions. This work of the Applications Laboratories culminated with the publication of a catalog of 700 spectra in 1962 and 1963. These spectra represented a carefully selected variety of chemical compounds and were meant to help chemists develop a feel for the interpretation of NMR data. Varian gave this catalog to its customers. It also made it available to other laboratories at a nominal cost.[36]

To educate chemists about the power of NMR as an analytical tool, the group innovated further by organizing an annual workshop series on NMR spectroscopy. Varian Associates held its first workshop in October 1957. The workshop lasted four days and attracted more than 100 American and foreign scientists from industry, government agencies, universities, and research foundations. Convinced of the value of this workshop for disseminating the techniques of NMR, Varian Associates organized three more workshops in Palo Alto. To encourage European adoption of the instrument, a similar workshop was organized in Zurich during the 1960s and continued for a decade. Conducted like an academic course, with lectures and laboratory sessions, this workshop became a major vehicle for informing chemists about NMR instrumentation. The workshop was divided into three parts. The first part was an introductory day of lectures on NMR instrumentation followed by hands-on laboratory instruction. The second and third parts of the course were devoted to more advanced topics. As an example, in one section Shoolery offered step-by-step instruction in the use of NMR as a tool for structural analysis, the difficult business of translating spectral lines into chemical structures.[37]

These marketing programs proved very effective in creating a market for Varian's instruments. The wide acceptance of Varian's spectrometers was also greatly facilitated by the federal government's growing investments in chemical research and chemical instrumentation in the first half of the 1960s. During this period, the National Science Foundation, the Atomic Energy Commission, and the National Institutes of Health made substantial investments in chemistry, offering grants specifically for the acquisition of chemical instruments. For example, the National Science Foundation gave its first two chemical instrumentation grants in 1959. Three years later, it awarded 21 such grants. The Atomic Energy

Commission and the National Institutes of Health administered similar programs. As a result of these programs, chemistry departments at American universities greatly increased their investments in chemical instruments in the first half of the 1960s. The top 125 PhD-granting chemistry departments in the US had spent $14 million on instrumentation between 1954 and 1959. In the next 5 years, they invested $36 million in scientific instrumentation. The European market for chemical instrumentation also grew rapidly in the 1960s, as France, Germany, and the United Kingdom tried to close the perceived scientific gap with the United States and invested heavily in scientific research and education (Committee for the Survey of Chemistry 1965; Stine 1992).

Varian benefited from these public investments in science. It saw its sales of NMR instruments go from $1,163,000 in 1957 to $11,045,000 in 1965. Much of these sales came from the A-60 spectrometer. The company also introduced high-end NMR spectrometers to the market in the early 1960s. These machines were designed for chemical laboratories engaged in fundamental research. By the mid 1960s, Varian Associates controlled roughly 70 percent of the worldwide market for NMR spectrometers. (Japanese and European firms owned the rest.) The company was thus a significant force in the discipline of chemistry as NMR spectroscopic techniques became increasingly used in chemical laboratories. In 1964, 18 percent of all papers published in US chemical journals relied on results obtained from NMR spectroscopic techniques. NMR spectroscopy had become an indispensable tool in chemistry.[38]

To complement the development of NMR spectrometers at Varian, Ginzton acquired several instrumentation and vacuum companies in the mid 1960s. His goal was again to reduce Varian's dependence on the Department.of Defense. He bought firms that strengthened Varian's market position in vacuum equipment and chemical instruments. In 1964, Varian purchased Mikros, a vacuum company based in Portland, Oregon. Mikros produced vacuum oil pumps and oil-pump-based vacuum systems. These products extended Varian's product offerings into the low-vacuum market. Building on Varian's strength in NMR spectroscopy, Ginzton acquired firms making other types of chemical instruments. In 1965, he bought Wilkens Instrument, a California-based manufacturer of gas chromatographs used to separate organic compounds. He also purchased a small German firm, which made mass spectrometers. In 1966, Varian further enlarged its instrument portfolio by merging with Applied Physics Corporation. Applied Physics was an important manufacturer of UV spectrometers, used in organic chemistry.

These acquisitions significantly strengthened Varian's instrument line and made the company one of the leading suppliers of chemical instruments in the United States.[39]

By the mid 1960s, Varian was a different company than it had been at the beginning of the "McNamara depression." It was now oriented toward commercial markets. In 1959, more than 90 percent of Varian's sales had come from the military sector. Eight years later, military sales accounted for only 40 percent of the firm's total revenues. The rest came from civilian customers. Varian's diversification policy, its acquisition of instrumentation businesses, and its merger with Eitel-McCullough also greatly changed the firm's product line. In 1959, Varian was a microwave tube firm with a small instrumentation business. Seven years later, Varian was a science-based conglomerate, manufacturing vacuum pumps, evaporators for semiconductor manufacturing, medical and scientific instruments, and a wide range of specialty vacuum tubes. Because of its re-orientation toward commercial markets, Varian expanded rapidly. Its sales grew from $38.1 million in 1959 to $145.1 million in 1966. By that time, Varian was one of the 500 largest corporations in the United States. But Varian's diversification efforts did not come without a price. In the mid 1960s, the firm became more bureaucratic and lost some of its entrepreneurial impulse. Varian, which had once prided itself for its egalitarianism and its sharing of information with all employees, became much more hierarchical. The share of the company owned by its employees also shrank. In 1965, employees owned only 30 percent of the company's stock. In other words, by the mid 1960s Varian Associates was not the engineering cooperative it had once been. It was more of a conventional electronics corporation.[40]

Silicon for Industry

While Varian Associates diversified into commercial systems, Fairchild Semiconductor, the main manufacturer of silicon devices in the Santa Clara Valley, built markets for its transistor products in the civilian sector. The instability of the military demand for silicon transistors during the "McNamara depression" encouraged Fairchild's reorientation toward commercial markets. But this re-orientation was also shaped by a capitalistic urge to exploit new market opportunities. In the first half of the 1960s, Robert Noyce, Fairchild Semiconductor's general manager, and Thomas Bay, his head of marketing and sales, increasingly understood that the firm's high-performance planar silicon transistors were in growing demand among commercial computer and consumer electronics corporations. As a result, they gradually re-oriented the firm toward the civilian sector, before deciding in 1963 to concentrate Fairchild's resources on the computer and consumer electronics markets.[41]

In 1960, the marketing and sales department discovered that recent regulations of the Federal Communications Commission were opening up a market for high-performance and high-reliability silicon components in television broadcasting. After much tergiversation, the FCC decided at this time to fully exploit the ultra-high-frequency (UHF) portion of the electromagnetic spectrum. The VHF bands then in use could not accommodate the increasing number of television broadcasting stations. The FCC decided to take advantage of the higher frequencies in order to meet the growing demand for television stations. This led the FCC to require all television monitors to have the capability of receiving UHF channels. This ordnance had the unintended consequence of creating a market for silicon transistors. Because TV tuners using conventional radio tubes could not handle the higher frequencies, television manufacturers became increasingly interested in planar silicon transistors. Unlike standard vacuum tubes, planar transistors could easily operate in the UHF range. They had the added benefit, unlike germanium transistors, of being dependable and temperature resistant, a characteristic that was important to the reliability-oriented television industry.[42]

Similarly, Fairchild's salesmen identified a new market opportunity in the commercial computer industry. In the early 1960s, commercial computer firms became increasingly interested in silicon devices.[43] By that time, computer architectures had largely stabilized, and manufacturers of commercial computers increasingly competed by improving the speed and reliability of logic circuits. As the performance of these circuits depended to a large degree on electronic components, computer manufacturers that had previously used germanium transistors became increasingly interested in planar devices. Planar transistors could withstand higher temperatures than germanium transistors. They also had equal or greater frequency characteristics.[44]

Control Data Corporation (CDC), which made high-end computers for computationally demanding applications, is a case in point. In 1961, Seymour Cray and his group at CDC's laboratory in Chippewa Falls, Wisconsin, decided to use planar silicon transistors for the prototype of their new computer, the 6600. The 6600 was a supercomputer meant to compete directly with IBM's machines in the scientific computer market. At first, Cray had designed his computer around standard packaging techniques. But he soon ran into substantial technical difficulties. Early design experiments revealed noise and oscillation problems in the back panel wiring which were deemed unacceptable for a system of that size. The switching speeds required were so high that wave lengths present in the signals were shorter than the wire lengths in the back panels. In other words, the physical size of the computer had become a limitation on speed.[45]

To shorten the wiring paths, Cray and his group made drastic changes in packaging configurations. The firm replaced traditional building blocks with a new package, nicknamed the "cordwood package." This package had much higher function densities. It was made of two small printed circuit boards back to back with all components mounted internally. Up to 64 transistors could be assembled in each module. This was a tenfold density improvement over previous packaging techniques at CDC. While these packages solved the computer's noise and oscillation problems, they imposed severe temperature constraints on transistors. The germanium transistors that the CDC group used in its cordwood packages did not withstand these high temperatures. As a result, in 1961 the engineering team designing the Control Data 6600 looked at planar silicon devices, which were more temperature resistant than germanium transistors.[46]

Noyce and Bay were quick to understand that the computer and consumer electronics markets had great potential. The computer market was especially promising. Because of the growing requirements for computers in science, industry, and government, the commercial computer industry exploded in the early 1960s. Computer shipments swelled from 1,000 in 1957 to 4,500 in 1962. By 1967, the total production of computers had reached 18,700. Because computers such as the Control Data 6600 used more than half a million transistors, the commercial computer industry represented a very large market indeed, one that potentially dwarfed the military sector. The consumer electronics industry offered a very sizable and fast expanding market as well. The total production of television sets grew from 5 to 7 million between 1958 and 1963. There was additional demand for silicon devices in other consumer electronics products, such as transistor radios and stereo systems. These growing markets were supported by the rise of consumer culture. In the 1950s and the first half of the 1960s, Americans used their new affluence to buy household appliances and entertainment devices. In this consumerist age, radios and televisions were seen as basic necessities. Vast markets for advanced electronic components opened up (Flamm 1988).

To exploit these opportunities, Noyce and Bay restructured Fairchild's sales and marketing organization. Up to 1960, Fairchild had had a relatively small and military-oriented marketing and sales force. Noyce and Bay greatly strengthened the sales and marketing organization in the early and mid 1960s. Bay recruited a group of bright and aggressive salesmen and marketing specialists. Among them were Jerry Sanders and Floyd Kvamme, each of whom would rise to a leadership position in the semiconductor industry on the San Francisco Peninsula. These newcomers transformed Fairchild's sales and marketing department into one of the most effective in the semiconductor industry. To complement his internal sales force, Bay established a network of distributor representatives in 1961. The internal sales force focused on the larger accounts, while the distributors served the smaller customers. Noyce and Bay also gave the sales and marketing department strong technical backing by transferring the applications engineering laboratory from R&D to marketing in 1961. Previously, applications engineering had concentrated on characterizing and evaluating new devices for military applications. After the restructuring, applications engineering increasingly gave technical support to sales and marketing and acted as a technical interface with commercial customers.[47]

Bay also reconfigured the sales and marketing department in the early and mid 1960s. To better understand the needs of commercial users, Fairchild Semiconductor restructured product marketing as well as the regional sales offices along market lines in 1963. "When [the company] really started getting into the commercial markets," Bay recalled, "we [the management group] reorganized to not only regions but market area. For instance, on the West Coast, we had a computer guy who handled Scientific Data Systems and GE Phoenix, and we had a consumer guy. And then we had product marketing organized similarly: we had a product marketing group for each of those markets to keep our finger on what the [users] were looking for and so what products we needed to continue to grow in various market segments."[48] Two years later, Fairchild, in an innovative move, restructured its entire sales and marketing department around four markets: computer, consumer, industrial, and military. Enjoying substantial autonomy, each market division had its own sales, product marketing, applications engineering, and product planning groups and its own legal office to draft production contracts with customers.[49]

In conjunction with the building of a strong commercial marketing and sales department, Noyce, Bay, and Gordon Moore (the head of R&D at Fairchild Semiconductor) gained a better understanding of commercial requirements by hiring system engineers. In 1962, Moore established a high-speed memory engineering department in the research laboratory to develop a memory system. He staffed it with engineers from GE's computer division. A year later, Moore established a digital systems department in the research laboratory under Rex Rice, a computer engineer from IBM, to learn about new computer architectures and packaging techniques. Management also recruited a large number of circuit and system engineers with experience in the computer and consumer electronics industries. They were assigned to beef up the device development, marketing, and applications engineering departments. Indeed, most engineers in the applications engineering laboratory had previously worked at computer and consumer electronics firms before joining Fairchild. A large contingent of applications engineers had been schooled in computer technology at IBM and at Philco's Western Development Laboratory. Others came from Zenith and various consumer electronics companies. As Fairchild brought in system engineers from user sectors and built a strong sales and marketing department, the firm gained a solid system and marketing expertise in the computer and consumer electronics industries. Because of its strong system expertise, Fairchild Semiconductor knew nearly as much as system firms about system design.[50]

Exploiting this system competence as well as the strong interface with commercial users, Fairchild's device engineers developed high-performance transistors and diodes that met the complex circuit and system requirements of computer and consumer electronics manufacturers. To do so, they collaborated closely with in-house applications engineers. These applications engineers identified the needs of commercial and military manufacturers by relying on their circuit and system expertise and by conferring with customers. Based on these requirements, they determined the devices' characteristics. "We sought to understand the circuit and system requirements of the customer," reminisced John Hulme, the head of the applications engineering laboratory, "and from there define the product. We would start by asking: What does the system need to do? How can we do it with what we have?"[51] Having identified the circuit and system needs of the users, applications engineers, in collaboration with the research and development laboratory, designed components that fit the firm's manufacturing processes and met the system needs of commercial users.[52]

Following this approach to component design, Fairchild developed a broad line of system-oriented products in the early and mid 1960s. In

1960 the firm had eight military-type transistors on the market; in 1964 it manufactured 130 different transistors, many of which had commercial applications. Each transistor was meant to fit a particular circuit or system niche. For example, Fairchild designed, among others, transistors for saturated switching circuits, non-saturating switching circuits, and amplifier and oscillator circuits. Particularly noteworthy was an automatic gain control transistor, which a group of research engineers designed in 1962. Automatic gain control (AGC) is a feature critical to the design of intermediate frequency (IF) modules widely used in radio and television applications. Until then, engineers designing radios and television sets could obtain this electronic function through the use of vacuum tubes. Fairchild engineers were the first ones to design an AGC silicon transistor. The design of this transistor made all transistor IF modules practical and thereby improved the reliability and manufacturability of consumer electronic products. This transistor gave the firm a substantial competitive advantage in the consumer electronics market. Going one step further in its efforts to meet the circuit and system needs of commercial users, Fairchild customized its transistors for individual computer manufacturers. For example, the device development group at Fairchild designed new switching transistors for Control Data's line of supercomputers.[53]

Fairchild Semiconductor seeded a market for its products by developing new applications around its diodes and transistors and by giving these circuit and system designs to its customers at no cost. Its goal was to broaden markets for existing products and to develop markets for new products. The applications engineering laboratory, again, was central to this effort. It concentrated on writing applications notes and designing prototypes of commercial equipment around the firm's diodes and transistors. Fairchild's applications notes were meant to show commercial customers how their problems could be solved better with silicon devices than with vacuum tubes, germanium transistors, and electromechanical components. The applications notes described the functions commercial manufacturers could perform with Fairchild's diodes and transistors and the systems they could make with the firm's components. They also explained in great detail how these circuits and systems could be designed. They included circuit diagrams, board layouts, and, in some cases, a mathematical analysis of the circuit or a short discussion of the system's mechanical design. A note on a "transistor stereo FM multiplex adapter," published in 1962, offers a good example of these applications notes. It showed how to build a new kind of adapter for multiplex and

Figure 5.3
Fairchild's applications engineering laboratory, with applications notes in the
foreground, 1965. Courtesy of Fairchild Semiconductor and Stanford University
Archives.

monophonic FM reception with Fairchild's diodes and transistors.
Besides a comparative treatment of various demodulation techniques
and their respective advantages and drawbacks, the note presented a dia-
gram of the FM adapter and a printed circuit board layout in actual size.
It also explained how to integrate the new device into stereo equipment.
Besides producing detailed applications notes, Fairchild Semiconductor's marketing group published a newsletter from 1961 on. *Fan
Out* introduced the firm's new devices and offered short descriptions of
possible applications for them to its customers.[54]

With their creative designs and deft use of silicon components, the
applications notes sold Fairchild's devices. They became instant engineering classics and were widely used by circuit and system engineers in
the commercial sector. "People would design right out of the applications
notes," recalled Hulme. "They would take the applications note and
would essentially copy it."[55] Fairchild's applications notes were used
extensively by computer and consumer electronics firms in the United

Figure 5.4
A color television monitor designed at Fairchild, 1965. Courtesy of Fairchild
Semiconductor and Stanford University Archives.

States. They were also copied by radio firms in Asia. It has been reported
that Sony and most Hong Kong electronics firms closely modeled their
transistor radios after Fairchild's notes. Indeed, Fairchild's application
laboratory served as the de facto engineering department of many Hong
Kong's radio manufacturers, introducing a stream of radio designs which
these firms later produced and marketed.[56]

Going beyond the writing of applications notes, engineers in the applications laboratory built and demonstrated prototypes to convince commercial manufacturers of the system potential of Fairchild's devices. They
designed and constructed a wide variety of electronics-based devices and
systems such as burglar alarms, calculators, and radio transmitters. For
example, in the early 1960s they developed a signaling system for the
Ford Thunderbird to demonstrate that it could be built more easily and
at lower cost with silicon devices than with electromechanical components. "We wanted that business," recalled a former marketing manager
at Fairchild, "and we wanted to show that we could do that electronically.
So we bought a replacement rear fender for a Thunderbird and we had
this mounted in the conference room. These Ford executives came in
and we made it do its thing, turn right, turn left based on electronics and,

of course, it was a tiny package. That's how we tried to get people interested in our components."[57]

Similarly, the applications engineering laboratory designed one of the first transistorized television sets in 1962 to demonstrate the feasibility of using transistors in TV. Along with notes on IF amplifiers and other television circuits, Fairchild's applications engineers built a television prototype which was smaller, more reliable, and had better temperature characteristics than TV sets then available on the market. With the assistance of the firm's sales and marketing department, they later demonstrated it with the corresponding transistor kit to television manufacturers such as Zenith and General Electric. The Fairchild design vastly exceeded conventional television performance. It was more reliable than television sets on the market. It was also cheaper to produce as it was based on a modular design. As a result, a major manufacturer of television sets adopted it in 1963 and introduced it to the market in the following year. By 1965, Fairchild's television set was mass produced by General Electric, Zenith, and Sylvania. Fairchild Semiconductor dominated the television transistor business well into the late 1960s.[58]

Mass Production

Along with the design of system-oriented products and the development of new applications, Fairchild Semiconductor drastically reduced prices in order to build a large commercial market for its products. Understanding that the adoption of silicon components in the consumer electronics and computer industries depended to a large degree on price cuts and increases in production volume, Noyce, Bay, and Moore pursued a strategy of drastic price reductions. Not unlike Texas Instruments, which a few years earlier had cut prices to create a commercial market for germanium devices, Fairchild gradually reduced its prices on silicon transistors in the early 1960s. In the spring of 1963, the company announced a dramatic price cut on some of its planar transistors. The firm reduced prices on these products from $5 to 25 cents. By 1964, the price for these devices had fallen to 10 cents. Sometimes, Fairchild Semiconductor was selling components below manufacturing costs with the expectation that profitability would be obtained by reaching higher production volumes (Freund 1971).[59]

These price reductions and the related growth in production volumes were made possible by a thorough restructuring of the manufacturing department. Central to this reorganization was Charles Sporck, a mechanical engineer who joined Fairchild as production manager in the fall of 1959. Over the next few years, Noyce gave Sporck the mandate of reducing production costs and moving to mass-production volumes. This may have been the most difficult manufacturing problem in the US industry at the time. Producing advanced silicon transistors in small quantities was already challenging. The manufacturing processes were poorly understood. They were also very hard to control. As a result, manufacturing yields, the percentage of good transistors coming out of the line, were low. Forcing these fragile and unstable manufacturing systems to produce millions or tens of millions of transistors was an even more difficult task. But Sporck was the right man for the job. He was forceful, ambitious, and aggressive, and he had significant mass-production experience. Before moving to Fairchild, Sporck had worked as production supervisor at General Electric, where he had directed the production line for power capacitors and manufactured millions of power capacitors per year. Because of his GE experience, Sporck had also built contacts with other mass-production engineers whom he could rely on to transform Fairchild Semiconductor into a mass-production company.[60]

In the early 1960s, capitalizing on these contacts, Sporck recruited mass-production engineers from General Electric to scale up production and reduce manufacturing volumes. He hired additional mass-production specialists from the Ford Motor Company and other firms. In the first half of the 1960s, Sporck and his group introduced a new mindset characterized by cost consciousness and an emphasis on output. They also gradually reshaped the organization and the techniques of production at Fairchild. They transformed a small craft-oriented manufacturer into the first mass producer of planar diodes and transistors. To do so, they introduced new organizational forms and production-control techniques from Ford and General Electric. These men also scaled up production and drastically reduced manufacturing costs through yield improvement, throughput enlargement, and the establishment of offshore assembly plants in the Far East. This massive restructuring of the production function did not go without conflict. Eugene Kleiner, a co-founder of Fairchild Semiconductor, and other engineers with a background in specialty production, resisted Sporck's aggressive policies. The historical record is largely silent on these conflicts, but it is clear that the specialty production-oriented engineers lost the fight. Noyce demoted Kleiner from his position as head of industrial engineering and sent him to oversee the facilities of the research and development laboratory. In 1961, Kleiner left Fairchild Semiconductor. He later founded Edex, a manufacturer of teaching machines, and became an active investor in small electronics firms on the Peninsula. He formalized these activities by establishing Kleiner Perkins, a venture capital partnership, in 1972.[61]

To increase production volumes and reduce manufacturing costs, Sporck restructured the organization of production at Fairchild Semiconductor along the General Electric model. Until then manufacturing at Fairchild had been organized around process steps and had been characterized by parallel organizations for production and process engineering. In the early 1960s, Sporck moved the company in a new direction by reorganizing the firm's production department along product lines. These product lines were built around families of products, which used similar manufacturing processes such as small geometry NPNs, large geometry NPNs, PNPs, and radio-frequency transistors. Product lines were responsible for processing devices from wafer start to shipping and for lowering manufacturing costs and improving production yields. Directed by a product engineer, each product line had a production group and an engineering group. The production group was responsible for processing the devices to specifications and schedule. The engineering group maintained, troubleshot, and documented the process. These product lines had their own diffusion and assembly areas, making them, in effect, mini-factories within the larger plant. While under the previous organization authority and responsibility had been diluted and relations between engineering and manufacturing had often been antagonistic, the new structure centralized power and authority in the hands of the product engineer and helped build groups which addressed problems as teams instead of as adversaries. Product lines also made for more efficient manufacturing as the equipment was used continuously and the engineering and production staff could gain experience in the manufacture of a few products.[62]

The mass-production engineers emphasized production control. Sporck and the engineers whom he brought from General Electric centralized and strengthened the production planning function. They also introduced new production and inventory control techniques from General Electric to plan, schedule, and expedite the flow of parts and materials through the factory. In particular, they brought in new techniques for controlling wafer-in-process inventories. This was a major challenge, as the number of wafers increased dramatically in the early and mid 1960s. More generally, Fairchild Semiconductor's manufacturing engineers set up an elaborate system to track the progression of wafers, dice, and transistor packages within the fabrication area and the assembly lines. This was a task of substantial proportion as Fairchild processed thousands of different components, some counting in the millions, at any given time in its factories by the mid 1960s.[63]

The new management techniques and organizational structures brought by Sporck and his associates enabled these men to enlarge production throughputs and to improve yields in the wafer-fabrication areas. In the first half of the 1960s, Sporck and his mass-production engineers maximized throughput by redesigning production equipment and manufacturing processes. They continually increased the number of wafers in each batch. For instance, epitaxial reactors, used for the deposition of semiconductor films over the silicon crystal, treated 16 silicon wafers per load in 1962. In 1964 they were able to process 26 wafers per load. Sporck and his group also used ever larger silicon wafers. Whereas their predecessors had used ¾-inch ingots, these men progressively enlarged the wafers' diameter to two inches, leading to a sevenfold increase in wafer surface areas. The increase in batch and wafer size made possible a substantial increase in production volumes. It also had the intended benefit of maximizing the use of capital equipment and therefore reducing per unit costs.[64]

Along with continuous volume increases, Fairchild's manufacturing engineers tightened their control of production processes as a way of improving yields. Yields were the largest cost determinant in the wafer-fabrication area. Yields, or the percentage of fully operational units emerging from the production process, were low in the early 1960s. They averaged 30 percent for planar transistors. To improve yields, the product line groups tightened their control of complex and unforgiving processes. They achieved this through experimentation, trial and error, and by systematically applying statistical process control techniques to manufacturing problems. These efforts were backed up by a large process research and engineering program in the research laboratory. Engineers and physicists in R&D systematically analyzed failure mechanisms. They investigated the physics and chemistry underlying silicon processing and developed new manufacturing processes. Fairchild's engineers concentrated at first on photomasking and solid-state diffusion. These processes were the main determinants of yield loss in the early 1960s. Sources of defects in diffusion and photolithography such as misalignment in successive photoengraving steps were increasingly identified and remedied. As a result, yields increased rapidly. They plateaued in the mid 1960s, however, leaving Fairchild's process-engineering groups to look for other failure modes. After they identified dust particles as a major yield determinant, they tightened the cleanliness standards in the fabrication area.[65]

In conjunction with their program of yield improvement and throughput enlargement in wafer fabrication, Sporck and the mass-production engineers devoted considerable attention to assembly operations. This was the second part of the manufacturing process, where labor, rather than yield, was the largest cost determinant. At first, they used motion studies to improve the productivity of each work station. These studies enabled them to maximize the output of individual stations in the assembly process. Die attach, the process by which the die or chip is bonded to the package, is a good example. Fairchild's industrial engineers more than doubled the productivity of die attach stations between 1961 and 1964. In 1961 operators attached 90 silicon dice per hour; in 1964 the average output reached 220 dice per hour. Fairchild's engineers achieved this increase in productivity by small and seemingly trifling adjustments in work stations. For instance, they brought the packages closer to the die plate. They mounted a triangular plastic piece on the microscope to help the operator position herself better in relation to the microscope. And finally, they added an electronic mechanism, which better controlled the temperature of the bonding material.[66]

Because these improvements in productivity were not sufficient to reach commercial prices, Sporck delocalized Fairchild Semiconductor's assembly operations to low-labor-cost areas. Sporck's primary goal was to reduce labor costs and overhead expenses. But he was also interested in minimizing Fairchild's exposure to unionization efforts. In the early 1960s, unions sought to organize the company's diode plant in San Rafael and nearly won an election there. As a result, Sporck and Noyce, who viewed unionization as a death threat to the company, vowed to shift assembly operations to areas where labor unions were weaker. To do so, Fairchild Semiconductor built a new plant in Portland, Maine in 1962. Wages for unskilled workers in Maine were only half of those in the Bay Area. The workforce in Portland was also strongly anti-union. A year later, Sporck and Noyce established a new subsidiary in Hong Kong to take advantage of the Territory's cheap labor. Fairchild was the first US semiconductor firm to do so. According to most accounts, the idea of setting up an assembly plant in Hong Kong came from Noyce. Noyce had invested in a small radio firm in Hong Kong, which alerted him to the low cost of direct labor in the area. Noyce and Sporck also considered Asian labor to be diligent and easily disciplined.[67]

Noyce's and Sporck's decision to establish an assembly plant in Hong Kong met substantial opposition. Fairchild Camera's board of directors worried about Hong Kong's political instability and its proximity to communist China. Engineers at Fairchild Semiconductor also doubted that such a complex product could be produced in East Asia. But the labor-cost differential between Hong Kong and California was so great that this opposition quickly quieted down. While the hourly rate for assemblers on the San Francisco Peninsula was around $2.80, operators in Hong Kong earned only 25 cents per hour. Hong Kong also offered inexpensive engineers eager to work in electronics. "There was a lot of engineering talent around," Sporck later reminisced. "Many guys had gotten degrees in the United States and went back to Hong Kong and they had nothing to do. So we had the opportunity of putting tremendously talented people to work on mundane things, but to them that was not so mundane compared to what they were working on normally."[68] In 1963, the standard wage for an engineer in Hong Kong was only $150 a month—less than one-fourth of the cost of employing an engineer in Mountain View. Supervisors and department heads earned up to $300 a month. In other words, Fairchild Semiconductor could hire a mid-level manager in Hong Kong for the cost of an operator in Santa Clara Valley.[69]

Sporck and his men took over an old Hong Kong factory that had made rubber shoes and converted it into a transistor assembly plant. The plant was cramped and looked like a sweat shop. It was directed by an American engineer, Norman Peterson, and employed young female operators and a crew of technologists who had been trained in the US and Taiwan. Fairchild increased production rapidly. In its first year of operation, the Hong Kong plant assembled 120 million transistors, at very low cost. To push costs even lower, in 1964 Fairchild introduced a new packaging technique. Until that time, transistors were packaged in "headers" or small metallic boxes. These headers were expensive. To dispense with the metal headers, the research and development laboratory in Palo Alto developed a plastic encapsulation technique. Under this process, operators placed the transistor die on a small ceramic disk and

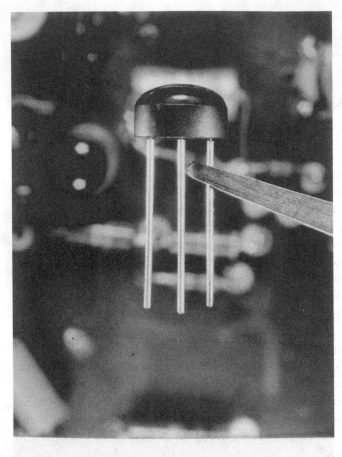

EPOXY 2N3638

Figure 5.5
An epoxy transistor, 1964. Courtesy of Fairchild Semiconductor and Stanford
University Archives.

covered it with a drop of epoxy. The encapsulation required little skill
from the operator and could be easily performed by workers in Hong
Kong. Along with offshore assembly, the epoxy package enabled
Fairchild to dramatically reduce its prices on planar silicon transistors
and thereby match those of radio tubes and germanium transistors.[70]

Exploiting the low labor costs to the fullest, in a few years Fairchild
shifted a large share of its assembly and testing operations to Hong

Kong. At first, the Hong Kong plant specialized in transistor assembly. It rapidly gained other functions such as die testing, final testing, and wafer sorting. By 1966, Fairchild Semiconductor employed more than 4,000 operators, technicians, and engineers and was Hong Kong's largest electronics employer. As the Hong Kong plant rapidly ran out of capacity, Fairchild established another factory in South Korea in 1965. The building of a network of factories in East Asia was an important business innovation. It gave Fairchild a substantial competitive advantage over Motorola and Texas Instruments that continued to do most of their assembly in the United States. Offshore assembly also permitted Fairchild to dramatically reduce its costs and expand the commercial markets for silicon transistors, especially in consumer electronics. As a result of growing sales in the commercial market, Fairchild Semiconductor became less of a military contractor and more of a civilian manufacturer. In 1964, a third of Fairchild's revenues came from computing, consumer electronics, and other civilian industries. By 1966, nearly half of Fairchild's sales were unrelated to defense. The growing commercial demand also enabled the firm to grow rapidly in the first half of the 1960s. Fairchild Semiconductor's sales grew from $27 million in 1961 to $90 million in 1965.[71]

Conclusion

In the early 1960s, the Department of Defense, which had sustained the growth of the electronics cluster on the San Francisco Peninsula, forced local firms to consolidate in order to survive. Paradoxically, it also led them to reduce their dependence on the military sector and diversify into commercial markets. Another significant factor in this diversification process was the emergence of new markets in the civilian sector. To grow these commercial markets, Fairchild Semiconductor and Varian Associates developed innovative marketing techniques to sell these products and they often were actively engaged in their customers' product engineering efforts. For example, Varian's chemists and engineers sold NMR spectrometers by convincing chemists of the usefulness of their machines. They accomplished this by helping them solve their chemical problems with NMR techniques. Similarly, Fairchild's engineers wrote applications notes to assist their customers. They also engineered novel electronic products such as a transistorized television set to convince users of the potential of silicon devices. They gave these system designs to customers at no cost. In addition, for Fairchild to meet the volume and price requirements of commercial users, they acquired expertise in the mass production of silicon transistors. The firm's management hired engineers from mass-production industries and adopted their techniques for their manufacturing requirements.

These moves enabled Fairchild Semiconductor and Varian Associates to build significant commercial businesses in the first half of the 1960s. Watkins-Johnson, Hewlett-Packard, and the vacuum tube division of Litton Industries substantially increased their commercial sales at this time as well. In doing so, they transformed the electronics manufacturing district, which until then had been overwhelmingly oriented toward the military sector, into a cluster that served a mix of both commercial and military customers. This process led to the renewal of the local entrepreneurial and managerial class. Many of the entrepreneurs who had spearheaded the rise of the district in the 1940s and the 1950s, including William Eitel, Jack McCullough, and Myrl Stearns, retired and left the industry. Charles Sporck rose to prominence by directing the re-orientation of Fairchild toward commercial markets. Sporck and the mid-level managers who worked for him later played a central role in the growth of the integrated circuit business in Northern California.

The re-orientation toward the civilian sector led local entrepreneurs and their engineering and managerial staffs to master new technical and marketing capabilities. They learned how to create markets for their products in the commercial sector, notably by giving technical support to the users and helping them solve their engineering and research problems. In addition to this marketing know-how, they learned how to make complex devices in mass quantities. This mass-production expertise was new to the area. With the exception of Eitel-McCullough, which had produced power-grid tubes in large volumes during World War II, the Peninsula's electronic component firms had been adept at manufacturing components in small quantities. In the first half of the 1960s, Fairchild's managers and manufacturing engineers innovated by producing their devices by the tens of millions every year. This major shift involved the adoption of new scheduling techniques from the electrical industry and learning how to control high-volume manufacturing processes. At the same time, Fairchild's management innovated by setting up and running transnational manufacturing systems. They delocalized much of the firm's assembly operations to Hong Kong and made this plant an essential component in their production organization. The manufacturing and marketing expertise that semiconductor managers and engineers acquired at Fairchild in the first half of the 1960s was later crucial for the expansion of the integrated circuit business on the San Francisco Peninsula. This know-how enabled Fairchild and other local semiconductor firms to open up markets for integrated circuits in the commercial sector. It also allowed them to manufacture microcircuits in mass quantities in the second half of the 1960s.

6

Miniaturization

In the first half of the 1960s, the San Francisco Peninsula became the main center for the design and production of integrated circuits in the United States. The expansion of the microcircuit business in the Santa Clara Valley was remarkable indeed. In 1961, Fairchild Semiconductor, the firm that pioneered these electronic components, sold only $500,000 worth of microcircuits. Five years later the microcircuit revenues of local corporations reached $60 million and Peninsula firms controlled more than half of the sales of microcircuits in the United States. The rapid growth of the integrated circuit business revitalized the manufacturing district, which had suffered from the "McNamara depression" and the decline of the microwave tube industry in the early 1960s. Reversing this downward trend, integrated electronics created thousands of jobs and gave rise to new corporations.[1] How did the microcircuit cluster emerge on the Peninsula? How did local firms develop integrated circuit technology and grow markets for it? How did the Santa Clara Valley become the main center for integrated circuit manufacturing in the United States?

The growth of the microcircuit business and the evolution of integrated circuit technology can be largely explained by conflicts at Fairchild Semiconductor, competition among local firms, and the changing demand for miniaturized electronic circuits. In a book on the innovation process, Clayton Christensen has pointed out that "well managed" corporations often fail to take advantage of their revolutionary innovations. Because these innovations do not meet the needs of existing customers and do not have the performance requirements of mainstream markets, these firms invest little in their further development. As a result, the engineers who created these innovations leave their employer and start new businesses that commercialize their work (Christensen 1997).

This familiar pattern in American high technology was central to the rise of the integrated circuit business on the San Francisco Peninsula. In the early 1960s, many managers at Fairchild Semiconductor had little interest in the new technology. Others were frankly hostile to it. In their view, microcircuits were a disruptive technology that made Fairchild compete with its own customers. These men also derided the fact that integrated circuits seriously underperformed conventional circuits made of discrete devices and that they did not meet the speed requirements of most semiconductor users. In contrast, the Fairchild engineers who had developed the first planar integrated circuits were enthusiastic about the new technology. They saw an immediate need for microcircuits in the military sector. Since the mid 1950s, the Department of Defense had promoted the miniaturization of avionics equipment as a way to reduce their weight and thereby increase the payloads of aircraft and missiles. In their view, the growing miniaturization of electronic circuits would make entirely new systems possible and revolutionize the electronics industry. Because of Fairchild management's lack of interest in integrated circuitry, these engineers left the firm. They started their own integrated circuit ventures on the basis of Fairchild's manufacturing processes. In 1961, Jay Last and Jean Hoerni set up Amelco, the semiconductor subsidiary of Teledyne. A few months later, a group that included David Allison, Lionel Kattner, and David James established Signetics. To their great surprise, these groups discovered that there was little demand for microcircuits, even in the military sector. To survive, Amelco complemented its custom circuit business with transistors and transistor assemblies. Taking a user-oriented approach, Signetics' engineers developed digital circuits of a logic configuration (diode transistor logic) that was familiar to most system firms. In spite of this important product innovation, Signetics sold few circuits at first. It nearly went bankrupt. Only a cash infusion from Corning Glass enabled Signetics to survive. When the Department of Defense strongly encouraged its system contractors to use microcircuits in their designs in the spring of 1963, Signetics was the primary beneficiary of the market expansion that ensued. It became the leading US producer of microcircuits.

The growing demand for Signetics' circuits did not go unnoticed at Fairchild. By 1963, the firm's managers had finally understood the full potential of the new technology and poured substantial resources into the development of microcircuit products. In order to capture a large share of the market for integrated circuits and claim for themselves the leadership position held by Signetics, Fairchild's engineers developed products that exactly matched the specifications of Signetics' integrated circuits. Moreover, Fairchild sold their products at less than half of Signetics' price.

This course of action had a devastating effect on Signetics. It also led to the rapid growth of Fairchild's microcircuit business in the military sector. To grow commercial markets for its products, Fairchild later developed a new semiconductor package, the dual-in-line package. This package lowered the cost of assembling integrated-circuit-based systems and made microcircuits attractive to computer manufacturers. To further strengthen its position in integrated circuitry, Fairchild also moved into the manufacture of analog circuits. By 1966, Fairchild was the largest integrated circuit maker on the Peninsula, ahead of Amelco and Signetics (which had significant sales in the military and computer markets).

Technological Enthusiasm

Fairchild Semiconductor was the source of planar integrated circuits and of the microcircuit business. It was a research group at Fairchild Semiconductor that developed the first planar integrated circuit and the complex processes for making it. Many of the engineers who expanded the integrated circuit business on the Peninsula first encountered the new technology at Fairchild. In 1959, Jay Last, one of Fairchild Semiconductor's founders, established a new micrologic section in the research laboratory. The goal of this section was to make miniature circuits. Building an integrated circuit, Last and his group discovered, was extremely difficult because of the close tolerances that were required. It also presented the challenge of electrically isolating the transistors, diodes, and resistors on the same die. To electrically insulate the individual components, Lionel Kattner and Isy Haas, two research engineers working with Last, devised a new process. They were helped by David Allison, a diffusion expert and the head of the transistor development section at Fairchild. With this new process, they isolated devices on the same chip by diffusing isolation regions from the top and the back of the silicon wafer. Using this innovative process, Kattner and Haas made the first planar integrated circuit in September 1960.

In the next year, Kattner and other members of the microelectronics group designed an entire family of circuits for digital applications. These circuits were of the DCTL (direct-coupled transistor logic) configuration. Digital designers can use different basic circuit schemes to implement logic functions—using different types of circuit elements (transistors, diodes, capacitors, and resistors) and configuring them in various ways in order to produce pulses or differences in voltage. These pulses represent the basic zeroes and ones of digital design. Jay Last and his group chose the DCTL configuration at the urging of Robert Norman. Norman had invented this configuration at Sperry Gyroscope and was now the head of the laboratory's device evaluation group at Fairchild. The main advantage of DCTL was that it required only transistors and resistors. This made it easier to fabricate than other logic configurations in use at the time. Fairchild showed its first integrated circuit at a news conference in New York in March 1961. There, the firm announced that it would introduce five other members of its "micrologic" family later in the year. Fairchild had put on the market the first planar integrated circuits. (TI announced its first line of integrated circuits in October 1961.) Fairchild sold $500,000 worth of integrated circuits in 1961.[2]

Despite Fairchild's apparent success, the company rapidly lost many of the people who made the integrated circuit possible—in large part because of stark disagreements over integrated circuits and their future. Before the announcement of the first integrated circuit, the micrologic group lost its leader, Jay Last. Last and another Fairchild founder, Jean Hoerni, left the company in January 1961. Hoerni was the head of the physics section of the research laboratory and had invented the planar process. These men quit the firm they had founded for a variety of reasons. Critical to Last's decision to leave were his misgivings about the future of integrated circuits at Fairchild Semiconductor. Last encountered substantial opposition to the development of microcircuits within the company. Tom Bay, the marketing and sales manager, was concerned that the development of integrated circuits would lead Fairchild to compete with its own customers. By selling integrated circuits, the company would compete with firms that bought transistors from Fairchild to produce their own circuits made of discrete components. This would undermine the firm's transistor and diode sales, its chief source of revenues. At a staff meeting in the fall of 1960, Bay accused Last of "pissing away a million dollars" on integrated circuits and sought to have the project discontinued. Gordon Moore, who headed the R&D department, was not supportive either. He was skeptical about integrated circuits and did not believe in their future commercial importance. "My single biggest mistake at Fairchild," Moore later reminisced, "was not appreciating how big the business for integrated circuits could be. Most of us working in the laboratory at the time of the completion of the first family of integrated circuits did not realize that we had barely scratched the surface of a technology that would be so important. It was just another product completed, leaving us looking around for a new device to make, wondering, 'What's next?'" (Moore and Davis 2001) In late 1960, Moore re-oriented the laboratory toward new fields of inquiry such as superconducting films and the development of transducers and microwave devices. Last, who believed in integrated circuits and was interested in realizing their potential, felt that Fairchild was not the best place for fostering the development of microcircuits.[3]

Last and Hoerni had other reasons also for their disenchantment with Fairchild Semiconductor. The firm had grown rapidly, and their positions within it had changed. In the early days of the firm, each founder had equal status and was equally involved in the decision-making process. But by 1960, Noyce and Moore, who had cultivated and nurtured their relationships with Fairchild Camera and Instrument (Fairchild Semiconductor's parent company), had emerged as the firm's acknowledged leaders. Hoerni and Last, each of whom headed a section in the R&D laboratory, saw their influence increasingly limited to R&D. These men had also sold their interest in Fairchild Semiconductor to Fairchild Camera in 1959. As a result, they had lost their ownership in the firm. In addition, Hoerni felt that he was not appreciated by his new employer. More than 30 years later, he complained that John Carter, Fairchild Camera's president, had never acknowledged him for the development of the planar process.[4]

Last's and Hoerni's disenchantment and Last's misgivings about the future of integrated circuits at Fairchild made them particularly receptive to an offer from their friend Arthur Rock (who was then at Hayden Stone) to start a semiconductor laboratory for Teledyne. Teledyne was a recent spinoff of Litton Industries, the military conglomerate that Tex Thornton had built around Charles Litton's microwave tube business in the 1950s. It had been established by Henry Singleton and George Kozmetsky, two former vice-presidents at Litton Industries, in the summer of 1960, with financing by Rock. Teledyne's founders and their backers at Hayden Stone had two goals: first, to build a large military electronics system business specializing in avionics, inertial guidance, and communication systems; second, to construct a conglomerate using many of the techniques that Thornton and his associates had pioneered at Litton Industries in the second half of the 1950s. (See chapter 2.)

To differentiate themselves and compete with Litton, Singleton and Kozmetsky wanted to base their company around the new technology of microelectronics. Until then, Litton Industries had stayed away from the semiconductor field. In 1957 it had turned down Arthur Rock's offer of investing in the group that later started Fairchild Semiconductor. Singleton and Kozmetsky thought that this was a serious mistake. They believed that the future of electronics would be so bound up with the new technologies of microelectronics and integrated circuits that no system manufacturer would succeed unless he mastered the manufacture of advanced component technologies. They further predicted that the classical electrical manufacturing process was about to undergo a dramatic change in direction. Instead of buying electronic components from specialty houses, manufacturers would be increasingly forced to make their own circuit elements. Having an internal semiconductor capability was also important for marketing reasons. Singleton and Kozmetsky were interested in building a one-stop operation that would sell everything from silicon to missile subsystems to the Department of Defense. In other words, integrated circuits and other microelectronic components would give them a competitive edge—technically and business wise.[5] Hence their interest in Jean Hoerni and Jay Last. In order to build a microelectronics-based system company, Singleton and Kozmetsky needed people who could build an in-house semiconductor capability. Rock and Alfred Coyle, a Hayden Stone senior partner, enticed Last and Hoerni to set up a semiconductor laboratory for the new firm. At Rock's urging, Last and Hoerni drove to Southern California to meet with the entrepreneurs at the end of December 1960. Singleton made them what they thought was an attractive offer. He asked them to establish and co-manage a device-development laboratory and a small-scale production facility for Teledyne. The goal of this particular semiconductor laboratory would be to develop and produce integrated circuits and other advanced devices for the electronics system division that Singleton and Kozmetsky had recently organized. Finally, Singleton assured them that the laboratory would be autonomous. He offered them and their future associates generous options of Teledyne stock based on the Litton model.[6]

Hoerni and Last accepted Singleton's offer and submitted their resignations to Fairchild Camera in January 1961. They set up Amelco Semiconductor as Teledyne's semiconductor operation in the following month. (By mid 1961 they were vice-presidents of Teledyne.) At Last's and Hoerni's insistence, the operation was located in Mountain View rather than in Los Angeles—these men liked living in the Bay Area and had no interest in moving to Southern California. They also wanted to take advantage of the Peninsula's rich labor pool and supply network. No other location in the United States had more engineers, technicians, and operators skilled in the manufacture of diffused silicon devices. There were also many suppliers of chemicals and, increasingly, semiconductor processing equipment in the area. Sheldon Roberts and Eugene Kleiner, two other Fairchild Semiconductor founders who had become similarly disillusioned with the company, followed Hoerni and Last to Amelco a few weeks later. Roberts joined Amelco's technical staff, while Kleiner, acting as a consultant, built the service departments. Haas, who had developed the first planar micrologic circuit at Fairchild, quickly followed Last in order to continue his work with him on integrated circuits.[7]

The departure of Hoerni and Last was a considerable setback for Fairchild Semiconductor. These men had made major technological contributions to the firm and helped it establish itself as a key manufacturer of diffused silicon transistors. Many engineers and managers at Fairchild recalled having been shaken by their departure. But it was in the laboratory that the impact of their forming Amelco was the most pronounced. Their departure deeply affected the morale of the research staff, especially in the physics and micrologic sections. It created a managerial vacuum. Last's and Hoerni's establishment of a new semiconductor operation also set an example for other research engineers to follow.[8] And they did. Less than 3 months after Amelco Semiconductor's formation another group of research engineers at Fairchild made preliminary steps toward the formation of Signetics, a new integrated circuit company. This group was young and ambitious. It was made up of four physicists and engineers who had worked with Last and Hoerni in the R&D laboratory. Two members of the group, Lionel Kattner and David Allison, had developed the first integrated circuits at Fairchild. David James, a free-spirited Briton, was a talented experimental physicist with a PhD from the University of Bristol. He had collaborated closely with Hoerni in the physics section, where he worked on the planar process and headed a project on the growth of epitaxial films. After Hoerni's departure, he succeeded him as section head. The fourth member of the group was Mark Weissenstern, an Israeli with a master's degree in electrical engineering from MIT. In 1960, Weissenstern directed the development of the "cheapie," an inexpensive planar transistor. He was also involved in packaging studies in the device development section and was working on plastic encapsulation.[9]

These men were interested in setting up their own company for a variety of reasons. They were discontented with Fairchild Semiconductor. Allison was disturbed by the fact that, in spite of his many contributions to the company, the founders of Fairchild had not shared the proceeds of the firm's sale to Fairchild Camera with him. According to Allison, in 1959 the founding group voted on whether to share their gains with "key employees." Some, especially Last, pressed for it. But the majority elected not to do so. Kattner and Weissenstern were equally disaffected. Moore gradually discontinued Weissenstern's "cheapie" project on the grounds that the anticipated cost of the transistor was not appreciably lower than the one of Fairchild's standard products. Although the micrologic project had been a resounding success, Kattner was alienated by the fact that Moore never directly recognized him for the development of the first planar integrated circuit. It was also clear to James that, after his promotion to head of the physics section, he would not progress much more at Fairchild. Starting a semiconductor company would enable these men to advance their careers and get equity for their work.[10]

Integrated circuits seemed to offer the perfect vehicle for the formation of a new semiconductor company. It was an open field that Fairchild Semiconductor was not fully exploiting. No other semiconductor corporation, including Texas Instruments, had yet been able to produce integrated circuits. This was an opportunity to exploit. James, Kattner, Allison, and Weissenstern were also enthusiastic about the technological and economic potential of microcircuits. They saw this as a revolutionary technology that would soon replace transistors and other discrete components. Integrated circuits, they predicted, would have a substantial industrial impact. Microcircuits would find their ways into the daily life of the "average citizen" by the end of the decade. They envisioned a variety of applications such as miniature radios and "compact automatic computers that would automatically direct a housewife's cooking, washing, baking, and other daily chores."[11] They also believed that microelectronics would become one of the largest industries in the United States.[12]

In the spring of 1961, James, Allison, Kattner, and Weissenstern asked Texas Instruments (Fairchild's chief competitor) and some New York investment banks to finance the future venture. Kattner, who had worked at Texas Instruments, approached his former employer through a friend and neighbor who worked as a TI sales representative. Kattner revealed his plans and inquired whether TI would be interested in supporting the establishment of an integrated circuit company on the San Francisco Peninsula. This feeler elicited considerable interest at TI's headquarters in Dallas. TI had announced the first integrated circuit (the mesa circuit, developed by Jack Kilby) in 1958, but had not yet been able to bring it to production. Kilby's circuit could not be manufactured in quantity. Only the planar process would make integrated circuit products feasible. Texas Instruments, however, had not yet fully mastered this process. (TI would introduce its first planar transistor 6 months later.) Now that Fairchild had announced its line of integrated circuits, TI was desperate to commercialize microcircuits as well. Investing in James' and Kattner's venture would give TI access to Fairchild's process technology. This investment would also enable TI to commercialize planar integrated circuits. Following up on Kattner's inquiry, Mark Shepherd, TI's vice-president, immediately flew to California to meet with the group. After a weekend of discussions, Shepherd declared his interest in funding the new venture. The deal, however, fell through in the next few weeks as Shepherd and Patrick Haggerty, TI's president, realized that supporting two competing integrated circuit groups in Texas and in California would create substantial managerial difficulties. They also surmised that Fairchild Semiconductor would sue TI for the theft of trade secrets—a case TI probably would lose.[13]

While the four entrepreneurs negotiated with TI, they also contacted investment banks and brokerage firms in New York. The group knew that Hayden Stone had arranged the financing of Fairchild Semiconductor. It was of course impossible for them to contact the same bank. But other Wall Street firms financed new electronics ventures. At first, James sent a letter outlining their plans to Merrill Lynch. But Merrill Lynch had recently left the business. It forwarded the letter to Lehman Brothers. Within days of TI's rejection, James received a phone call from Warren Hellman, an associate at Lehman Brothers, declaring his interest in the proposal. In June 1961, Hellman flew James and Kattner to New York to meet with the bank's partners. In collaboration with Lehman, the group drafted a business plan. The new venture was to make custom integrated circuits. In other words, it would design circuits in collaboration with its customers, put the designs into silicon, and manufacture them in the desired quantity. This market had yet to be tapped. The group felt that there would be a substantial demand for such circuits. In August, the bank informed James that they would support the new venture. Lehman Brothers was not new to the electronics business. It had financed the purchase of Litton Industries by Thornton almost a decade earlier and had made enormous profits on this transaction. Lehman Brothers saw similar opportunities in integrated circuits. It was interested in bringing the new venture public in a few years.[14]

To finance the startup, Lehman Brothers organized an investment syndicate. In addition to Lehman, this syndicate included individual investors such as the actor Bob Hope and banks in New York and San Francisco. These included some of the better known names in US finance: Goldman Sachs, Lazard Frères, White Weld and Co., J. Barth and Co., and Milbank and Co.. These investors put $1 million into the new venture, 85 percent of it in non-convertible debentures and the rest in preferred stock. In return, they received approximately 80 percent of the firm's shares. The founders got the remaining 20 percent. James, Kattner, Allison, and Weissenstern left Fairchild and incorporated the new venture, Signetics, in August 1961. The group was soon joined by Orville Baker and Jack Yelverton. Baker was a brilliant circuit engineer who had worked at IBM before joining the device evaluation section at Fairchild. Yelverton, who had an MBA from New York University, had a solid background in business administration. At Fairchild, he had worked as the head of personnel and chief recruiter. He helped formulate the firm's personnel policies, notably regarding management-employee relations. He had also headed administration at the diode plant in San Rafael. Unlike Last's and Hoerni's amicable departure, the group's resignation was viewed by Noyce as a defection. Noyce, who had left Shockley Semiconductor in a similar fashion a few years earlier, was incensed at seeing some of his employees start a competing operation with the support of Lehman Brothers. These feelings would color Fairchild Semiconductor's relations with the new firm and play a signifi-cant role in Noyce's subsequent attempt to crush the new venture.[15]

Microelectronics and System Design

When Fairchild Semiconductor was concentrating on the manufacture of transistors, the two new startups, Signetics and Amelco Semi-conductor, focused on integrated circuits and other microelectronic products from the start. These new ventures started with the idea of mak-ing custom circuits, but their paths rapidly diverged. Signetics focused on standard circuits. Amelco evolved into a broad-based semiconductor manufacturer that also engineered micro-electronic components for Teledyne, its parent company. Amelco began with a serious handicap. From the beginning, the new venture was undercapitalized. Teledyne had just been established and lacked capital itself. It invested roughly one million dollars in the semiconductor laboratory from February 1961 to September 1962. (Some of these monies came from the Davis and Rock venture capital partnership that bought a stake in Teledyne in the spring of 1961.) Teledyne's investment was decidedly insufficient in an industry that was becoming increasingly capital intensive. Fairchild Camera had invested several million dollars in Fairchild Semiconductor during the same period of time a few years earlier. Moreover, the monies came to Amelco in small chunks rather than in a large lump sum. This dearth of capital put substantial constraints on Amelco's development. It also ensured that the semiconductor operation would permanently be in a precarious financial position. In other words, Teledyne wanted to build a semiconductor capability on the cheap.

Last and Hoerni made the most of the little money they had. They staffed the laboratory with some of their former associates at Fairchild. In addition to Roberts and Haas, many of their research engineers at Amelco had previously worked at Fairchild. Others came from Bell Telephone Laboratories, from Texas Instruments, and from optical firms on the San Francisco Peninsula. These men were offered options of Teledyne stock. Most technicians and operators were hired locally as well. Last, Hoerni, and their recruits spent much of the first year building a state of the art integrated circuit facility. To cut costs, they bought diffusion ovens from Electroglas, a new semiconductor equipment manufacturer located in Redwood City. The photo-masking equipment was developed internally. Last designed it in collaboration with Robert Lewis, an optical engineer. Since the making of integrated circuits depended on extremely accurate super-imposable photographic masks, Last developed a complex optical system to make these special masks. To obtain the high precision required, he reduced vibrations to a minimum—by installing the camera on a 20-foot-long camera bed anchored on a granite block. This system was substantially more advanced and could produce more precise masks than the one used at Fairchild Semiconductor at the time.[16]

Along with the building of the processing facilities, the group worked on microelectronic components and semiconductor devices for Teledyne's system divisions. Their goal was to engineer special components that would enable Teledyne to get military contracts for the development of new electronics systems. Hoerni engineered high-performance transistors for Teledyne's tracking receivers and airborne digital computers. Last and Haas developed miniature circuits in support of Teledyne's proposals for inertial guidance systems. They worked on two types of miniature circuits: integrated circuits and hybrid circuits. Hybrid circuits, also called thin-film circuits, represented a parallel approach to the miniaturization of electronic circuits. In this approach, the passive elements—resistors and capacitors—were made by the vacuum deposition of metals on a glass or ceramic substrate. Active elements such as transistors and diodes were mounted in chip form on the substrate and connected to the thin film elements by thermal compression bonds. Teledyne's system engineers approached Last and Haas with their own circuits. Haas and Last decided which technology to use based on what was best for the particular task at hand. Among their first projects was the engineering of an accelerometer for one of Teledyne's inertial guidance system development proposals. They were also involved in the development of a miniature precision gyroscope and associated electronic elements for NASA.[17]

To supplement Teledyne's funding, Last and Hoerni looked for their own military development contracts. This was a reversal of the policy they had previously followed at Fairchild Semiconductor. At Fairchild, Last and Hoerni had turned down offers of development monies from the Department of Defense. Now that they worked for a military-oriented systems firm and needed DoD contracts to support their research projects, they approached the agencies they had turned down earlier. Last made overtures to the Signal Corps Engineering Laboratories at Fort Monmouth and the Aeronautical Systems Division at the Wright-Patterson Air Force Base. These organizations were supporting extensive R&D programs on integrated circuits at Motorola and Texas Instruments. In order to generate interest from these organizations, Last organized visits by military representatives to Amelco. The research staff submitted a variety of proposals on integrated circuit packaging, the development of high-efficiency transistors for integrated circuits, and optical interconnection of semiconductor devices to these and other agencies in 1961 and early 1962.[18]

Amelco's exclusive orientation toward servicing the needs of the system company was short lived, however. In the spring of 1962, Hoerni and Last expanded the laboratory's activities to include the production and sale of semiconductors for the open market. This shift was brought about by severe financial pressures. The laboratory received fewer military R&D contracts than they had hoped. To Singleton's chagrin, the systems company also won very few contract competitions, especially for inertial guidance systems. Military cutbacks during the "McNamara depression" might partially explain Teledyne's poor showing. The field of inertial guidance was also dominated by Litton Industries and by the Instrumentation Laboratory at MIT, which did not leave much room for a newcomer. As a result, Teledyne was starved for money. To reduce their cash outlays, Singleton and Kozmetsky told Amelco Semiconductor that it had to support itself through sales. However, they did not relieve the semiconductor operation from its obligations toward the systems company. Far from it. They expected Amelco Semiconductor to be self-supporting and at the same time provide engineering support to the systems company.[19]

To generate outside sales and put the semiconductor operation on a self-sustaining basis, Hoerni and Last transformed the semiconductor laboratory into a full-fledged business. "Our aim in the next six months," Last wrote to Singleton in May 1962, "is to come as close as possible to breaking even on a profit and loss basis for the rest of this fiscal year and to build up the facilities and the nucleus of trained personnel capable of generating sales of $5 million in fiscal year 1963."[20] To build the company, Last and Hoerni hired sales and marketing specialists. They also organized a manufacturing department by recruiting engineers and foremen from Fairchild Semiconductor and Texas Instruments. Operators were hired away from local firms. To house the expanded operation, Last and Hoerni built a 55,000-square-foot plant. During these transformations, the operation's administrative structure changed. Until then, Amelco Semiconductor had been co-managed by Hoerni and Last. As the company expanded, Hoerni became general manager. Last increasingly concentrated on the R&D side of the business, especially microelectronic components.[21]

Last, Hoerni, and their group also engineered a series of products for external customers. The composition of this product line was shaped by Last's and Hoerni's technical interests, as well as competitive constraints and market opportunities. Hoerni was intrigued with a new type of transistor, the field-effect transistor. Last was enthusiastic about microelectronics. Last and Hoerni also sought to limit the fields in which they competed with Fairchild Semiconductor. This led them to emphasize the development of specialty components for niche markets. The state of the market for semiconductors further shaped Amelco's product line. The demand for discrete components was much larger than the one for microcircuits in early 1962. With the exception of Teledyne, system firms had little interest in using integrated circuits in their products. Microcircuits were a new, untried, and still largely experimental technology. At that time, most systems houses designed and produced their own electronic circuits. Buying these from an outside vendor would cut into their profits. Circuit engineers also saw microcircuits as a threat to their livelihood and resisted their use. There were other reasons as well for their resistance. They did not know how to evaluate integrated circuits. They could not apply the standard techniques used in the evaluation of transistors to these new components. Because the market for microcircuits was still in its infancy, Last and Hoerni realized they could not build a viable company on integrated circuits alone. As a result, they put together a product line made of both discrete devices and microelectronic components.[22]

Because of these different interests, constraints, and opportunities, Amelco's product line was highly diversified. Amelco commercialized the high-quality bipolar transistors Hoerni had designed for Teledyne's computers and communication equipment. The firm introduced more bipolar transistors in late 1962 and in 1963. Hoerni also developed field-effect transistors—unipolar devices with performance characteristics similar to those of vacuum tubes. Using the work he had done on field-effect transistors at Fairchild, Hoerni applied the planar process to the manufacture of these devices. Amelco introduced field-effect transistors to the market in May 1962. These devices were used in analog computers, electronic switches, and amplifiers.[23]

The group developed and commercialized a variety of miniature circuits as well. Among these were special assemblies made of individual transistors, diodes, resistors, and capacitors. These discrete components were mounted close together in a single package. Amelco's engineers developed numerous types of special assemblies, with a particular focus on linear circuits and especially operational amplifiers. An operational amplifier is a particular kind of linear or analog circuit. (Analog or linear circuits process the magnitude of an electronic signal, while digital circuits deal with discontinuous "0" and "1" signals.) It is a basic building block in electronic design. Using a variant of the Fairchild Semiconductor manufacturing process, Amelco Semiconductor also developed and commercialized digital integrated circuits. The firm announced two such circuits in May 1962. One year later, the firm introduced a family of digital circuits called optimized microcircuits ("OMICs"). This family was designed by Haas and other members of the Amelco engineering group. They used the same logic configuration as the one they had employed in Fairchild's first micrologic family, the DCTLs. Haas and Last saw a number of advantages in the DCTL configuration. Fairchild had already opened a market for it and there were technical and economic benefits as well. DCTL was the easiest logic configuration to integrate on a silicon chip. It required a small number of component types (resistors and transistors) and appeared easiest to manufacture. Through clever design and processing techniques, Haas engineered a family of circuits that had better electrical characteristics than the one he had helped design at Fairchild 2 years earlier. In addition, Haas and his group innovated by making linear integrated circuits, including operational amplifiers.[24]

Amelco also sold custom hybrid circuits and integrated circuits. These circuits were designed and made in the R&D lab to meet customers' specific requirements. One of Amelco's first custom orders was from United Aircraft. In the summer of 1962, United Aircraft asked Amelco to engineer two prototypes of miniaturized cardiac monitors. These monitors were intended to measure the cardiac activity of astronauts and transmit the signals to a recorder within a range of 50 feet. To meet this special order, Last and his group developed a complex hybrid circuit made of tunnel diodes, transistors, thin-film resistors, and capacitors. This order was followed by others from United Aircraft and other firms. In 1963, Last wrote the following to a Navy contracting officer:

Amelco is actively engaged at the present time in producing packaged circuits using thin film resistors and capacitors. Although details of the circuits cannot be discussed in detail due to proprietary restrictions imposed by our customers, we can make a few general remarks concerning those circuits. We have delivered an 80 megacycle transmitter completely contained in a package 0.350" × 0.150" × 0.060" with the small size made possible by the use of arrays of thin film resistors and capacitors. We have delivered a variety of differential amplifier circuits and linear amplifiers using resistance networks [made by thin-film techniques]. Our approach to the thin film circuitry problem has been favorably received by a large number of customers and many prototype circuits are in various stages of fabrication and sales negotiation.[25]

The group also devoted substantial efforts to the design and the small-scale fabrication of custom integrated circuits. Little is known about these projects because many of them were highly classified. In 1962, the Amelco engineers designed and fabricated a microcircuit for the electronic locks of the atomic bombs for the Sandia Laboratory. They also worked on highly secret circuits for Westinghouse. Other customers included Honeywell and the Cubic Corporation, a computer startup financed by Hayden Stone.[26]

Though the R&D group did substantial work for outside customers, its primary focus and responsibility was to act as the internal component design operation for the systems company. It developed custom integrated circuits and hybrid circuits to support proposals for system development contracts. Teledyne became increasingly successful at this game in 1963 and 1964. This brought subcontracts for circuit development and small-quantity production to Amelco. For example, in 1963 Amelco received subcontracts from Teledyne to work on microcircuits for the Lawrence Livermore Laboratory.[27] Amelco's and Teledyne's greatest success, however, was a series of contracts for the development of the Integrated Helicopter Avionics System (IHAS). The IHAS program was one of the most ambitious avionics programs of the first half of the 1960s.

The Department of Defense set up this program in the context of America's growing involvement in the civil war in South Vietnam. To bolster the South Vietnamese government, the Kennedy administration sent a growing contingent of military advisers—more than 16,000 of them by the end of 1963. The military increasingly relied on helicopters, but found that the helicopters they had did not meet the requirements of guerilla warfare in Southeast Asia. They needed helicopters that could fly in close formation, at low altitude, over any type of terrain, under conditions of little or no visibility. These requirements necessitated the development of an automatic avionics system that would allow helicopters to fly close to the forest canopy with little input from the pilots. The military contacted more than ten firms to submit preliminary proposals for this system in 1963. Teledyne was one of them. In collaboration with Last's group, Teledyne's system engineers proposed a fully integrated avionics system. This system was controlled by a digital differential analyzer, a computer that did analog computations by digital techniques. The computer was to be composed of integrated circuits. In the spring of 1963, Haas and other engineers in the R&D section designed more than ten microcircuits for the computer.[28]

Teledyne's innovative proposal led to their winning a military development contract for this system in the summer of 1963. Only two other firms, Nortronics and Texas Instruments, were awarded such contracts. In the following year, Last, Haas, and the microelectronics group concentrated most of their efforts on the engineering of microcircuits for the IHAS. They also developed a new package for this system (called MEMA, for Microelectronic Modular Assemblies). It was a ceramic package. A MEMA contained up to 40 integrated circuits. The integrated circuit chips were attached to a metallized ceramic base. To interconnect the integrated circuits in the same package, the group developed miniature printed circuit boards. Twenty of these modules were used in the computer's central processing unit. These innovations contributed to the Department of Defense's decision to award a $25 million sole-source contract to Teledyne for the further development and pilot production of the avionics system in 1965. This was a huge boost for Amelco. It was also the largest contract that Teledyne had ever received.[29]

Amelco's emergence as a supplier of semiconductors and developer of microelectronic parts for Teledyne was not a smooth process. Last and Hoerni had to shift their priorities and projects constantly to build a viable business and at the same time attend to the needs of the systems company. Last recalls having to go back and forth from device develop-

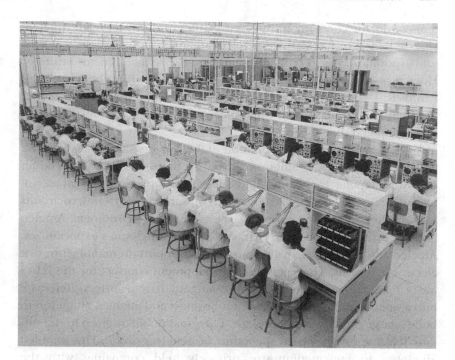

Figure 6.1
Amelco's manufacturing plant, circa 1963. Courtesy of Jay Last.

ment for external customers one week to meeting Teledyne's requirements the next. When Amelco Semiconductor did not meet the
demands of Teledyne's system engineers rapidly enough, these men
asked Henry Singleton to put pressure on the semiconductor group.
This was a difficult, almost untenable, position to be in. Amelco also
experienced substantial setbacks in its move to external markets. These
challenges were typical of the semiconductor business. Amelco experienced the usual poor manufacturing yields. Its devices had reliability
limitations. It also had difficulties in coordinating sales and manufacturing. These issues were compounded by the fact that Hoerni had little
business experience and did not pay enough attention to cash flow and
profits. As a result, Amelco Semiconductor was losing money in 1963.
This prompted Kozmetsky to remove Hoerni from the position of general manager. In a face-saving measure, he offered him the position of
Director of Corporate R&D at Teledyne in June 1963. When Hoerni
declined the job and insisted that he be fired, Kozmetsky replaced
Hoerni with Jim Battey, an experienced executive from Clevite, a

semiconductor company in the East. After his dismissal from Amelco, Hoerni joined Union Carbide as a consultant. Union Carbide, a chemical corporation, was interested in entering the semiconductor business. Hoerni established a semiconductor division for them on the Peninsula—Union Carbide Electronics. This operation specialized in the manufacture of special assemblies and field-effect transistors and thus competed directly with Amelco.[30]

Amelco Semiconductor's situation stabilized. Its sales grew from $2 million in 1963 to $3.6 million in 1964. At that time, its main product lines were bipolar transistors, field-effect transistors, and special assemblies. The firm also sold $396,000 worth of standard OMIC microcircuits and another $367,000 in custom circuits to outside customers. Amelco became profitable that same year, vindicating Singleton's expectations. Amelco also helped Teledyne emerge as an important manufacturer of electronic systems. The sole-source development contract for the IHAS was a major breakthrough for the company. It transformed Teledyne into a significant force in the avionics business and enabled Singleton to build a large military conglomerate. The great publicity given to the various IHAS contracts raised the value of Teledyne stock. This enabled Singleton to buy medium-size privately held companies with the company's stock. Following Litton's recipe, Teledyne acquired 21 small specialty companies between 1963 and 1965. These firms included manufacturers of aircraft fittings, telemetry equipment, radar systems, and microwave tubes. One of these acquisitions was MEC, a manufacturer of microwave tubes on the San Francisco Peninsula. As a result of these acquisitions and the rapid expansion of the system business, Teledyne's sales went from $4.9 million in 1961 to $86 million in 1965.[31]

Making User-Oriented Integrated Circuits

The group that established Signetics made strategic decisions that were different from those that Last and Hoerni made. While Amelco Semiconductor evolved into a broad-based manufacturer of specialty transistors and miniaturized circuits, Signetics concentrated exclusively on the manufacture of integrated circuits. In the weeks that followed the firm's establishment, the founders revisited their original business plan and debated which fields the company should venture into. David Allison proposed that the corporation make bipolar transistors as well as integrated circuits. They knew how to design and manufacture transistors very well. Allison argued that the market for microcircuits was very small, which made it risky to concentrate on integrated circuits alone. Other members of the group felt differently. They reasoned that, with the little capital they had, they could not compete with Fairchild in the low-cost, high-volume transistor market. Instead they felt that they should concentrate their resources on microcircuits and attack that market. David James, who had been nominated president by Lehman Brothers, decided to focus Signetics on microcircuits alone. This was a significant decision because it put the firm on a different trajectory than Amelco Semiconductor. Signetics was the first corporation to specialize in integrated circuits in the United States.[32]

Implementing their original business plan, the group at Signetics set out to produce custom circuits. Their goal was to help their system customers design electronic circuits. They also intended to put these circuits in silicon and manufacture them in the requested quantities. "We intend to be a job shop," James declared to a *Business Week* reporter in April 1962, "and we don't think that we will have too much trouble getting people to come to us early in the design stage of equipment development once we prove that we can produce."[33]

To engineer and produce custom circuits, Signetics mobilized the same resources as Amelco. They hired some engineers from Fairchild Semiconductor and Hewlett-Packard, and they recruited others from Texas Instruments. All these men were offered stock options when they joined Signetics. The founding group also used local machine shops and equipment manufacturers to equip their plant. They procured the diffusion furnaces from Electroglas and employed the services of the local industry for the fabrication of high-precision assembly equipment. They also bought their photo-masking equipment from Electromask, a newly established company in Los Angeles.[34] To set up the plant and build a functioning organization, the core management group organized themselves in ways that were reminiscent of early Fairchild. Each member of the group took on a set of technical tasks and had functional responsibilities as well. They assigned among themselves the setting up of processes such as diffusion and photolithography. In addition to these tasks, each member of the group had managerial responsibilities. David Allison was in charge of technical development. Orville Baker headed circuit design. Lionel Kattner directed manufacturing. Mark Weissenstern took on quality control, testing, and, for a time, marketing. Jack Yelverton set up the firm's administrative services, introducing many of the management-employee-relations techniques he had helped develop at Fairchild Semiconductor.[35]

To interact with the customers and develop custom circuits, these men organized the firm in an innovative way. At Fairchild, product engineering was mostly done in the R&D labs at that time. This made it sometimes difficult to interface development work with market demands, especially as the firm grew. Because of its custom orientation, Signetics needed to be much more aware of and open to customer inputs. As a result, the Signetics group decided not to set up an independent R&D laboratory. All product engineering was to be done in the "technical development department." This department was made up of two sections: device development (under Allison) and circuit development (under Baker). Baker's group was closely tied to marketing. It also collaborated with the device development group. This new organization enabled the close coupling of process and design engineers with marketing specialists.[36]

In the firm's first 6 months of operation, Baker, Weissenstern, and other members of the group went on a fact-finding mission to a number of electronic system corporations, hoping to drum up custom business. To their surprise, they discovered that the demand for custom integrated circuits was more limited than they had originally thought. Very few system firms were interested in having anyone engineer and manufacture specific circuits. Those rare corporations that were interested wanted circuits that required transistors and resistors that Signetics could not produce with its existing processes. (Signetics was developing its own variant of the Fairchild process for making integrated circuits.) As a result of these field studies, the founding group reexamined the premise on which they had founded the company. They decided to produce a family of standard integrated circuits in addition to custom circuits. The idea was that this standard family would bring in much-needed revenues and also act as a form of advertisement for Signetics' custom circuit capability. The founding group approached the development of standard microcircuits in an original and innovative fashion. Unlike the micrologic group at Fairchild, which had given little attention to the needs of customers when it developed its first integrated circuits, Signetics integrated the users' requirements into their designs. This was consistent with Signetics' orientation toward custom circuits and customer focus.[37]

To meet the needs of potential system users, Baker employed a different logic configuration than the RTL/DCTL logic used by Fairchild, Amelco, and Texas Instruments.[38] Baker utilized diode transistor logic. This was a form of logic that Baker knew well. He had worked on DTL circuits made of discrete components at IBM. In his view, DTL had a

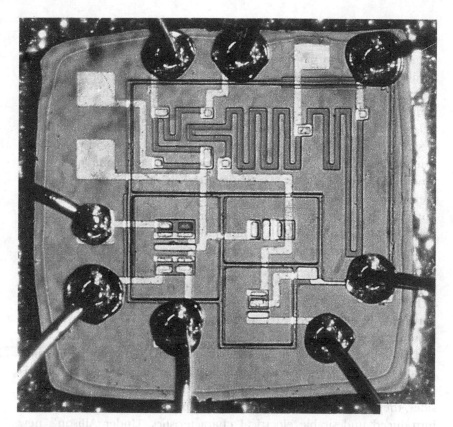

Figure 6.2
Signetics' DTL gate, 1962. Courtesy of Lionel Kattner.

number of advantages over RTL and DCTL. Unlike Fairchild's DCTL and RTL products, DTL circuits had good noise immunity. In other words, external electrical signals did not interfere with their functioning. This was important since the end user, the military, had an unyielding interest in reliability. Furthermore, DTL was an established and widely used form of logic. Most design engineers did not know and were not interested in learning about DCTL and RTL. Since many military electronics firms were already using DTL circuits made of discrete components, they knew how to evaluate them and design them into their systems. They could easily adjust their testing equipment to integrated forms of DTL circuits. In short, DTL was a good candidate for integration because it was reliable and familiar to most circuit and system engineers. Baker later visited engineers at system firms and talked with them about

DTL circuits. The response was overwhelmingly positive. They would buy DTL circuits if these were on the market.[39]

In the winter and the spring of 1962, Baker and Allison developed a family of DTL circuits. Baker began with a DTL gate and later engineered six circuits that used different combinations of this gate. These included gates, flip-flops, and binary elements. These were all the circuits needed for the design of the central processing unit of a computer. Baker designed these circuits so as to make them easy to use and place on a printed circuit board. These circuits were remarkably complex for that time. For example, the binary element was made of twenty circuit elements: two transistors, ten diodes, eight resistors, and two thin-film capacitors. Making these circuits was particularly difficult, not only because of their complexity and use of thin films but also because of their logic configuration. They were made of a greater variety of elements than RTL circuits. They also required tighter device tolerances, more exacting process control, and new ways of isolating devices on the same silicon chip.[40]

To overcome these difficulties, Allison developed new processes. In particular, he developed a triple-diffusion process. (Texas Instruments' first microcircuit family also used triple diffusion.) At Fairchild, Kattner and Haas had previously isolated devices on the same chip by diffusing isolation regions from the top and the back of the silicon wafer. This process had serious limitations. Since the wafers were diffused from both sides, they were very thin. As a result, they broke easily. The process also introduced undesirable electrical characteristics. Under Allison's new scheme, the wafers were triple diffused. At first, Allison diffused a deep N-type well in a P-type wafer to make the collector; he then diffused the base and the emitter in this well. James and Allison also developed ways of integrating thin-film resistors and capacitors as well as diffused active elements in the same monolithic substrate. Signetics introduced its first DTL gate to the market at the Institute of Radio Engineers show in March 1962. They announced the whole DTL family in July of the same year. This was a remarkable accomplishment. They had designed an innovative line of circuits in less than a year. They had also integrated complex circuits that no other firm had put into silicon before.

Almost at the same time as the introduction of the DTL family, Signetics was running out of money, having expended most of its capital. A second round of funding was required. Lehman Brothers and the other investors, who had recently been battered by a decline in the stock market, were not inclined to invest more money in the startup. This left the founders to their own devices. They approached potential investors,

Figure 6.3
Signetics' founders greeting Corning executives at the San Francisco airport, circa 1963. Left to right: Lionel Kattner, Amory Houghton (chairman of Corning), David James, Malcolm Hunt (vice-president of Corning). Courtesy of Lionel Kattner.

including Litton Industries. They were supported in this endeavor by Joseph Van Poppelen, an ambitious businessman whom they had recently hired as vice-president of sales and marketing. Van Poppelen had previously headed sales at Motorola. When none of his contacts bore fruit, the situation became critical. It was at that time that Lehman Brothers advised the founding group at Signetics that Corning Glass Works was interested in buying a controlling interest. Corning, a long-time client of Lehman Brothers, was an old, established corporation with a long tradition of technological innovation. Corning was no stranger to the electronics field as it produced television bulbs as well as specialty glass used by vacuum tube manufacturers. Corning had recently diversified into the production of resistors, capacitors, and hybrid circuits. Malcolm Hunt, the head of their electronic division, was interested in Signetics for a variety of reasons. He feared that integrated circuits would soon displace discrete components such as the ones he was producing. He wanted to complement his line of thin-film circuits with digital integrated circuits. Hunt was eager to establish a foothold in this new and promising industry.[41]

Intent on buying into Signetics, Hunt paid a visit and courted the Signetics group. He assured them that Corning would support Signetics financially and technically and leave it completely autonomous. Van Poppelen and the management group voted as to whether they would ally themselves with Corning. Five voted in favor of the deal. Van Poppelen and Kattner dissented on the grounds that Corning was not a good fit for an integrated circuit company. In their view, Corning was a staid Eastern firm, ignorant about semiconductors and antithetic to the ways and culture of the electronics industry on the West Coast. Their view later proved to be correct. In November 1962, following the vote, the founders and original investors signed an agreement with Corning. Under the terms of the agreement, Corning invested $1,700,000 in the semiconductor firm. In return, they got 51 percent of the company's stock and a majority on the company's board. Hunt became Signetics' chairman while David James remained president. The agreement also included an innovative arrangement whereby Signetics' original stockholders could sell their shares to Corning. The value of the stock would be set by an independent appraiser at the end of every fiscal year.[42]

Corning's cash kept the firm afloat. It also financed an aggressive sales and marketing campaign, that advertised Signetics' custom circuit capability and its family of DTL circuits. In late 1962 and 1963, Van Poppelen built up the marketing and sales department by hiring some of his former associates at Motorola. They publicized their DTL circuits through seminars and articles in the trade and engineering press. They also sought to build a custom business by developing "preFEBs" and "variFEBs." PreFEBs were made of groups of transistors, diodes, resistors, or capacitors on the same silicon chip. Van Poppelen and his group devised these preFEBs to help engineers at system houses get a feel for and familiarize themselves with integrated circuits. These engineers could also design their own circuits by arranging these elements. VariFEBs offered a different approach to custom circuit design. VariFEBs were integrated circuit dice. These dice were fully processed with the exception of the aluminum interconnection pattern. They were identical to those used by Signetics for its standard circuits. Under this scheme, the customers would choose their own pattern and ask Signetics to deposit it on the dice.[43]

While developing these new products, Van Poppelen also sought to get Signetics' circuits officially approved by the Department of Defense. Under the procurement system in force since the late 1940s, the Army, the Navy, and the Air Force independently certified suppliers of electronic components. It was only after these suppliers had been certified that system contractors could use their products. Van Poppelen and Hunt actively promoted Signetics' circuits to the officers in charge of system development programs at the Department of Defense. They concentrated their efforts on Arthur Lowell, a powerful Marine colonel who was in charge of the development and procurement programs of the Navy's Bureau of Weapons. Lowell was one of the main champions of integrated circuits in the military in the early 1960s. Winning him over was essential. To do so, Van Poppelen and Hunt cultivated their relationships with Lowell by making frequent visits to Washington. This involved a lot of dining and drinking. The campaign was a success as Signetics was approved as a supplier by the Navy in 1963.[44]

In spite of these efforts and Signetics' product innovations, the firm had great difficulty generating sales. "It was like shouting in the void," James later recalled. "Nobody was buying except the Army and the Navy and their purchases were very small, disappointingly small for quite a while."[45] Signetics sold circuits to a few military contractors such as Honeywell, Lockheed, RCA, and Martin Marietta, but these orders were mostly in sample quantities. Signetics was no more successful with custom circuits. The only significant custom contract that the company received in 1963 was for the development of high-speed circuits for the National Cash Register Company. System firms were not buying. Firms that did so purchased their circuits from established manufacturers rather than from untried startups such as Signetics. Autonetics and the Instrumentation Laboratory at MIT, the two organizations that committed themselves to microcircuits in 1962, bought their components from Fairchild and Texas Instruments. Autonetics awarded a large contract to TI for the development of a family of 22 planar microcircuits. These were to be used in the guidance and control system of the Minuteman II, the successor to the first Minuteman missile. The Instrumentation Laboratory designed the guidance computer for the Apollo rocket around a circuit from Fairchild's micrologic family. The Instrumentation Laboratory chose this circuit because Fairchild was the only firm able to deliver microcircuits in quantity at that time. The fact that Robert Noyce (Fairchild Semiconductor's general manager) had graduated from MIT might have figured in this decision as well. As sales languished, Signetics incurred heavy losses. In desperation, the management, who had shunned military R&D contracts until that time, began to actively pursue those contracts. Van Poppelen hired a military sales specialist from Motorola, whose sole responsibility was to ferret out R&D contracts from the Department of Defense. As a result of these efforts, Signetics received a development contract for new packaging techniques. In 1963 it got a $159,000 contract to develop a high-speed DTL family for the Signal Corps Engineering Laboratories.[46]

Signetics' fortunes began to change in the spring and summer of 1963. Starting around that time, Signetics received a surge of orders for its DTL circuits because of a significant change in Department of Defense policy. In April 1963, James Bridges, the Director for Research and Engineering at the DoD, sent a memorandum to the military services urging them to embrace microelectronics. Like silicon components a few years before, microelectronics would enable the military to procure more reliable weapon systems. Increased reliability would enable the military to perform missions successfully, reduce maintenance costs, and ultimately expand the use of electronics in military systems. "This gain in reliability, coupled with reduction in size, weight, and power requirements, and probable cost savings," wrote Bridges, "makes it imperative that we encourage the earliest practicable application of microelectronics to military electronic equipment and systems. It is recognized that the currently proven technology in microelectronics generally limits its application in developments for service use to digital circuits such as are employed extensively in computers and other data processing and handling devices. In initiating new developments for these types of equipment, it is suggested that they be examined carefully to determine the applicability of the new technology."[47] This directive had a profound effect on military procurement policies. Until that time, the military services had supported the development of integrated circuits. But with the exception of the Navy's Bureau of Weapons, the services had been reluctant to use them in their systems. In an about face, they now embraced the new technology. Military agencies let it be known that system proposals had to incorporate integrated circuits or hybrid circuits in order to be funded. This led managers at system firms to put substantial pressure on design teams to incorporate integrated circuits in their new systems. Not surprisingly, military contractors and especially avionics firms became much more receptive to integrated circuits. In other words, the Department of Defense created a large market for integrated circuits in the military sector that went well beyond the Apollo and Minuteman II programs.[48]

Signetics was the main beneficiary of the creation of this market. The demand for DTL circuits grew rapidly and became much greater than the demand for RTL and DCTL components. As Baker had foreseen, system firms chose DTL circuits because of their electrical performance and noise margins. DTL circuits were also easy to use and their logic configuration was familiar to most digital circuit engineers. Signetics circuits were being designed into most new electronics systems at that time.[49]

Now that Signetics had a market, it encountered a second hurdle: it could not fill orders for its circuits. The firm experienced severe yield problems. In the summer and fall of 1963, yields on flip-flops and binary elements were on the order of 1 percent. Those for gates reached 10 percent. These low yields were caused by photoresist problems and dislocations in the silicon crystal. Signetics also ran into substantial difficulties in fabricating circuits with thin-film capacitors. These capacitors were made by growing a thin oxide layer between aluminum and silicon. The oxide often developed pinholes, which led to device failure. Signetics' devices also had the "purple plague," a problem that it shared with Amelco, Fairchild, and other manufacturers of integrated circuits. The aluminum film that interconnected the different elements would take on a purple color and later dissolve. This also led to the failure of the device. Under Kattner's direction, Signetics' engineers devoted considerable effort to solving these difficulties, improving yields, and standardizing the process.[50]

These sales and manufacturing troubles had important consequences for the running of the company. They caused a rift between the founders and their main investor, Corning. In light of Signetics' persistent problems and its inability to meet its sales and production forecasts, Malcolm Hunt and other Corning executives lost their confidence in Signetics' management and gradually became more involved in the company's affairs. At first, Hunt sent a financial specialist and a production expert, Don Liddie, to Signetics. Liddie worked as an assistant to David James, Signetics' president. He also organized the production-control department and introduced more discipline in manufacturing. Hunt later intervened at the highest levels of the company. Judging James ineffective, he demoted him from president to vice-president of research and development in the fall of 1963. Van Poppelen, the vice-president for marketing, became executive vice-president and general manager. He ran the corporation for Corning. Corning also forced Weissenstern, one of the founders, to leave the company, and bought out his interest. These moves led to growing tensions between the remaining founders on the one hand and Corning and Van Poppelen on the other.[51]

In spite of this managerial instability and the growing tensions between Corning and Signetics' founders, the company's sales revenues grew rapidly in the fall of 1963 and in much of 1964. After years of losses, Signetics became profitable. This surge in revenues and profits can be attributed to substantial improvements in yields. The firm's efforts to clean up and standardize the process finally bore fruit. David Allison also made an important innovation in processing by eliminating one of the main causes of device failure: dislocations in the silicon crystal. Silicon wafers would often undergo a lot of stress when introduced into high-temperature diffusion furnaces. The high temperatures would cause dislocations in the crystal. Because these dislocations would run across the transistors, they would destroy them. Before putting the wafers in, Allison lowered the temperature of the furnace to 800°C from the usual 1,200°C. Later he raised the temperature back to 1,200°C. Dislocations in the crystal decreased dramatically, and the yields went up right away. This simple procedure was quickly adopted by other firms on the San Francisco Peninsula.

To meet the growing demand for standard circuits, Signetics hired more employees and enlarged its plant space. Its staff grew from 170 in mid 1963 to 350 in 1964. Signetics also expanded its facilities by renting another building in Mountain View in the summer of 1963. A few months later, Signetics leased an additional 14,000-square-foot facility specifically for R&D, reliability engineering, circuit design, equipment design, and the machine shop. Signetics also planned and started the construction of a large plant in Sunnyvale in 1964. By then, Signetics had become the largest supplier of integrated circuits on the Peninsula. Its microcircuit sales were substantially larger than Fairchild's or Amelco's.[52]

Dumping and Market Building

The emergence of DTL as a popular logic configuration and the rapid growth of Signetics did not go unnoticed at Fairchild Semiconductor. These successes sparked a major controversy between the marketing department and the integrated circuit development group about which course of action to pursue in integrated circuits. Prior to this, Fairchild's managers were not aggressively pushing the development and marketing of microcircuits. As they viewed the success of Signetics and received a growing number of inquiries regarding integrated circuits from their customers, they changed course. By early 1963, the firm had finally begun to believe in the future of integrated circuits. Gordon Moore, the head of

R&D, was now convinced that integrated circuits had a future. His opinion was reinforced by the marketing department.

To position Fairchild favorably in the integrated circuit field, Noyce and other top managers resolved to devote significantly greater financial and engineering resources to the development of integrated circuits and related processing techniques than they had done in the past. The firm also created a new section in the R&D department devoted solely to digital systems research. To head this section, Noyce and Moore recruited Rex Rice, a former IBM engineer. The goal of this new section was to develop new computer architectures and packaging techniques. It was also meant to help the company evaluate which digital functions should be integrated in a silicon chip.[53]

Although there was a general agreement about the future importance of microcircuits, the various groups at Fairchild fought ferociously over which direction to pursue. The marketing department thought that the growing market for integrated circuits would go DTL. Marketing and sales were having considerable difficulty selling DCTL circuits. DTL elicited much greater interest among customers. Taking its cue from the market place, the marketing department advocated that Fairchild produce and commercialize DTL circuits in addition to its DCTL and RTL lines. On the other side of the debate was a group of circuit engineers in the R&D laboratory. These engineers had replaced Last, Allison, Kattner, and Haas after their departure to Amelco and Signetics. The microcircuit group was led by Robert Norman and Phil Ferguson. Norman had invented DCTL at Sperry Gyroscope in the mid 1950s and had helped to develop the first family of integrated circuits at Fairchild. Ferguson was the head of the device development section (where he had replaced Allison). These men argued vehemently that Fairchild should concentrate its resources on DCTL and RTL circuits. Norman maintained that these logic configurations offered superior characteristics to DTL. According to Norman, they were faster and could be produced at higher yields. The group headed by Norman and Ferguson was then developing a new family of low-power RTL circuits with financing from the National Security Agency.[54] The NSA contract supported the development of integrated circuits for the agency's spy satellites. Norman and Ferguson felt that these circuits would compete easily with DTL and find a large market in aerospace programs beyond the NSA's immediate needs. Noyce and Moore sided with the marketing department. They decided that Fairchild would develop DTL circuits to compete directly with Signetics and they had the power to enforce their decision.[55]

This decision enraged Norman, Ferguson, and many other circuit engineers. In the summer of 1963 they left Fairchild to start their own semiconductor company, General Micro Electronics (GME). These men were joined by Howard Bobb (a Fairchild salesman with close ties to the NSA) and Lowell (the newly retired head of avionics at the Bureau of Weapons, who had championed the use of integrated circuits in naval systems in the early 1960s). The new startup was supported by Pyle National, a Midwestern manufacturer of electrical cables. Like Corning, Pyle National was concerned that integrated circuits would make its component business obsolete. The founding group at GME set out to produce and commercialize the family of low-power circuits they had engineered at Fairchild. They also obtained contracts from the NSA contracts for further development of RTL circuits.[56]

To replace Ferguson, Noyce asked Pierre Lamond, a tough French engineer who had made a name for himself by overseeing the production of the switching transistor for Control Data, to direct the device development section. Robert Seeds, a talented circuit engineer from IBM, was appointed the head of the integrated circuit development group. These men directed Fairchild's much-enlarged microcircuit development program. They gave considerable emphasis to the development of DTL integrated circuits. Since Signetics' DTL family was rapidly becoming the standard in the market, Lamond and Seeds elected to copy it. Given the nature of the market, this decision was a highly rational choice. However, it did not fit with Fairchild's tradition of industry leadership. Not only were they emulating Signetics' circuits, but five other semiconductor corporations decided to second source Signetics' DTL line around the same time. Like these other firms, Seeds and his group engineered DTL circuits fulfilling the same electronic functions and having the same pin configuration as Signetics'. What was unusual about Fairchild's family was its processing, which relied on epitaxy, a complex process that the company had employed solely for its fast switching transistors and a few low-power RTL microcircuits. This process enabled Fairchild to design DTL products that performed faster and better than Signetics'.[57]

In June 1964, when the R&D laboratory at Fairchild completed the first elements in its DTL microcircuit family, the market for these circuits was thoroughly dominated by Signetics. This left little space for other players. It was in this context that Robert Noyce, Thomas Bay, and Charles Sporck (the manufacturing manager who had played a big part in the firm's expansion into commercial transistors a few years earlier) determined to dump Fairchild's devices on the market. They decided to introduce Fairchild's DTL circuits at less than half the price of Signetics' products. This price represented roughly half of what Fairchild antici- pated would be the cost of manufacturing the devices. In so doing, these men had two goals: to establish Fairchild as the dominant manufacturer of DTL circuits and to drive Signetics out of the market. It was not the first time that Noyce had attempted to put Signetics out of business. A few months earlier, he had unsuccessfully attempted to lure Allison back to Fairchild in the hope that the departure of its expert process and device development engineer would deal a lethal blow to Signetics. Now, Noyce sought to grab Signetics' only market. Around the same time, Fairchild Semiconductor also went after General Micro Electronics. It sued GME for the theft of trade secrets, expecting that this would finally cripple the firm. For Fairchild's upper managers, Signetics and GME were danger- ous competitors that had to be destroyed. Noyce also intended these actions to convey a strong message to Fairchild's engineers and managers who might be tempted to defect and start their own business.[58]

Noyce's pricing tactics also had another goal: to stimulate the demand for integrated circuits and thereby create a market for them in the com- mercial sector by making them cheaper than equivalent circuits made of individual diodes and transistors. The company had used this technique to great effect in the transistor business a few years earlier. Low-cost assembly in Hong Kong had enabled Fairchild to slash its prices for tran- sistors. This in turn had enabled the firm to create markets for them in the computer and consumer electronics industries. Noyce knew that low- ering prices for integrated circuits would have a similar effect. Users would adopt them because they were less costly than circuits made of dis- crete devices. In other words, price reductions would greatly enlarge the demand for microcircuits, especially among commercially oriented firms that were more price conscious than military contractors.

Not surprisingly, Fairchild's drastic pricing of its DTL line had a destructive effect on Signetics. In order to take advantage of the price offered by Fairchild, many Signetics customers canceled their orders. To save what remained of the firm's backlog, Van Poppelen, Signetics' exec- utive vice-president, along with Baker, Yelverton, and the three remain- ing founders, decided to match Fairchild's prices. This meant that, after only a few profitable months, Signetics would return to the red and incur heavy losses. The corporation made this decision in the hope that it would keep some of its customers and maintain some cash flow. But this move was not well received at Corning. Hunt and other Corning

managers saw Signetics' decision as additional proof of poor management. Determined to solve the firm's perceived managerial problems
once and for all, Hunt demoted Van Poppelen to marketing vice-
president in September 1964. Jim Riley, a seasoned manager from
Corning's electronic component division, took over as Signetics' president and general manager. Riley placed Corning engineers and managers in the most responsible positions at Signetics. As an example,
Liddie became the head of manufacturing and reorganized it along
Corning lines. To stem Signetics' heavy losses, Riley took radical cost-
cutting measures. In the fall of 1964, he laid off nearly half of Signetics'
workforce and slashed its product-development expenses. The firm concentrated its efforts on stabilizing and standardizing the process in order
to improve yields and reduce manufacturing costs.[59]

By winter, Signetics needed some urgent refinancing. Corning used
Signetics' dire condition to increase its share of ownership. At the same
time, Corning's managers would humiliate the founders and punish
them financially. In late December, a few days before the firm had to
close its doors for lack of funds, Corning's management convened a
meeting with Kattner and Allison in Corning, New York. The treasurer
offered to bring in additional capital by buying a new issue of stock at 5
cents per share. This would give Corning an 82 percent equity position
in Signetics. The offer would dilute the founders' stock and the stock
options granted to employees to almost nothing. Kattner and Allison
turned this proposition down. In last minute negotiations, Corning and
Signetics' representatives agreed that new shares be issued at the price of
50 cents a share. Subsequently, an independent evaluator determined
that the real worth per share was in the neighborhood of $3.50.[60]

Corning's hard bargaining tactics and the growing presence of
Corning people at Signetics further alienated the founders. In the spring
of 1965, disillusioned and disheartened, James and Kattner left the firm
they had established. Kattner joined Amelco as assistant general manager. Under his watch, in the next year and a half Amelco's integrated circuit sales grew from $1.2 million to more than $4 million (on an annual
basis). Amelco's total revenues reached $9.8 million in 1966. James
became the vice-president for research and development at Ultek, the
Varian Associates spinoff that made VacIon pumps and various types of
vacuum systems for semiconductor manufacturers. Yelverton started a
head-hunting agency in San Francisco, while Van Poppelen became general manager of ITT Semiconductor. By March 1965, Corning's takeover
was complete.[61]

Fairchild's price reduction also enabled it to capture a large share of the growing market for DTL circuits in the military and space sector and to displace Signetics as the largest producer of integrated circuits in the United States. In late 1964 and 1965, Fairchild saw the demand for its DTL circuits skyrocket, while its shipments of RTL and DCTL parts remained level. Many military and space system firms designed Fairchild's DTL circuits into their systems at that time. For example, Boeing adopted them for the prototype of the lunar orbiter. Sperry designed a new LORAN system around Fairchild's circuits. But while Fairchild was successful at selling DTL circuits, it was less successful at making them. When the firm announced its products' prices, it had yet to transfer them to production. Not unlike Signetics a year before, Fairchild ran into severe manufacturing problems with its DTL circuits. Several times in late 1964 and 1965, the firm "lost the process"—that is, yields plunged to zero. As a result, Fairchild was late in its deliveries. This caused serious production delays for its customers. Fairchild's manufacturing problems saved Signetics. Because of Fairchild's poor delivery record, some customers switched their purchases of DTL circuits back to Signetics. Strong sales, the cash infusion from Corning, and radical cost cutting enabled Signetics to survive. But Signetics had been permanently wounded, and it would never regain its leadership in the integrated circuit business.[62]

Once Fairchild Semiconductor improved its yields, it emerged as the leading manufacturer of DTL circuits, which were widely adopted by military contractors. Fairchild also made inroads in the commercial market, which until then had been closed to microcircuits. Because of the drastic price cut, DTL integrated circuits were now cheaper than equivalent circuits made of discrete components. In late 1964, a few commercial firms began to use Fairchild's circuits in their systems and subsystems. This was an important development. Until then, only military and space contractors had used integrated circuits. Prominent among the early commercial users was Scientific Data Systems (SDS), a manufacturer of minicomputers located in Southern California. In December 1964, the firm announced its use of Fairchild's DTL circuits in the analog-to-digital and digital-to-analog converter of its SDS-92, a new computer designed to control industrial processes.[63]

In 1965 and 1966, building on this foothold in computer applications, Fairchild's managers focused on the development of commercial markets for integrated circuits. They relied on the company's strengths in manufacturing and especially its recently acquired mass-production capability. Fairchild also made use of the applications engineering group's knowledge of the silicon process and the design of commercial electronic systems. Many of the men in the group had been hired to support Fairchild's push into commercial transistors a few years earlier. At Pierre Lamond's request, the applications engineers developed a series of DTL circuits that would meet the requirements of a variety of commercial users. (Lamond, who had directed the development of Fairchild's first DTL circuits in the R&D laboratory, had recently become the director of the manufacturing plant in Mountain View.) Applications engineers also identified users' circuit and system needs and, in collaboration with the R&D laboratory, designed integrated circuits that fit the firm's manufacturing processes and met the system needs of commercial customers. Their goal was to engineer "building blocks," namely circuits that designers would use to construct a wide range of electronic systems. They also developed increasingly complex circuits in order to meet new needs and open up new applications.[64]

In conjunction with the development of customer-oriented devices, the applications engineering group published numerous applications notes that explained how the customers could use these circuits in their systems. These applications notes were meant to show customers how their technical problems could be solved better with integrated circuits than with transistors, vacuum tubes, or electromechanical components. The applications notes described the functions commercial manufacturers could perform with Fairchild circuits and the systems that they could make with the firm's components. They also explained in great detail how these systems could be designed.

But the design of commercially oriented DTL products and the writing of applications notes did not suffice to build a large market for integrated circuits. Potential customers in a variety of commercial industries had to be further convinced. With the exception of SDS and a few other firms, many electronics corporations were at best lukewarm toward microcircuits. Like military contractors a few years earlier, commercial firms were reluctant to use a complex technology that they did not understand well, and buying integrated circuits from outside vendors (instead of making circuits composed of discrete devices in house) would cut into their profits. Commercial firms worried too about production costs. Because integrated circuit packages were expensive to solder to printed circuit boards, they increased the overall cost of system manufacturing.

To convince commercial firms of the great future of integrated circuits, Fairchild launched an aggressive marketing campaign in 1965.

This campaign included presentations at technical meetings and articles in the trade press by Robert Noyce, Gordon Moore, Victor Grinich, and other Fairchild managers. For example, in April 1965 an article by Moore titled "Cramming More Components onto Integrated Circuits" appeared in *Electronics*. In this article, Moore argued that microcircuits had become cheaper and more reliable than circuits made of discrete devices. Integrated circuits were also going to lower the price of digital electronics dramatically by incorporating more devices on a silicon chip. Moore noted that the number of transistors on silicon circuits had doubled every year since the development of the first microcircuit in 1960. He claimed that they would continue to do so in the foreseeable future (a statement later known as "Moore's Law"). Applying this observation, Moore predicted that the number of transistors on a silicon chip would increase from 50 in 1965 to 65,000 in 1975 (Moore 1965).[65] Moore also claimed that the exponential growth in the complexity of microcircuits and the corresponding decline in the cost of electronic functions would revolutionize electronic system engineering. "The future of integrated electronics," he contended, "is the future of electronics itself. The advantages of integration will bring about a proliferation of electronics. Integrated circuits will lead to such wonders as home computers—or at least terminals connected to a central computer—automatic controls for automobiles, and personal portable communications equipment. But the biggest potential lies in the production of big systems. In telephone communications, integrated circuits in digital filters will separate channels on multiplex equipment. Integrated circuits will also switch telephone circuits and perform data processing. Computers will be more powerful and will be organized in completely different ways. Machines similar to those today will be built at lower costs and with faster turnaround." (Moore 1965) Moore suggested that integrated circuits would unleash a wave of innovation in electronic system design. Firms that refused to use microcircuits condemned themselves to oblivion. The world belonged to those who exploited the enormous potential of integrated electronics (ibid.). In conjunction with this marketing campaign, Fairchild Semiconductor developed new integrated circuit packages that lowered the assembly cost of electronic systems. Packaging was indeed the chief obstacle to the widespread use of integrated circuits in the computer and consumer electronics industries. At that time, microcircuits were sold in flatpacks. Flatpacks were small rectangular glass and ceramic packages with close lead spacing. They could be stacked easily and had good thermal dissipation characteristics. But they were too cumbersome and expensive for volume production. (They had been developed for military applications in the early 1960s.) Moreover, flat-packs required the use of printed circuit boards with narrow traces, which were difficult and expensive to produce. Adding insult to injury, flatpacks were also difficult to solder to printed circuit boards. "Flatpacks," complained a Scientific Data Systems engineer in 1964, "are hard to handle and hard to insert on the circuit cards. The leads are too close and bend easily."[66] As a result, the use of flatpacks entailed sub-stantial labor costs. In order to produce integrated-circuit-based systems, firms had to employ very skilled and highly paid operators who would solder each individual flatpack by hand to the printed circuit boards. Few commercial firms were willing to incur these expenses.[67]

To stimulate the commercial demand for integrated circuits, engineers in the digital systems and device development sections of Fairchild's R&D department set out to develop a new package. They were interested in designing a package that would lend itself to "easy industry use in auto-mated insertion systems or very rapid hand insertion."[68] To facilitate circuit-board layout and to reduce the assembly costs of electronics sys-tems, Rex Rice, a computer engineer and the head of the digital systems laboratory, devised a new lead configuration for integrated circuit pack-ages. The new configuration had pins that were 100 rather than 50 mils apart (as they were in flatpacks). The pins were also in a line, rather than coming from each of the package's four sides as in the flatpack. Before putting the in-line package into production, Fairchild's marketing spe-cialists tested it on computer manufacturers, especially Control Data and General Electric. They discovered that customers did not like packages standing up in one line, but were willing to take two. As a result, Fairchild's device development group transformed the in-line prototype into a package with two parallel rows of pins.[69]

The end result was a radically new package, the dual in-line package (DIP). Whereas the flatpack was small and its leads spaced 50 mils apart, the DIP was larger and had wider lead spacing. This configuration facil-itated the use of cheap printed circuit boards. The DIP allowed for wire routing under the package and permitted a more efficient use of board space. Its in-line arrangement simplified the layout of printed circuit boards. The new package also was easier to handle and more adaptable to automated assembly. It plugged into the board and thereby could be soldered using mass-production techniques. As a result, dual in-line packages reduced assembly costs by a factor of 4. Fairchild Semi-

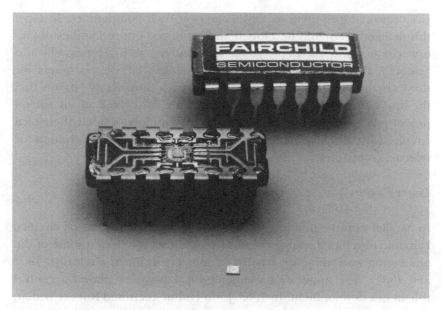

Figure 6.4
Capped and uncapped dual in-line packages with integrated circuit die, 1967.
Courtesy of Fairchild Semiconductor and Stanford University Archives.

conductor introduced the dual in-line package on its line of DTL circuits in 1965. The dual in-line package was a major innovation. It
opened up the commercial market for integrated circuits and played a
major role in the evolution of integrated electronics from innovation to
large-scale manufacture.[70]

In part because of deft marketing and the introduction of the DIP, the
demand for DTL and other digital integrated circuits increased significantly in 1965 and 1966. The overall market for digital circuits
expanded from $35 million in 1964 to $90 million in 1966. In 1964 semiconductor firms shipped very few integrated circuits to commercial
users, but by 1966 nearly half of the sales of integrated circuits in the
United States were to the commercial sector. Fairchild Semiconductor
was the prime beneficiary of this market expansion. In 1965 and 1966
it was the largest supplier of microcircuits to computer and other commercial users. (Texas Instruments and Motorola focused more on the
military market.) In addition to selling its line of DTL circuits, Fairchild
built a large custom circuit business. Computer manufacturers were
increasingly interested in custom circuits because they felt that custom

circuits would give them an advantage over their competitors. To meet this demand, Fairchild developed and produced entire families of custom circuits with flip-flops, gates, and buffers. IBM, Honeywell, and Scientific Data Systems were among Fairchild's largest customers for these circuits. They approached Fairchild with proprietary circuits that the firm's engineers transformed to meet the constraints of the silicon process. As more and more computer corporations wanted their own circuits, Fairchild became a major manufacturer of custom products for selected companies.[71]

Linear Circuits

In parallel with the building of a commercial market for digital circuits, Fairchild developed analog circuits (in doing so, it followed some of the pioneering work done by Amelco's engineers in linear circuits in 1962 and 1963). The applications engineering laboratory played a central role in this development. In 1963, John Hulme recruited Robert Widlar, a young, aggressive, and independent-minded engineer, to the applications laboratory. Hulme gave Widlar the job of designing an operational amplifier that would be compatible with Fairchild's manufacturing process. An operational amplifier is a particular kind of linear or analog circuit. It could be used for signal processing, signal generation, and feedback control applications. At the time when Hulme asked Widlar to design an operational amplifier, there were no good amplifier microcircuits available on the market. Texas Instruments had introduced operational amplifiers in 1962, but their performance was very limited. Amelco had also designed a monolithic operational amplifier for the IHAS program, but this circuit had not yet been commercialized. Hulme was determined that Fairchild take advantage of this interesting market opportunity.[72]

Building an operational amplifier with Fairchild's manufacturing process was no easy task. This process had been optimized for the making of digital circuits, namely circuits where the switching speed of transistors was essential. This process was now to be used for the production of microcircuits, in which other electrical characteristics, such as leakage currents and beta (current gain), were more important. To deal with the severe constraints this imposed on circuit design, Widlar made innovations in integrated circuit design (some of which he later patented). Among other innovations, he put active devices to work like passive components, used special configurations to make a low-beta transistor to behave like a high-beta unit, and created constant-current source replacements for the large resistors. Widlar also partnered closely with David Talbert, a process engineer in the manufacturing plant. At Widlar's urging, Talbert carefully tweaked and tightly controlled the manufacturing process in order to produce better transistors and resistors for analog circuits. Making linear circuits was much more difficult than producing digital circuits.[73]

Widlar's and Talbert's collaboration yielded two revolutionary products, the μA702 (μA standing for "micrologic amplifier") and the μA709. Fairchild introduced the 702 to the market in 1964. The 709, a circuit that improved substantially on the 702, followed in November 1965.[74] Widlar heavily publicized the 709. He gave a series of lectures and wrote papers and applications notes on the 709—thereby popularizing linear microcircuits and convincing system engineers that these circuits could be used. Military contractors rapidly adopted the 709. When the 709 became cheaper than similar circuits made of discrete components, it was adopted in numerous applications—from jet engines to process controls and a wide variety of instruments. In other words, the 709 became the standard operational amplifier. Raytheon, Amelco, and Union Carbide Electronics soon copied it and brought it to production. By 1967, more than a dozen US corporations sold this circuit. The 709 operational amplifier initiated the linear circuit business in the United States and gave Fairchild Semiconductor a prominent place in it.[75]

By 1967, Fairchild was the largest maker of integrated circuits on the San Francisco Peninsula. Having sold $35 million worth of microcircuits in 1966, it controlled about 30 percent of the US market for integrated circuits. It was particularly strong in DTL circuits, operational amplifiers, and custom digital products for mainframe computers. Signetics followed Fairchild's example and moved into the computer market. In 1966, Signetics had $12.6 million in sales and was the second-largest maker of integrated circuits in the area. (Amelco sold $9.8 million worth of semiconductors that year.) Fairchild's integrated circuit business was also highly profitable. As production volumes increased, the manufacturing staff gradually improved its control of the process and obtained better production yields. As a result, manufacturing costs decreased. In 1965 and the first half of 1966, Fairchild was probably the only firm making a substantial profit on integrated circuits in the United States. Fairchild's dumping strategy had succeeded.[76]

The rise of integrated circuits had significant effects on the Peninsula's electronics manufacturing complex and on Stanford University's electrical engineering program. Following Fairchild, Signetics, and Amelco in the new technology, Stanford and electronics systems firms rapidly learned how to design and make integrated circuits. Partially because of their close proximity to Fairchild and Signetics, these organizations became keenly aware of the potential of microcircuits. This propinquity also allowed them to acquire an expertise in the manufacture of microcircuits. At the forefront of this move into microelectronics was Hewlett-Packard. In 1965, Hewlett-Packard, a manufacturer of electronic measurement instruments, acquired the capability to make integrated circuits. At this time, Bill Hewlett and David Packard decided that their corporation would design and manufacture microcircuits for its own use. To do so, the firm hired local technicians as well as design and process engineers from semiconductor firms in the Santa Clara Valley. It also retrained hundreds of its own engineers in the new technology and built facilities for the production of integrated circuits. These substantial investments enabled the firm to introduce new integrated-circuit-based instruments to the market in the late 1960s.[77]

In parallel with Hewlett-Packard's venture into integrated circuitry, Stanford University incorporated the new technology into its curriculum and research program. This was not the first time that Stanford had sought to capture innovations coming from industry. It had done so with microwave tubes in the immediate postwar period and silicon devices in the mid 1950s. In 1964, John Linvill, the chairman of the electrical engineering department, recognized that microcircuits would revolutionize the engineering curriculum as much as transistors had revolutionized it 10 years earlier. As a result, he encouraged his faculty to develop a series of courses on integrated circuits. For example, in 1964 a junior faculty member developed a laboratory course that familiarized students with the complex processes used in the making of integrated circuits. He also established the Integrated Circuits Laboratory. This teaching laboratory was small and could fabricate simple circuits. A few years later, Linvill spearheaded the development of a research program on integrated circuits at Stanford. To obtain the advanced processing expertise that was necessary for such an endeavor, he hired a research engineer from Shockley Semiconductor and made him the Integrated Circuits Laboratory's chief engineer. With this influx of processing know-how, Stanford's Laboratory fabricated photo-transistor arrays and high-voltage integrated circuits in the late 1960s. The university was now a player in integrated circuit technology.[78]

Conclusion

Critical to the emergence and growth of the integrated circuit business on the San Francisco Peninsula was the cadre of engineers who developed the first integrated circuit at Fairchild Semiconductor. Because Fairchild's management showed limited interest in microcircuits, the group splintered and many engineers left the company to start new microcircuit operations. These men adopted different business models and pushed Fairchild's process and design technologies in a variety of directions. Jay Last and Jean Hoerni formed the semiconductor subsidiary of a startup system firm and made custom microcircuits for their parent company. Lionel Kattner, David James, and David Allison established Signetics, a stand-alone company that specialized in the making of standard digital circuits. This group devised a new approach to product engineering and made a major product innovation by developing a family of DTL circuits. When the Department of Defense forced its prime contractors to incorporate integrated circuits in their systems, these firms turned to Signetics' products and bought them in significant quantity.

It was only at this time that Fairchild's managers recognized the business potential of the new technology. To grow their integrated circuit sales, they asked their engineering staff to copy Signetics' products. They later commercialized these circuits at less than half of Signetics' price. This major price reduction enabled Fairchild to expand its microcircuit revenues and displace Signetics as the largest supplier of DTL circuits to the military sector. Price cuts, along with the development of the dual in-line package, stimulated the demand for microcircuits among computer manufacturers. Fairchild also innovated by introducing linear circuits to the market. As a result, Fairchild Semiconductor became the largest producer of integrated circuits in the United States, ahead of Signetics and Amelco.

The rise of integrated circuit manufacturing was a turning point in the history of the electronics cluster on the San Francisco Peninsula. The pioneering work on integrated circuits done at Fairchild, Amelco, and Signetics formed the basis for the extraordinary growth of the integrated circuit business on the Peninsula in the late 1960s and the early 1970s. By 1966, the main ingredients for this major expansion were in place. Firms on the Peninsula knew how to make many circuit types in quantity. They had strong circuit design and marketing capabilities. Many managers and engineers understood the microcircuit business well. The demand for integrated circuits was also growing fast. Military

contractors and computer manufacturers increasingly used microcircuits in their systems. New markets were emerging in instrumentation,
process control, consumer electronics, and other commercially oriented
industries. These markets promised to expand rapidly in the late 1960s
and the early 1970s. Fairchild, Signetics, and Amelco were well positioned to take advantage of the growth in the demand for integrated circuits. But paradoxically none of the Peninsula firms that pioneered
integrated electronics emerged as the primary beneficiaries of the
industry's expansion in the second half of the 1960s and the early 1970s.
The future belonged to a new wave of startups, including Intel, Intersil,
and National Semiconductor.

7

Valley of Silicon

The integrated circuit business exploded in the Santa Clara Valley in the late 1960s and the first half of the 1970s. Employment in the semiconductor industry grew from 6,000 in 1966 to roughly 19,000 in 1975. By that time, semiconductor corporations headquartered in Northern California employed tens of thousands of workers and engineers in other regions and in various Asian countries. Their sales volume also expanded significantly during this period. By the mid 1970s, local firms controlled a large share of the global market for integrated circuits. The San Francisco Peninsula had become the microelectronics center of the United States. And it was increasingly referred to as Silicon Valley, after the material used in the manufacture of microcircuits. It was in 1971 that Don Hoefler introduced the name "Silicon Valley" in a series of articles in *Electronic News*. This series on the rise of the semiconductor industry on the San Francisco Peninsula was entitled "Silicon Valley, USA." The name stuck. It was increasingly used to refer to the electronics cluster as a whole. How can one explain the growth of the integrated circuit business in the Santa Clara Valley? How did the electronics complex on the San Francisco Peninsula become the "Valley of Silicon"?

Most accounts of the rapid expansion of the integrated circuit industry in Northern California focus on Robert Noyce, Gordon Moore, and the formation of Intel. These narratives emphasize the conflicts that tore their first firm, Fairchild Semiconductor, apart and led them to start a new corporation. They also explore the ways in which Moore, Noyce, and their staff engineered advanced metal oxide semiconductor (MOS) processes for the making of high-density memory circuits and microprocessors.[1] Moore and Noyce clearly were major figures in the integrated circuit business, but they were not the only microcircuit innovator-entrepreneurs on the Peninsula. Jean Hoerni, Charles Sporck, and Pierre Lamond also played important roles in the industry's rapid expansion.

More important, because these narratives focus almost exclusively on Moore and Noyce, they ignore the larger social and economic forces that drove Intel and sustained the rise of the semiconductor industry as a whole.

Much of the growth of the microcircuit business on the Peninsula in the late 1960s and the early 1970s can be explained by a wave of entrepreneurship the magnitude of which had never been seen before on the San Francisco Peninsula. Between 1966 and 1970, Moore, Noyce, Hoerni, Sporck, and other entrepreneurs formed or reorganized more than 30 microcircuit corporations. These men were motivated primarily by peer competition, technological enthusiasm, and the potential financial rewards of the microcircuit business. To build new microcircuit operations, they used manufacturing processes that had been developed at Fairchild, Union Carbide Electronics, and other semiconductor corporations. They also raided these corporations' engineering and managerial staffs and their pools of skilled technicians and operators. In other words, the entrepreneurs, who had often held positions of authority at these firms, partially dismantled their former organizations to build new ones. (This predatory behavior and the older firms' focus on traditional technologies and markets contributed to the older firms' relative decline during this period.)

In their startups, the new entrepreneurs concentrated on the development of MOS manufacturing processes and the engineering of ever more powerful integrated circuits. They opened up large markets for these circuits in a variety of commercial sectors, including the watch industry, consumer electronics, instrumentation, telecommunication, and the automotive industry. They did so by substantially lowering the cost of electronic functions and by helping their customers to design systems based on integrated circuits. Some firms went as far as directing the product engineering efforts of their customers in order to sell their circuits. Thereby, the entrepreneurs and their corporations transformed integrated circuits into what the economists Timothy Bresnahan and Manuel Trajtenberg call a "general purpose technology"—a technology characterized by its pervasive use in a wide range of industrial sectors and by its technological dynamism. Integrated circuits (henceforth abbreviated "ICs") allowed users to develop entirely new products, introduce these products more rapidly to the market, and at the same time increase the productivity of their engineering staffs. By the mid 1970s, integrated circuits manufactured in Silicon Valley had become essential to many sectors of the US economy (Bresnahan and Trajtenberg 1995; Rosenberg 1963).

Opportunities

In the late 1960s, engineers and managers left Fairchild and other semi-conductor operations en masse to start new IC ventures on the Peninsula.[2] The entrepreneurs sought to take advantage of two opportunities in integrated electronics at that time: the fast growing demand for microcircuits in a variety of industrial sectors and the emergence of the MOS technology. In the mid 1960s, the demand for integrated circuits expanded significantly. The US market for microcircuits went from $41 million in 1964 to $80 million in 1965 and then to $120 million in 1966. This large increase was partially fueled by the precipitous decline in the price of integrated circuits. As a result of these reductions, microcircuits became cheaper than equivalent circuits that were made of discrete components. Microcircuits were also increasingly powerful. They could fulfill increasingly complex electronic functions. In 1966, the Electronics Industry Association and consulting organizations such as the Stanford Research Institute predicted that the US market for integrated circuits would grow to $420 million in 1970. They also forecast that the market for microcircuits in Western Europe would reach the $200 million mark at this time. These enormous markets opened up considerable opportunities for new startup firms. The new entrepreneurs were eager to exploit them.[3]

A number of entrepreneurs were also interested in taking advantage of the MOS technology, which differed from the bipolar technology that had been used in the manufacture of integrated circuits. The microcircuits marketed by Fairchild, Signetics, and Amelco in the first half of the 1960s were bipolar devices. These devices depended on the bulk of the silicon crystal for transistor action. In contrast, MOS transistors were field-effect devices. The surface of the silicon crystal was critical for their functioning. An MOS transistor had three electrodes: a source, a gate, and a drain. The gate was made of a highly conductive material (often metal) and was insulated by a layer of silicon oxide from the rest of the transistor. When one applied a voltage (charge) to the gate, the charge attracted electrons in the P substrate. This created a current from the source to the drain. It was possible to control the current flowing from the source to the drain by changing the charge of the gate. If there were a lot of positive charges on the gate, the gate would attract many electrons and the current from source to drain would increase. Reducing the charge on the gate decreased the flow of electrons between the two electrodes.

MOS technology had been pioneered at RCA and the Bell Telephone Laboratories before being promoted by Fairchild and one of its early

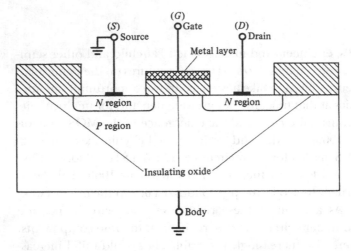

Figure 7.1
Cross-section view of an *n*-channel MOS transistor. Source: Grinich and Jackson 1975. Used with permission of McGraw-Hill.

spinoffs, General Micro Electronics. In the early 1960s, Fairchild's R&D laboratory developed MOS transistors and integrated circuits. It later investigated the structure of MOS transistors and examined the surface of the silicon crystal, especially the interface between the silicon crystal and the silicon oxide layer. This research program put the MOS structure on a firm scientific footing. General Micro Electronics also entered the fray in the early 1960s. In 1963, it hired MOS engineers from Fairchild and started a MOS circuit operation. The group at GME developed a MOS process and designed MOS circuits for NASA and the National Security Agency. It also made a family of MOS circuits for Victor Comptometer, a manufacturer of calculating machines (Bassett 2002).

In 1966, MOS process technology was still experimental. Like many new processes, the MOS manufacturing techniques were unstable and the yields were atrocious. Fairchild's and GME's MOS circuits were not very successful either. Their electrical characteristics were unpredictable. MOS circuits had the additional drawback of being slower than bipolar circuits. But in spite of these disadvantages MOS had a lot of promise. With MOS one could pack more transistors on a silicon chip than with bipolar techniques. In other words, MOS technology helped further miniaturize electronic circuits and lowered the cost of these circuits. MOS circuits also required less power than their bipolar counterparts. These characteristics made MOS components attractive for a wide variety of applications. Makers of aerospace systems and portable electronic equipment liked the fact that these circuits needed little power to operate. But the most promising market opportunities were in electronic systems that required complex circuits at low cost. Prominent among these were calculators, computer terminals, and computer memory systems. The new technology of MOS and the markets that it opened up were up for grabs. Many engineers were determined to take advantage of them (Bassett 2002). But these men felt that they could not exploit the emerging technical and market opportunities and get financial rewards for their work at the corporations where they were working. They were impelled to start new firms in order to take advantage of these opportunities. Fairchild, Signetics, and the other semiconductor outfits on the Peninsula focused on the military market, the computer market, and (to a lesser degree) the consumer electronics market. They had much less interest in emerging industrial applications. Although Fairchild et al. had active research programs in MOS, they focused most of their resources on bipolar technology. In other words, Fairchild and Signetics concentrated on incremental innovations to boost the performance of microcircuits for computer makers and weapon manufacturers—their main customers. The industrial market and MOS technology were peripheral to these core interests.

The entrepreneurs were also convinced that they would get little or no financial rewards from their work if they stayed at these corporations. They had not benefited financially from the development of integrated circuits and the creation of markets for them in the first half of the 1960s. The rewards had gone to upper management in the East. This was especially the case at Fairchild Camera and Instrument, Fairchild Semiconductor's parent company. Because of Fairchild Semiconductor's leadership position in integrated circuits, the value of Fairchild Camera's stock grew from $27 per share in January 1965 to $144 per share in October 1965. Richard Hodgson and John Carter, respectively Fairchild Camera's president and chairman, benefited handsomely from this speculative interest in Fairchild Camera stock. They had received large stock option packages in the late 1950s and first half of the 1960s. In 1965, they exercised their stock options and sold their stock, which enabled them to build fortunes. But the managers and engineers on the West Coast, who were primarily responsible for the appreciation of Fairchild Camera's stock, had few or no stock options and saw little of this new wealth. This led to considerable resentment on their part. When they asked for more options, John Carter turned down their requests. It increasingly became clear that they would not realize any substantial benefit from their work at Fairchild. The other parent companies in the East followed the same practice in that they offered few or no options to their semiconductor engineers and managers on the Peninsula. In this environment, the formation of new firms seemed to be the only way for silicon technologists to exploit the new business opportunities and reap the financial benefits of their work.

Starting a new semiconductor corporation became easier in the second half of the 1960s, and the industry that made equipment for the manufacture of semiconductors expanded on the Peninsula. New ventures such as Applied Materials commercialized pieces of equipment that previously had been developed and used only within Fairchild and other established semiconductor firms. The commercial availability of diffusion ovens, epitaxial reactors, and step and repeat cameras lowered the barriers to entry considerably in the integrated circuit business. New corporations could procure them immediately rather than having to devote significant engineering and financial resources to duplicate Fairchild's equipment.

Capital for such new ventures became more available. Different sources of funding emerged at that time. Managers and entrepreneurs in the semiconductor industry invested in startups—both locally and nationally.[4] But the real source of funding was the venture capital industry. Thirty new and reconstituted venture capital operations were formed on the Peninsula from 1968 to 1975. Among the most notable were the Mayfield Fund (1968), Arthur Rock and Associates (1969), and Kleiner-Perkins (1972). By the early 1970s, the Bay Area was one of the largest centers for venture capital in the United States. The growth of the venture capital industry encouraged the formation of new semiconductor startups by assuring potential entrepreneurs that investment backing was available. Venture capitalists also encouraged star engineers and managers to leave their firms and start new businesses. In short, the rise of the venture capital industry both facilitated and fostered entrepreneurship on the San Francisco Peninsula.[5]

Defections

The growth of the venture capital business, the tensions with the Eastern corporate staffs, and the emergence of new market and technical opportunities encouraged many experienced engineers and managers to leave existing firms and start their own corporations. Peer competition was another stimulus to defect. Entrepreneurs began launching these new enterprises because they wanted to show what they were able to do. They sought to outperform their rivals in the industry. All semiconductor businesses on the Peninsula were affected by this wave of competition and entrepreneurship. But the firm that was impacted the most was Fairchild. In a matter of years, Fairchild Semiconductor lost its group vice-president, two general managers, the director of research and development, the head of microcircuit manufacturing, and most of the engineers and mid-level managers who had transformed the company into a major producer of integrated circuits. To add to this exodus, the engineers working on MOS in the research laboratory also left Fairchild. In the period 1967–1970, defectors from Fairchild reorganized or established ten integrated circuit corporations.

A group of manufacturing engineers and designers of linear circuits associated with Charles Sporck, Pierre Lamond, and Robert Widlar were important figures in this entrepreneurial wave. These men moved to Molectro, then reorganized that firm and its parent company, National Semiconductor. Molectro, a Fairchild spinoff that had filed for bankruptcy protection, made bipolar integrated circuits. Molectro was later bought by National Semiconductor, a maker of silicon transistors located in Danbury, Connecticut. To revive Molectro, National Semiconductor's board recruited David Talbert and Robert Widlar, the engineers who had pioneered the development of linear circuits at Fairchild. In December 1965, National Semiconductor offered Widlar and Talbert large stock option packages (20,000 shares at $5 a share) and asked them to build a linear circuit business at Molectro. This was an enticing offer. Talbert and Widlar had no stock options at Fairchild and felt that the company had not properly rewarded or recognized their contributions. Less than a year after joining Molectro, Talbert and Widlar plotted to replace their bosses at National Semiconductor, whom they deemed incompetent, with a team of Fairchild managers. In a bold move, they turned to Charles Sporck, the general manager of the semiconductor division at Fairchild, and Pierre Lamond, the head of IC manufacturing, and suggested that they take over the company (Rotsky 1988).[6]

The timing could not have been better. Sporck and Lamond were interested in leaving Fairchild. They were becoming increasingly frustrated with John Carter and other Fairchild Camera's top managers. After his initial investment in Fairchild Semiconductor, Carter had transformed Fairchild Camera into a conglomerate. Between 1960 and 1965 he had bought 14 specialty-oriented businesses. The most notable acquisition was DuMont, an ailing manufacturer of cathode ray tubes, oscilloscopes, and testing equipment. Many of these firms were losing money at the time of their acquisition. Carter expected that he would turn them around and transform them into great successes. But Sporck, Lamond, and other managers in the semiconductor division objected to Carter's conglomerate-building strategy. In their judgment the new acquisitions were marginal companies. They were also resentful of the fact that Carter used the profits of the semiconductor division to shore up these money-losing operations. What infuriated the semiconductor people the most was the feeling that they were being treated as second-class citizens within the company and that their efforts were not getting the recognition they deserved. In 1966, two-thirds of Fairchild Camera's sales and 110 percent of its profits came from the semiconductor division. But this division got short shrift from Carter. And Carter gave only meager stock options to Sporck, Lamond, and their staff. These men saw little of the new wealth they were generating. They resolved to grab it.[7]

In 1966, Lamond began looking actively for opportunities outside of the company. A "headhunter" put him in contact with Plessey, a British electrical manufacturer that was looking for an experienced manager to run its struggling semiconductor division. Lamond, who was interested in the position, proposed to move the division's headquarters to California and build a global, West Coast-based semiconductor business for Plessey. To do so, Lamond assembled a team of promising mid-level managers from Fairchild: Floyd Kvamme, the marketing manager for integrated circuits; Roger Smullen, a manufacturing engineer; Fred Bialek, the head of Fairchild's offshore operations; and Ken Moyle, a process engineer who had recently moved to Hewlett-Packard. Lamond also recruited his own boss, Sporck, to head the new group. The group negotiated with Plessey for more than 6 months. But the discussions broke down over stock options. The New York law firm that Plessey had hired to conduct the negotiations balked at the group's requests for stock options.[8]

The collapse of the negotiations with Plessey led Sporck and Lamond to look carefully at Widlar's and Talbert's proposition regarding National Semiconductor. National Semiconductor was offering interesting challenges and opportunities for a group of ambitious managers. The firm was losing money and had an uninspiring product line. But it had two assets. Widlar and Talbert, the stars of linear circuits, were on the company's payroll. National Semiconductor's stock was also traded over the counter, which made it relatively easy to raise capital. Following up on Widlar's suggestion, Sporck contacted National Semiconductor's board of directors. He proposed that the group take over National Semiconductor and transform it into a major producer of integrated circuits. His plan was to have the group focus on military applications and the incipient industrial market. The board accepted this proposal, and in February 1967 it fired National Semiconductor's president and replaced him with Sporck. The board gave Sporck and his group considerable latitude in the management of the company. It also granted them substantial stock options. For instance, Sporck received options on 44,500 shares at $13.50 a share. The company also loaned money to each individual in the group to help them buy additional stock.[9]

The defection of Sporck and Lamond devastated Fairchild. Sporck had held Fairchild's semiconductor division together through the sheer force of his will. His departure weakened the firm severely. Sporck and Lamond also took some of Fairchild's most experienced production managers with them to National Semiconductor. These defections crippled IC manufacturing at Fairchild and resulted in a worsening of the firm's yield and delivery difficulties. To replace Sporck, Noyce (now the vice-president in charge of Fairchild Camera's West Coast operations) chose Thomas Bay, the first head of sales and marketing at Fairchild Semiconductor. This was a decision that Noyce later came to regret. Bay proved unable to stem the firm's deterioration. In 1967 and 1968, Motorola and Texas Instruments increased their share of the market for integrated circuits. Motorola copied Fairchild's DTL circuits and undercut them in the market. At the same time, Texas Instruments introduced new digital circuits with transistor-transistor logic (TTL), a different logic configuration. The TTL circuits, which were faster than Fairchild's DTL circuits, became popular and further eroded Fairchild's microcircuit sales. Compounding these setbacks was the downturn in the semiconductor business in 1967 and 1968. In the summer of 1967, Fairchild Semiconductor became unprofitable. This was the first time the division had lost money since 1958, when it had introduced the first double-diffused silicon transistor.[10]

The troubles of the semiconductor division led to a major crisis at Fairchild Camera. Since Fairchild Camera was highly dependent on the semiconductor division's profits, the corporation as a whole went into the red when these profits disappeared. Losses amounted to $7.6 million in 1967. These dismal financial results brought about considerable turmoil at Fairchild's corporate headquarters on Long Island. Sherman Fairchild, Fairchild Camera's founder and largest stockholder, lost confidence in John Carter and asked him to resign. He replaced him with

Richard Hodgson. Hodgson then proceeded to divest some of the most unprofitable operations, such as DuMont's oscilloscope business. But within a few months, a steep decline in the value of Fairchild Camera's stock cost him his job.[11] To replace Hodgson, Sherman Fairchild nominated a management committee made up of Noyce and two of Fairchild's trusted business associates. While this committee ran the company's day-to-day affairs, Sherman Fairchild looked for a new chief executive officer. Noyce, who had hoped to become the next CEO, quickly realized that he was not the favored candidate. Sherman Fairchild and most managers at Fairchild Camera saw him as "too soft" for the job.[12]

Unable to get the top job, Noyce prepared his exit. He discreetly planned the formation of a new company (later to be called Intel) with Gordon Moore, the head of the R&D laboratory at Fairchild. Together they had identified memory circuits and the new technology of MOS as the new business opportunity, and they were eager to exploit it. They knew that they could easily raise capital to build a new semiconductor company. The venture capitalist Arthur Rock, who had long been encouraging Noyce to start his own business, suggested that he could raise money for it. Not surprisingly, Noyce and Moore turned to Rock for investment capital. Noyce and Moore each put $250,000 into the venture. Rock raised an additional $2.5 million. Among the investors were Hodgson and five founders of Fairchild Semiconductor: Eugene Kleiner, Jay Last, Jean Hoerni, Sheldon Roberts, and Julius Blank.[13]

The departure of Noyce and Moore was even more devastating to Fairchild Semiconductor than that of Sporck and Lamond. The engineers and managers at Fairchild were aghast. For many, Noyce and Moore were the corporation, and the corporation was abandoning them. Their departure encouraged others at Fairchild to start new integrated circuit businesses. The choice of the new chief executive officer accelerated this entrepreneurial flowering. To direct Fairchild Camera, Sherman Fairchild chose Lester Hogan, the head of Motorola's semiconductor division. This nomination was not well received at Fairchild. Motorola was Fairchild's arch-rival. In the first months of his tenure at Fairchild, Hogan proceeded to hire more than 100 managers from Motorola. He replaced all of the current top managers in the semiconductor division with his own men. Many mid-level managers were also displaced. These dislocations fueled the entrepreneurial spirits of Fairchild's employees. Eight groups of Fairchild managers and engineers, many of them fired by Hogan and his minions, started new integrated circuit ventures on the Peninsula in the late 1960s. For example,

Jerry Sanders, after losing his job as head of sales and marketing at Fairchild, started American Micro Devices (AMD). Other startups focused on semiconductor memories (Computer Microtechnology, Advanced Memory Systems) or precision linear circuit products (Precision Monolithics). These corporations were funded by a mix of venture capital, investment bankers, and fellow semiconductor entrepreneurs. Noyce, for example, invested in AMD.[14]

In addition to this wave of Fairchild-generated entrepreneurship, engineers and managers working at other semiconductor corporations on the Peninsula started new integrated circuit firms in the area. Amelco, Signetics, General Micro Electronics, and Union Carbide Electronics were the main sources of these spinoffs. But Hewlett-Packard, Lockheed, IBM, and a New York-based company, General Instruments, also generated new microcircuit enterprises in the area. The lure of new technologies and markets, the availability of venture financing, and tensions within these firms led to the formation of new semiconductor businesses. For example, former GME employees established five separate corporations on the Peninsula in the late 1960s. Key to the formation of these firms was the sale of GME to Philco-Ford in 1966. Philco proceeded to cancel the outstanding GME stock options and to move the company to Philadelphia. These decisions led some of the best engineers to leave the firm and start their own businesses. These men formed MOS-oriented startups such as American Micro-Systems (1966), Electronic Arrays (1967), Nortec (1968), and Integrated Systems Technology (1968) (Sporck 2001).

A similar process unfolded at Union Carbide Electronics. Union Carbide Electronics was a recent spinoff of Amelco. In 1963, after being fired from Amelco, Jean Hoerni had started a semiconductor division for Union Carbide. Hoerni's alliance with Union Carbide lasted only 3 years. He quickly grew frustrated of dealing with a parent company on the East Coast that did not understand the semiconductor business. Adding to this frustration was the fact that Union Carbide refused to give him any stock options. As a result, Hoerni started a new integrated circuit venture, Intersil (for "international silicon"). The new company would focus on advanced MOS circuits for novel applications: calculators and electronic watches. Hoerni, a French Swiss, had followed the development of electronic watches with interest. He had ties to the large Swiss watch firm Société Suisse d'Industrie Horlogère (SSIH), which had funded the development of one of the first electronic watches. In 1967, SSIH asked Hoerni to make integrated circuits for its watches. Along with Olivetti, SSIH financed the new venture. Intersil was probably the first San

Francisco Peninsula startup to be partially funded by foreign investors. The rest of the capital came from Hoerni and Rock. Intersil opened for business in July 1967.[15]

Social Innovations

What set the men who founded Intel, Intersil, National Semiconductor, and American Micro-Systems apart from previous electronics entrepreneurs on the Peninsula was the fact that their primary goal was to bring their corporations public within 3–5 years of their founding. In essence, they wanted to organize an initial public offering and list their firms on the American Stock Exchange or the New York Stock Exchange. By listing their firms on a stock exchange, the entrepreneurs would reap substantial financial rewards—something they had not been able to do at Fairchild, General Micro Electronics, and Union Carbide Electronics. Also, an early IPO would help them meet the requirements of their financial backers. One of the major requirements of venture capitalists was to have fast returns on their investments. The venture capitalists invested with a 5-year time horizon (Freund 1971).

These financial goals had a profound effect on the ways in which the new integrated circuit startups were managed. In taking their firms public, entrepreneurs had to take the expectations of Wall Street into account early. The investment bankers, who sold the initial stock offerings to the public, had two main criteria: technological innovation and rapid sales growth. The firms that met these criteria had a good chance to become public and see the value of their stock grow. (Investment bankers and stock analysts did not emphasize profits as much in their decisions.) The demands of Wall Street led the entrepreneurs to emphasize process and product innovation and to accelerate the growth of their sales. This close alliance with Wall Street was new to the electronic manufacturing complex on the San Francisco Peninsula. Eitel, Litton, and Varian, who had built the vacuum tube industries on the Peninsula, were hostile to Wall Street financiers and had brought their firms public only after decades of work. With the reorganization of National Semiconductor and the formation of Intel, Intersil, and American Micro-Systems, a new conception of the firm emerged in the area. When the older entrepreneurs, especially the Varian brothers, conceived the firm as a social institution, Hoerni, Noyce, Sporck, and their backers thought of it as a vector for financial gain, market penetration, and technological innovation. In other words, they did not build their organizations to last but to develop and market new products and sell semiconductor securities. Now the capitalistic impulse was at the core of the electronics complex in the Santa Clara Valley (Freund 1971). These goals led the integrated circuit entrepreneurs to devise new human resource and management-employee relations strategies. To innovate and grow fast, the new firms needed to recruit the most talented engineers and managers to the business. In order to attract and retain these people, the entrepreneurs and their venture capital backers gave them large stock option packages. (Signetics' management adopted a similar policy.) The entrepreneurs offered stock options to a wide range of employees, including all the engineers, all the managers, and many of the foremen. Sporck, Noyce, Moore, and Hoerni reasoned that stock options would attract skilled engineers and motivate them to work hard on complex technologies. Stock options had the added benefits of involving employees in the entrepreneurial process and its financial rewards and aligning their interests with those of the entrepreneurs and the venture capitalists.

The wide distribution of stock options at National Semiconductor and the new Fairchild spinoffs was an important and radical social innovation. This practice reinforced and stabilized entrepreneurial corporatism, a mode of management-employee relations that had emerged tentatively at Fairchild Semiconductor in the early 1960s. This form of corporatism was different from the ones developed at Litton Industries and Varian Associates in the 1940s and the 1950s. Like Charles Litton and the Varian brothers, Noyce and Sporck sought to avoid the emergence of a sharp divide between employers and employees. They cultivated communication between managers and employees and delegated as much authority as possible throughout the organization. But the nature of the corporatism they practiced was different. Unlike Litton, the semiconductor entrepreneurs were not paternalistic. Rather, their goal was to make engineers feel and behave like entrepreneurs.[16]

The microcircuit entrepreneurs also innovated organizationally. Their goal, again, was to introduce new products quickly to the market. At Fairchild Semiconductor, product development had been distinct from manufacturing. Engineers in the R&D laboratory developed new products and processes, which were later transferred to the production department. The manufacturing organization also had an engineering group in the applications laboratory, which developed follow-up products. By 1966, most upper managers at Fairchild had realized that this organizational setup was inefficient. It led to turf battles between research and manufacturing over product engineering. Circuits developed in the research laboratory did not transfer well into manufacturing. The manufacturing lines used different equipment and processes than the laboratory, which often made transfers from R&D difficult. In some cases, products developed in the laboratory were never put into production at Fairchild, but were exploited at other firms on the Peninsula (Sporck 2001).[17]

When Sporck, Moore, and Noyce left Fairchild, they were determined to remedy this state of affairs and to tightly integrate the development and manufacturing functions within their new firms. To do so, they decided not to establish independent R&D laboratories in their new ventures. Product and process engineers would work in the manufacturing plant and use the same equipment as the production groups. For example, Sporck and Lamond set up National Semiconductor so that the design engineering teams reported directly to plant management. Each team included circuit-design and process engineers, and each focused on a specific product line. Under this scheme, engineers developed new products and processes directly on the manufacturing line, using existing equipment and interacting daily with the people who were going to manufacture their circuits. The design engineering groups were also responsible for a product from its initial design stage through its production. They were expected to help solve any problems that might appear over the course of the product's life. This new organization enabled the integrated circuit startups to introduce new products to the market faster than Fairchild Semiconductor (Fields 1969a; Murray 1972).

Developing the Industrial Market

Using these new managerial techniques, the microcircuit entrepreneurs went in two different business and technical directions. Some entrepreneurs took advantage of the opportunities offered by the new MOS technology. They stabilized MOS processes, and they designed memory circuits and watch circuits. Other entrepreneurs focused on developing the "industrial market." This was a nascent market, but its potential seemed to be unlimited. The industrial market at that time included everything except the military, mainframe computer, and consumer electronics markets. In other words, there were a wide range of potential applications for microcircuits—for example, calculating machines, computer peripherals, process control equipment, communication systems, machine tools, automotive products, and scientific and medical instruments. Any firm that used gears, springs, levers, electromechanical devices, or transistors in its products was a potential "industrial" customer. The industrial market represented a large and untapped territory for integrated circuits.[18]

Charles Sporck, Pierre Lamond, and their associates at National Semiconductor were among the first entrepreneurs to identify the industrial market as a significant opportunity, and they decided to go after it. In order to transform National Semiconductor into an important supplier of microcircuits to this market, Sporck and Lamond had to reshape the firm. They moved its headquarters from Connecticut to the Santa Clara Valley, the cradle of knowledge about the manufacture of integrated circuits. They expanded the firm's microcircuit operations on the Peninsula. They built new manufacturing facilities and raided Fairchild's professional staff. In the winter of 1967, Sporck and his associates hired more than 35 engineers and managers from Fairchild. Among these recruits were Fairchild's best manufacturing engineers and offshore production managers and Donald Valentine, its head of sales and marketing. Fairchild's pool of experienced operators and technicians was not immune to Sporck's and Lamond's raids. These workers brought much-needed skills to National Semiconductor's microcircuit operation. They had mastered the complex techniques required for the production of integrated circuits and were able to start production at National Semiconductor quickly.[19]

To finance this venture, Sporck and Lamond extracted as much cash as they could from National Semiconductor's transistor plant in Connecticut. They immediately recognized that this transistor plant employed too many people for the sales revenues it generated. They were resolved to reduce its workforce in order to lower labor costs and generate cash for their plans for integrated circuits. They were brutal in their execution of their strategy. After taking over National Semiconductor, Lamond, Fred Bialek, and Roger Smullen immediately flew to Connecticut. In a mere few days, they had fired more than half of the plant's employees—over 300 operators, technicians, and engineers. The mass firings had the immediate effect of making the plant highly productive. Fred Bialek, the factory's new director, later boasted that within 3 months of Sporck's takeover the plant was shipping 3 times as many transistors with only 40 percent of the original workforce. Higher productivity and lower labor costs generated substantial profits for the corporation—profits that Sporck and Lamond immediately reinvested in the integrated circuit business on the West Coast.[20]

The management group at National Semiconductor financed that company's foray into integrated circuits in other ways—for example, by selling National Semiconductor's transistor process to a semiconductor startup in Spain. This brought them 500,000 pesetas, which the group had to smuggle out of the country in suitcases because the Spanish government imposed restrictions on the use of foreign currency at that time. Sporck and Lamond employed more orthodox ways of raising capital. Taking advantage of the fact that National Semiconductor's shares were traded on the over the counter market, the group sold $1.5 million worth of National Semiconductor common stock in a private placement in July 1967. A few months later, Sporck and Lamond obtained a bank credit line of $2 million. But their most successful financial move may have been the clever accounting scheme that they had devised. National Semiconductor's managers put considerable pressure on their customers to pay their bills rapidly. National Semiconductor expected payments within 30 days of purchase. To give the customers a strong incentive to honor their financial commitments earlier, they gave a 2 percent discount to those who paid their bills within 10 days. On the other hand, National Semiconductor paid its own suppliers on a 90-day schedule. This difference inflated National Semiconductor's cash position and financed its expansion into integrated circuits.[21]

Sporck and Lamond devoted a large share of their newly acquired financial and engineering resources to the development of integrated circuit products. They emphasized product engineering and gave less weight to the development of new manufacturing processes. Rather than innovate new processes, engineers at National Semiconductor tended to improve on processes engineered at Fairchild or at Texas Instruments. Since industrial users required standard circuits, the engineering teams focused on standard designs to supply this market. They focused mainly on the development of circuits that could be used as building blocks in a variety of systems. These circuits could be entirely new products or improvements on circuits already marketed by other firms. In order to reduce market risks and to meet the needs of a wide variety of industrial users, engineers at National Semiconductor developed a broad product line that included bipolar linear circuits, bipolar digital circuits, and MOS microcircuits.[22]

At first the group focused on linear circuits. National Semiconductor was in a unique position in this arena. Robert Widlar and Dave Talbert, the two main linear circuit engineers in the United States, were working at National Semiconductor. Before Sporck's takeover of National

Semiconductor, Widlar and Talbert had started a linear circuit line and designed a voltage regulator, a device that could smooth out variations in the flow of an electric current. This was the first monolithic voltage regulator on the market, a circuit that many in the industry thought impossible to make in silicon. In the late 1960s, much to the chagrin of Sporck and Lamond, Widlar and Talbert continued to call the shots in linear circuits. Widlar added to their displeasure by behaving as he pleased. Knowing that Sporck and Lamond were unable to replace him, Widlar drank and gave free rein to his irreverent and obnoxious self. Among his many pranks, he once brought a goat to mow National Semiconductor's lawn. On another occasion, he destroyed the company's paging system with firecrackers. He also threatened his co-workers with an axe and defied management as much as he could (Pease 1991; Rotsky 1991).[23] Widlar and his group also developed a long series of innovative linear circuits. In 1967, Widlar designed the LM101 operational amplifier, an improved version of the operational amplifier he had designed a few years earlier at Fairchild Semiconductor. The new circuit had better electrical characteristics than his first design. Among its other advantages, the LM101 was short-circuit-proof and required fewer external components. This circuit became the "hottest" linear circuit product on the market. A host of other linear circuits such as voltage regulators and radio-frequency amplifiers quickly followed in the next few years. By 1970, National Semiconductor had the industry's largest line of linear circuit products.[24]

Under Lamond's direction, National Semiconductor complemented its line of linear circuit products with MOS microcircuits and digital bipolar circuits. An engineering group developed a family of MOS memory circuits for industrial applications. But more important, Lamond actively promoted digital bipolar circuits. In order to establish National Semiconductor as a major supplier of digital circuits, Lamond used the tactics he had employed earlier at Fairchild. When Signetics had built a substantial business in DTL circuits in 1963 and 1964, Fairchild had copied Signetics' microcircuits and drastically lowered prices on these circuits in order to get market share. Lamond utilized the same tactic at National Semiconductor. But this time the target was Texas Instruments. In March 1967, Texas Instruments marketed a family of microcircuits with a TTL configuration. Lamond and Kvamme decided to copy TI's circuits. These circuits were fast and promised to become the main products in the digital market. Signetics, Amelco, and Fairchild (under Lester Hogan) made the same choice and second sourced TI's circuits in 1968 and 1969.[25]

Whereas Fairchild and Signetics focused on the simplest circuits in TI's TTL line, Lamond and Kvamme decided to copy and improve upon its most complex circuits—those with from 30 to 50 transistors. These circuits represented only a small share of the TTL market (13 percent in sales and 4 percent in volume). But they were more difficult to obtain from Texas Instruments. They were also more expensive, which allowed National Semiconductor to get more revenue per wafer. And they were produced in lower volumes, which meant that they could not overtax the firm's wafer-fabrication facility. To reverse engineer these circuits, Lamond turned to the technologists who had developed them at Texas Instruments. He recruited Jeff Kalb, the architect of TI's TTL family, and Robert Schwartz, the lead process engineer for these circuits. In October 1967, these men developed a complex dual binary flip-flop for National Semiconductor. National Semiconductor's version of this circuit was faster and more reliable than TI's. In subsequent years, Kalb and Schwartz re-engineered the other complex circuits in TI's line for National Semiconductor. They also introduced parts that were proprietary to the firm. By the spring of 1970, National Semiconductor had 31 complex circuits in its TTL line.[26]

In tandem with the engineering of a broad product line, the managerial group at National Semiconductor was creative in developing new markets for its products. In the industrial market, National Semiconductor faced an unusual challenge. Most potential customers had a limited understanding of integrated circuits and semiconductor electronics. At the same time, these customers numbered in the tens of thousands. This contrasted with the more traditional military, mainframe, and consumer electronics markets, which were dominated by a few large users. In other words, National Semiconductor had to contact a large number of potential customers and educate them about the potential of integrated circuitry.[27] These were challenges that Don Valentine, the cold and effective head of sales and marketing at National Semiconductor, had to address. Instead of building an internal sales force, which was standard practice, Valentine decided to outsource the sales function. He did this by setting up a network of manufacturing representatives—independent salesmen who sold National Semiconductor's circuits on commission and were paid 5–7 percent of the customer order. Valentine built this network by contacting the best salesmen in the semiconductor industry, men he knew from his days at Fairchild. He proposed to help them set up their own business. Valentine assured them that National Semiconductor would give them the exclusivity of its integrated circuit business in a particular sales area and at the same time help them find electronic products from other firms that the representatives would sell. Many entrepreneurially oriented salesmen liked Valentine's proposition and started their own organizations. This diverse network enabled National Semiconductor to create a large sales force at low cost. There were no salaries to pay and no overhead. The reps cost money only when they sold National Semiconductor's products. Because of the fairly large commissions paid by National Semiconductor, the representatives had a strong incentive to sell National Semiconductor's integrated circuits. The use of manufacturing representatives enabled National Semiconductor to reach many potential customers in the industrial market.[28]

Valentine coupled this sales force with an active applications engineering effort. Because the reps were not always knowledgeable about IC technology, Valentine deployed applications engineers to reinforce sales. These "field applications engineers" had several functions. They helped the sales reps close their sales. They also assisted the customers technically. Many industrial users knew little about integrated circuits and had never used them before. To overcome this obstacle, the field applications engineers "educated" customers about National Semiconductor's microcircuits. They often helped the users design their products around National Semiconductor's circuits. It was not an uncommon practice for field applications engineers to act as de facto project managers in the customers' plants. Many of the field applications engineers had held managerial positions before. These skills helped them to oversee the development of new products by the customers' engineering teams. In conjunction with this field application effort, National Semiconductor published its own applications notes. These notes discussed the main characteristics of the firm's circuits. They explained how the circuits could be used and described the types of systems one could make with them. National Semiconductor's notes were widely read by the users. National Semiconductor's applications engineering efforts were a major factor in opening up the industrial market for microcircuits.[29]

To further educate customers about the potential of integrated circuits, Valentine and Kvamme organized numerous seminars in the United States and in Europe. These seminars attracted overflow audiences. A typical example of these seminars was the "IC Seminar" that National Semiconductor offered in Los Angeles in June 1968. This seminar was widely advertised in the trade press and promised attendees that they would "get smarter, invent better, and be one-up on practically everybody." "In the case you missed [Robert Widlar and Floyd Kvamme] in [the previous seminar in] Paris," proclaimed the advertisement in *Electronics*, "we've got them again for a seminar of what's happening with Linear and MOS. Bob Widlar, who designed more than half the world's linear circuits, will pursue needed ideas and inventive applications. Floyd Kvamme, who has answers to micro-circuit questions that nobody's even thought of yet, will eloquently reveal some new *how to do it* wisdom on MOS." At the seminar in Los Angeles, Widlar, Kvamme, and other National Semiconductor engineers discussed National Semiconductor's new circuits and their applications. They also talked about the products that National Semiconductor had in the pipeline. At the end of the seminar, they held an "IC clinic" in which members of the audience discussed design problems with National Semiconductor's engineers.[30]

In parallel with the development of new products and marketing techniques, Charles Sporck and his group made incremental innovations in manufacturing. When Sporck directed production at Fairchild, he pioneered the assembly of transistors and integrated circuits in Asia. In 1962, Sporck oversaw the establishment of a large assembly plant in Hong Kong. The move to Hong Kong enabled Fairchild to shift part of its assembly operations to a location with low labor costs. As a result, Fairchild was able to produce inexpensive transistors for the consumer electronics market. At National Semiconductor, Sporck and Lamond went one step further in this direction. They decided to assemble all their microcircuits overseas. Sporck and Lamond further innovated by establishing their first semiconductor assembly plant in Singapore. Labor costs in Hong Kong were increasing. Singapore's labor rates were substantially lower than Hong Kong's. The Singaporean government also granted substantial tax breaks to US firms and made sure that the workforce was tractable. National Semiconductor gained a substantial cost advantage by locating in Singapore. This move enabled the firm to produce its circuits at lower cost than Fairchild Semiconductor and Texas Instruments, which was still assembling a substantial share of its products in the United States.[31]

The shrewd tactics of Sporck, Lamond, and Valentine yielded remarkable results. Partly because of their market-building techniques, the industrial demand for integrated circuits grew substantially in the late 1960s. National Semiconductor captured a large share of this market. Its sales grew from $7.2 million in 1966 to $41.9 million in 1970 and employment at National Semiconductor expanded from 300 to 2,800 during this period. In the spring of 1970, National Semiconductor sold microcircuits to 1,250 industrial firms. National Semiconductor had an especially broad customer base in linear circuits. Users of analog circuits ranged from makers of transaction terminals to producers of communication equipment and scientific instruments. For example, Hewlett-Packard was a large customer of National Semiconductor's linear circuits. National Semiconductor's TTL circuits found a large market among makers of calculators and computer peripherals, including IBM. National Semiconductor also supplied its MOS circuits to Japanese calculator firms. The success of these product lines boosted the value of National Semiconductor's stock. It also enabled Sporck to list the company on the American stock exchange in October 1970. Taking advantage of the rise in the value of National Semiconductor's shares, Sporck, Lamond, Widlar, Talbert, and their associates exercised their stock options and sold some of their newly acquired shares on the market. For instance, it is widely believed that Widlar sold his stock holdings for $1 million dollars in 1970, and that he then retired in his early thirties. National Semiconductor was also able to raise additional funds by the initial public offering. With this influx of capital, National Semiconductor was poised to become a major player in the integrated circuit business in the United States.[32]

Metal Oxide Semiconductors

While Sporck, Lamond, and Valentine built the industrial market for TTL and linear integrated circuits, other entrepreneurs concentrated on developing the emerging MOS technology.[33] These entrepreneurs included some of the stars of the semiconductor industry: Robert Noyce, Gordon Moore, and Jean Hoerni. But they were not the only ones. Other engineers and managers who had worked on MOS at GME, at Fairchild, and at other silicon firms wanted to make a go of it too. This entrepreneurial group was interested in MOS because the technology permitted the further miniaturization of electronic circuits. MOS also promised to open up new applications for integrated circuits. To exploit MOS technology, these men raided the engineering staffs of Fairchild, GME, and Union Carbide Electronics. They appropriated the MOS manufacturing techniques that these firms had recently developed and proceeded to improve upon them. The MOS processes engineered by Fairchild and GME were still experimental. They yielded circuits with poor and often unstable electrical characteristics. In brief, they did not easily lend themselves to commercial production. To create a profitable business, the entrepreneurs had to build reproducible and economical MOS manufacturing processes. They devoted a large share of their financial and

engineering resources to process development. The net result of these
efforts was a proliferation of MOS manufacturing processes on the
Peninsula. Each firm developed a different variant of MOS process technology, and sometimes several variants, in order to make circuits with the
right electrical characteristics for different market requirements.

In tandem with the heavy process-development effort, the entrepreneurs oversaw the engineering of new IC products. To use their limited
engineering resources in the most efficient fashion, they concentrated
on one or two kinds of MOS circuits. A few engineer-entrepreneurs pioneered the development of MOS microcircuits for electronic watches.
Others focused on logic circuits for electronic calculators, striving to sat-
isfy the market demands caused by the emergence of many Japanese
companies into this field. But the majority of MOS entrepreneurs worked
on memory circuits. Memory circuits are microcircuits that store information in binary form. The market for memory circuits held a lot of
promise. IBM had recently decided to use memory circuits in one of its
mainframe computers. IBM's managers expected that the rapid price
declines typical of semiconductor devices would soon make memory circuits cheaper than ferrite core memories. (Ferrite core memories were
until then the preferred method of information storage in computers
and other digital systems.) This decision created a market for memory
circuits in the computer industry, especially for IBM's competitors
Honeywell and Burroughs. Now that IBM was using memory circuits,
these firms had to adopt them. By refusing to do so, they were concerned, they would appear holding on an increasingly obsolescent technology. This would ultimately cut into their sales. As a result, the market
opportunities for memory circuits seeméd almost limitless. Observers at
the time predicted that the demand for semiconductor memories would
grow to $500 million a year in the 1970s.[34]

The entrepreneurs competed ferociously in this and other markets
segments. Being first to introduce new MOS circuits or new types of MOS
circuits to the market was critical. The entrepreneurs were well aware
that the first IC product brought to market often became a de facto standard.[35] As a result, the first mover reaped substantial financial benefits
from its engineering efforts. The firms that followed had a much lower
(and sometimes negative) return on their investments. Being first to
introduce a new product also meant that the startup would establish a
solid position in a market area and as a result would be better able to survive subsequent competition from much larger firms. In order to be the
first ones to the market, the entrepreneurs put enormous pressure on

their engineering staffs to perform. They also sought to undermine their competitors by stealing their most valued engineers. This practice often led the firms that had lost engineers to sue these men and their new employers for theft of trade secrets. In the late 1960s, trade-secret lawsuits in the Peninsula's semiconductor industry grew exponentially.

Beneath this atmosphere of intense and often ruthless competition was an undercurrent of information sharing. Engineers involved in MOS startups exchanged process and design data with engineers at other firms in an informal way. Most MOS engineers faced the same difficult task of developing and stabilizing complex manufacturing techniques. Any information that would help them solve the numerous process issues that they were encountering was welcome. This shared concern made the various engineering groups amenable to trading information. A web of previous associations also facilitated information sharing. Most of the MOS engineers had worked for other corporations before, notably GME and Fairchild Semiconductor. At these firms, they had built friendships with other engineers and managers who had often moved to new corporations themselves. These previous associations made it easy to contact engineers at other firms and ask for their advice on particularly difficult process or design questions (Lindgren 1969).

Bars also fostered the exchange of information among engineering groups. In the first half of the 1960s, engineers and managers at Fairchild and other silicon corporations on the Peninsula had developed the habit of meeting after work at a local bar. (The Wagon Wheel was a favorite.) At these bars, they would discuss the problems of the day. Bars were also where sales and marketing men met with the manufacturing guys to discuss order prices and delivery schedules. After leaving Fairchild, many of these engineers returned to these bars and discussed the business with their former associates. A lot of information flowed over beer and hard liquor, to the point that the management of many of the startups expressly forbid their engineers to go to the Wagon Wheel and other bars. The end result of these daily interactions was that design techniques and solutions to particularly difficult process problems moved from firm to firm. As a result, the MOS community on the Peninsula developed a repertoire of process "tricks" that were known only in the area. These tricks enabled them to solve their own process problems and obtain good manufacturing yields. In contrast, MOS firms located outside Northern California were not plugged into these networks and did not benefit from this shared knowledge. This put them at a distinct competitive disadvantage.[36]

To examine the ways in which the different entrepreneurs competed with each other and established positions in the emerging market for MOS circuits, I will concentrate on Noyce, Moore, and Hoerni. These entrepreneurs built some of the most successful startups on the Peninsula in the late 1960s. But they shaped their firms in strikingly different ways. They focused on different product and market opportunities. Moore and Noyce concentrated exclusively on standard memory circuits for computers and computer peripherals. Hoerni made custom circuits for a wide variety of applications, especially electronic watches. These different market orientations led them to develop different manufacturing processes. Noyce and Moore engineered their manufacturing processes for density (the number of transistors that can be crammed on a chip). Hoerni designed manufacturing processes that permitted the production of low-power circuits for timekeeping applications. The entrepreneurs also had different geographic orientations. Whereas Noyce and Moore targeted the US market, Hoerni's venture was international from the start.

At first, Hoerni focused on making custom circuits for European corporations. Hoerni, the scion of a Swiss banking family and the best-known European in solid-state electronics, could easily make contact with potential customers in Europe. The European market also offered interesting opportunities. European electronic firms needed advanced custom circuits that they could not procure locally, because European semiconductor firms had not yet fully mastered the production of complex MOS and bipolar circuits. US firms did not cultivate that market well either. Seizing this opportunity, Hoerni secured contracts for the development of integrated circuits from large European corporations in the months following the formation of Intersil. Some customers also took a minority position in the company in order to get a ready supply of integrated circuits if they needed to. Olivetti, an Italian maker of mechanical calculating machines, was interested in designing electronic calculators. It gave Hoerni a contract to make a calculator circuit. Hoerni and his engineers also developed MOS memory circuits for Siemens. More important, Hoerni got contracts from SSIH to engineer circuits for electronic watches. SSIH was developing electronic watches (especially at the Battelle Institute in Geneva) and needed integrated circuits for its watch prototypes. In late 1967 and early 1968, Hoerni and John Hall (a former Union Carbide Electronics employee who had followed Hoerni to Intersil) engineered both a bipolar and an MOS watch circuit for SSIH. Hoerni expected SSIH to give them substantial production contracts for these circuits. But these orders never materialized as SSIH decided not to bring its watch prototypes to production. Little is known about the reasons that prompted this decision, but it is likely that SSIH decided to postpone the commercialization of electronic watches for the same reasons as other Swiss corporations. Since the early 1960s, a group of Swiss watch firms around the Fédération Horlogére had invested heavily in the development of electronic watches. It had established a new organization for this purpose, the Centre Electronique Horloger, in Neuchâtel. By 1967, the Centre Electronique Horloger had engineered a working prototype of an electronic watch. But the firms that had funded the project were reluctant to commercialize it. In the late 1960s, these and other Swiss watchmakers experienced a surge of orders for mechanical watches. To meet these orders, they made massive investments in new production facilities and employee training programs. As a result, they were not eager to invest more resources in the building of an entirely new product line. The companies that had established the Centre Electronique Horloger worried that the electronic watches would hurt their business in the long run. Their prosperity and profits rested on mechanical technology. By commercializing electronic watches, they might release a product that would cannibalize their mechanical watch business in the long run. It is likely that SSIH made similar calculations at the same time. Like the firms involved with the Centre Electronique Horloger, SSIH decided to postpone the commercialization of its electronic watch. It was only in 1970 that the company introduced it to the market.[37]

The slow commercialization of the electronic watch put Hoerni in a difficult financial position. He could not count on production contracts from SSIH to support his business. Nor had Olivetti followed through with production orders for calculator circuits. These major setbacks led Hoerni to revise his strategy and do business with the Japanese watch industry. Don Rogers, Intersil's dynamic sales and marketing manager, was central to this radical change in strategy. Rogers encouraged Hoerni to do business with SSIH's arch-enemy, the Hattori Seiko Company of Japan. Hattori Seiko (known in the West as Seiko) was the leading manufacturer of watches in Japan and was a major competitor of SSIH in the high-end watch industry. Rogers, who had headed Fairchild Semiconductor's sales effort in Asia in the first half of the 1960s, was aware of Hattori's interest in electronic watches. By the mid 1960s, Hattori had developed a prototype of an electronic watch. But this watch used hybrid circuits instead of the more complex integrated circuits of the Swiss watches. As a result, the Seiko watch was not as accurate as the Swiss prototypes. Hattori needed integrated circuits, but the company was not able to make them internally. This opened up a potential market for Intersil and its expertise in the design and processing of watch microcircuits.[38]

In 1968, following up on Rogers' suggestion, Hoerni met with Soiji Hattori, the patriarch who ran Hattori Seiko. At this meeting, Hoerni offered to engineer integrated circuits for Hattori. In particular, he proposed to make a low-voltage complementary MOS (CMOS) circuit for their watch prototypes.[39] This was a technology that Hoerni had never used before. CMOS, first developed at Fairchild, was one of the more complex variants of MOS. CMOS circuits combined n-channel and p-channel transistors. The great advantage of CMOS was its low power requirements. CMOS derived this advantage from the fact that it dissipated significant power only during a change of state. One transistor in the complementary node was on while the other was off until the pulse arrived; then each of them changed state. The instant change was the only time that significant current flowed. As a result, CMOS circuits required little power. This made them ideal for battery-driven watches. But no organization had yet been able to produce CMOS circuits for watches. The Centre Electronique Horloger had worked on them in vain. RCA, one of the pioneers of CMOS technology, had recently commercialized a CMOS circuit for time-keeping applications. But this circuit operated at 15 volts, and the battery used in the electronic watch prototypes put out only 1.3 volt. Not surprisingly, the president of Hattori became interested in Hoerni's proposition. Having a CMOS circuit would enable Hattori to engineer and commercialize a superior electronic watch. It would also give the firm a competitive edge over the Swiss. Hattori gave a $75,000 circuit-development contract to Hoerni. The contract stipulated that the circuit had to be ready in 6 months. This was a short time to engineer a circuit of that complexity.[40]

On his return from Japan, Hoerni asked Hall, his chief engineer, to develop the circuit. Hall and Hoerni, making use of Hoerni's earlier work at Union Carbide Electronics, developed an innovative CMOS process. At Union Carbide, Hoerni had investigated the causes of instability in MOS devices and had developed techniques to stabilize the electrical characteristics of MOS circuits. Building on this work, in less than 6 months Hall and Hoerni engineered a working CMOS circuit that operated at 1.3 volt and met Hattori's other engineering specifications. This extraordinary technical feat enabled Hoerni to establish close relations with Hattori Seiko. In 1969, Hattori decided to use Hoerni's CMOS circuit in its electronic watches. Hattori later placed significant production contracts with Intersil. At the same time, the Japanese firm signed other microcircuit development contracts with Intersil and sent some of its best engineers to California. For example, a Seiko engineer designed a set of chips for an electronic calculator at Intersil in 1969. Seiko's close association with Intersil enabled the firm to commercialize the first CMOS watch in 1970. In doing so, Seiko was well ahead of the Swiss, who had recently introduced a bipolar-circuit-based watch to the market.[41]

Hoerni's close relations with Seiko created tension with his Swiss partners and investors. Intersil, a venture that the Société Suisse d'Horlogerie Industrielle had partially funded, was working with its most dangerous competitor! "The Swiss," Hoerni later recalled, "jumped up to the ceiling when they learned about the Seiko contract. And I remember one of these old guys at the SSIH, saying 'Well, Mr. Hoerni if we really thought that you needed these $75,000 we would have *given* them to you. But going to Japan, it is a national disgrace.'"[42] This conflict affected Intersil's relations with SSIH. It appears that SSIH did not give any new contracts to Intersil for the next few years. But Intersil's unique expertise in watch circuits led other Swiss firms to do business with Intersil. In 1970, Ebauches SA, one of the leading watch firms in Switzerland, asked Hoerni to produce some watch circuits.[43] These complex dealings with the Swiss and the Japanese enabled Hoerni to grow his firm. Intersil's sales reached $1.9 million in 1969 and doubled to $3.8 million the following year. Intersil also gained the reputation of being one of the technical powerhouses in the Peninsula's semiconductor industry. It could produce CMOS circuits that National Semiconductor, Fairchild, and the other MOS startups could not make.[44]

In the late 1960s, Noyce and Moore built another San Francisco Peninsula powerhouse: Intel. They oriented it toward the production of large-scale integration (LSI) memory circuits. LSI circuits were circuits with thousands of transistors. These complex circuits made possible the storage of a large number of memory bits. To manufacture these circuits, the entrepreneurs appropriated production techniques that had recently been developed at Fairchild Semiconductor. Several months before Noyce's and Moore's departure from Fairchild, two research engineers in the company's research laboratory, Thomas Klein and Federico Faggin, had devised new techniques for making MOS gates. (This work was based in part on earlier research done at the Bell Telephone Laboratories.) Instead of making the transistor gate out of metal which was the common

practice at the time, they used polycrystalline silicon. The use of polysili-
con had several advantages over metal gate techniques. "The silicon gate
process," Moore later recalled, "offered the advantage of self-registration
(meaning that one layer of the structure is automatically aligned with
those previously applied to the wafer). As a result, the devices could be
smaller and perform at higher frequencies." (Moore 1996a) In other
words, the silicon-gate process made possible the further miniaturization
of integrated circuits. It also boosted the circuits' speed by a factor of
2 or more.[45]

Noyce and Moore quickly realized the great potential of these new
techniques. At Intel, they focused much of the company's engineering
resources on the transformation of these laboratory techniques into a
reproducible and economic manufacturing process. Building a stable silicon-gate process posed a difficult technical problem, probably one of
the most challenging problems in the US semiconductor industry at the
time. Fairchild Semiconductor tried to solve it but did not quite succeed.
(The move of many Fairchild engineers skilled in MOS to Intel did not
help.) On the other hand, the engineers around Noyce and Moore at
Intel had strong incentives to make the process work. The survival of the
company and their own financial well being were at stake. "The concept
of the silicon gate device was lovely," an Intel manufacturing engineer
later recalled,

but the technique for putting the silicon on top of the oxide had never been
worked out. About this point in time, we were putting it down through evaporation. It would go down, it would look nice and we would sit there and watch the
wafer and pretty soon the surface would just start to roll up, like a sardine can.
We overcame the peeling problem by changing the process completely, going to
a totally different technique. We hired a guy who was from American Micro-
Systems (another MOS startup on the San Francisco Peninsula). He had grown
epitaxial layers accidentally on top of oxide in an epireactor, so he suggested we
try that and we did and it worked, so we were off and running in a new direction.
After that, things were solved painstakingly. Things did not come easy, they were
a little bit here, a little bit there kind of a thing.[46]

The development of the MOS process required a lot of good engineering. Over a period of a year or so, Moore and the MOS engineers carefully isolated the process defects, examined their causes, and developed
new methods that eliminated them. The end result of these efforts was
remarkable. Intel's new process allowed the making of twice as many
transistors per wafer as any other MOS process at that time. This gave
Intel an enormous competitive advantage over other manufacturers of

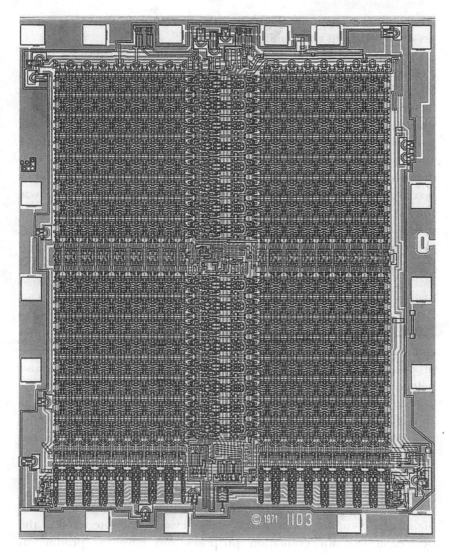

Figure 7.2
Intel's 1103 memory circuit. Courtesy of Intel Corporation.

semiconductor memories. It enabled the firm to produce denser memory circuits than its competitors (Real 1984).[47]

In conjunction with the development of the silicon-gate process, Intel's engineering staff designed highly integrated memory circuits. The first MOS circuit they designed was the 1101, a 256-bit static random-access memory (SRAM) chip. The circuit was compatible with TTL circuits, which made it useful in small systems such as printers. But the 1101 was too slow and too costly for computers (Sideris 1973). Within a year, Intel advertised a more integrated product, the 1103. This was a dynamic random-access memory (DRAM) chip in which the capacitive charge had to be refreshed continually. Crucial to the development of this circuit was Intel's close collaboration with Honeywell, the maker of mainframe computers. Alarmed by IBM's adoption of semiconductor memories, Honeywell decided to use DRAM circuits in future versions of its mainframe computers. As a result, it circulated a request for a 512-bit DRAM among semiconductor suppliers. Intel was one of eight companies to place a bid for this circuit. Intel won the bid and developed the 512-bit circuit. In parallel, the firm engineered two other circuits that had twice as many bits (1,024 bits) as the circuit requested by Honeywell. Intel could design these circuits because of its mastery of the silicon-gate process. Honeywell's engineers liked the new 1,024-bit designs and helped debug these circuits. They also influenced Intel's choice of bringing the more manufacturable circuit of the two to production. This circuit, often referred to as the 1103, had 4,000 transistors. This made it the most complex integrated circuit on the market at the time. It was also the first 1,024-bit MOS chip.[48]

The 1103 was well received in the market. It met the growing demand for high-density memory chips in electronic systems. Manufacturers of minicomputers and computer terminals bought it in large quantities. Mainframe computer makers such as Honeywell were interested in the device, at first for evaluation purposes and later for production machines. The rapid adoption of the 1103 in a wide variety of applications transformed this product into the "standard" 1,024-bit memory chip, the circuit that most electronic system engineers knew about and incorporated into their designs. In other words, with the 1103 the Intel team had succeeded in designing a new high-profile integrated circuit. The commercial success of the 1103 quickly inspired others to copy this chip. Intel licensed its design and process to MIL, a Canadian company, to meet the second-source requirements of its customers and also to generate cash to support its operations. But many of the circuits other firms introduced in the next few years were not authorized. The growing acceptance of the 1103 boosted Intel's sales. The company, which had sold $500,000 worth of circuits in 1969, had $4 million in revenues the following year. By that time, Intel had a clear leadership in the field of MOS memories (Sideris 1973).

As a result of these efforts at Intel, Intersil, and other firms, the San Francisco Peninsula emerged as the main technological and commercial center for MOS in the United States. In the late 1960s, startups on the Peninsula acquired a unique MOS-processing capability. They solved many of the technical problems of MOS and engineered a wide array of manufacturing processes. These various processes enabled local firms to produce a wide range of innovative circuits. Silicon Valley firms made watch chips and memory chips. They also manufactured logic circuits, with a special focus on circuits for electronic calculators. All these circuits opened up new fields of application for integrated electronics. As a result of these innovative processes and product designs, firms on the San Francisco Peninsula controlled roughly 60 percent of the US market for MOS circuits in 1970. They were also strong in overseas markets, and especially in Japan. The local MOS industry seemed poised for rapid expansion.[49]

Downturn

The semiconductor recession of 1970–71 momentarily halted the growth of National Semiconductor and the MOS startups. This downturn was partly caused by a steep decline in the demand for semiconductor components in the military/aerospace sector and in the computer industry. At that time, these markets were the largest markets for integrated circuits. They dwarfed the industrial and consumer electronics markets. In the early 1970s, the Department of Defense made substantial cutbacks in its procurement of integrated circuits and electronic systems. The computer industry also experienced a downturn of its own and was forced to retrench. This recession caused computer manufacturers to stop buying integrated circuits. In some cases they shipped the microcircuits they had already purchased back to the suppliers. This decline in the demand for microcircuits was intensified by the general economic malaise of the early 1970s. Because of Richard Nixon administration's restrictive monetary policy, the US economy experienced a severe economic recession in 1970 and 1971. Credit became tight and unemployment rose from 3.5 percent to 6 percent. This unfavorable economic climate further weakened the demand for electronic components.[50]

This crisis in the semiconductor industry was further aggravated by a general overcapacity to manufacture integrated circuits. As the industry had anticipated a boom in the demand for microcircuits, many semiconductor firms had made massive investments in expanding manufacturing capacity in the late 1960s. For example, Lester Hogan, Noyce's successor at Fairchild, had invested $35 million in the modernization of the company's wafer fabs and the construction of new production facilities. Sporck had also invested in new factories at National Semiconductor. In the spring of 1970, he opened a new fabrication facility in Santa Clara. With this new plant, National Semiconductor's manufacturing capacity had grown nearly fivefold. Texas Instruments, Motorola, and many East Coast corporations also had followed suit and substantially expanded their manufacturing facilities in the late 1960s. When the demand for integrated circuits declined, these plants became redundant. They added significantly to the fixed costs of many semiconductor manufacturers and negatively impacted their bottom line (Fields 1969a; Murray 1972).

Many firms experienced crises as they became saddled with unproductive investments and as their sales declined steeply. Most of the East Coast semiconductor operations went out of business. Sylvania, which had pioneered TTL circuits, closed its semiconductor division in 1971. Fairchild and Signetics were also adversely affected by the downturn. They had focused on the military and computer markets, which were the hardest hit by the recession. As a result, Fairchild's sales plummeted. The corporation also incurred heavy financial losses: $28 million over 2 years. To stay in business, Hogan, Fairchild's chief executive officer, laid off more than 40 percent of the company's workforce. In one year, Fairchild's worldwide payroll declined from 23,000 to 14,000 employees. In the early 1970s, Signetics also experienced a struggle for survival. Its sales fell from $41.3 million in 1969 to $31.8 million in 1971. It lost $18 million during this period. Like Fairchild, Signetics had to cut its workforce by 40 percent. By 1971, Fairchild and Signetics were shadows of their former selves.[51]

National Semiconductor and the new crop of IC startups fared better. They had focused on newer, more dynamic markets. But the downturn still had adverse effects on them. Qualidyne and a few other startups went out of business. Others such as Computer Micro-Technology were acquired by competitors. Intel and Intersil lost money in 1970. The downturn also had negative consequences on National Semiconductor. National Semiconductor had focused on bipolar circuits. This segment of the microcircuit market experienced the largest decline in the early

1970s. National Semiconductor had also invested heavily in new manufacturing facilities. As a result, the corporation had high fixed costs and its financial position was precarious. To absorb these costs and stay in business, Sporck and Lamond launched a price war. They slashed the price of TTL circuits from $1 to $0.15 per gate. These drastic price cuts enabled National Semiconductor to gain a large share of the TTL market at the expense of its competitors: Fairchild, Signetics, Amelco, and Sylvania.[52] An increase in orders for TTL circuits enabled National Semiconductor to utilize its new wafer plant to full capacity. With its plant churning out circuits at full speed, the company remained profitable throughout the recession. But National Semiconductor's sales declined from $41.9 million in 1970 to $38.5 million in 1971. The company that depended heavily on consistent on-time payments from its customers also experienced a severe cash flow crisis in the summer of 1971. Its customers were increasingly late in their payments. With no money left in the bank, Sporck and Lamond were unable to pay their employees. In order to avoid bankruptcy, Lamond was forced to look for a bank loan. This proved nearly impossible at a time of tight credit. Only the Bank of America was willing to lend money to the firm. It gave National Semiconductor a $6 million credit line, which saved the company.[53]

In spite of these difficult economic conditions, National Semiconductor and many MOS startups continued to invest heavily in the development of new products. National Semiconductor substantially added to its line of linear and logic circuits. In particular, it introduced a family of "tri-state" TTL circuits in order to penetrate the minicomputer market. Tri-state circuits (so called because they could switch to an off state in addition to the traditional binary output) were essential for the new computer architectures recently introduced at Hewlett-Packard and other minicomputer manufacturers. But the main focus at National Semiconductor was on semiconductor memories, as the market for memories looked promising. National Semiconductor devoted substantial resources to the development of memory products. In 1970 and 1971, it commercialized 15 memory circuits. These included "copies" of Intel's RAM products as well as proprietary read-only memories (ROMs). ROMs were used for storing permanent program instructions, while RAMs stored items only for the duration of running a program. Many of these memories were designed to be compatible with the company's tri-state TTL circuits.[54]

Noyce and Moore extended Intel's memory product line and introduced innovative circuits to the market. For example, Intel's engineers invented an electrically programmable read-only memory (EPROM) in 1970. In this circuit, information was electrically written bit by bit. It could then be completely erased by exposure to ultraviolet light. In parallel to this thrust into memory circuits, Intel designed the microprocessor, a general-purpose logic circuit. The microprocessor was a CPU on a chip that could be programmed for different applications. The idea of the microprocessor was not new. It was as old as the integrated circuit itself. In 1959 and 1960, the engineering group that had developed the first microcircuit at Fairchild had toyed with the idea of the "computer on a chip." But no firm had yet been able to realize this idea in silicon. Only Intel could. It had a process, the silicon-gate process, which made possible the making of high-density circuits (Sideris 1973; Moore 1996; Santo 1988).

In 1969, Busicom, a Japanese maker of scientific and business calculators, had asked Intel to make a set of twelve custom chips for them. Intel had accepted the custom contract, but rapidly redefined the deliverables. With Busicom's approval, Intel set out to make a set of four chips that included a microprocessor. Ted Hoff, an applications engineer at Intel, designed the microprocessor's architecture. Federico Faggin, who had co-developed the silicon-gate process at Fairchild and had recently joined Intel's engineering staff, did the logic and circuit design as well as the layout for the chip. Intel delivered the microprocessor (the 4004) to Busicom in the spring of 1971. Because Busicom encountered severe financial difficulties at this time, Intel was able to buy the rights from their Japanese customer to sell the chip for other applications than calculators. Intel put the microprocessor on the market in November 1971.[55]

When Intel expanded into microprocessors and EPROMs, Hoerni entered the business of linear circuits and computer memories. He hired some of the best linear circuit designers from Fairchild Semiconductor and started to make linear amplifiers. In 1970, Hoerni allied himself with a group of entrepreneurs composed of a former Fairchild sales manager and two managers at National Semiconductor, Ken Moyle and Roger Smullen, to start a new memory circuit company, Intersil Memory. This startup was partially owned by Intersil. Much of the funding for it came from Fred Adler, a venture capitalist based in New York. The group focused on the making of bipolar memory circuits. They developed bipolar random-access memory circuits as well as an inventive programmable read-only memory (PROM).[56]

These product-development efforts at Intel, Intersil, National Semiconductor, and other startups contrasted with the severe cutbacks Fairchild and Signetics were forced to make in their product engineering programs during this period. In 1970 and 1971, Fairchild and Signetics, which faced enormous losses and needed to limit their financial outlays in order to survive, slashed their product-development expenditures. Signetics cut back on its investments on MOS. The cuts in research and product development were even more pronounced at Fairchild. In his bid to save the company, Hogan reduced the R&D laboratory to one-fifth of its former size. He also made major cutbacks in the development of new products. In spite of its development of the silicon-gate process, Fairchild had not developed a successful MOS product line. The financial cutbacks and heavy managerial turnover in the MOS group put Fairchild's product development back even further. As a result, Fairchild abandoned the MOS business, especially the MOS memory business, to National Semiconductor and the startups. It also had few new products to offer when the demand for high-density and high-quality products grew in late 1971 and 1972 (Hoefler 1971b; Rhea 1972).

The recession of 1970–71 was a turning point in the history of the integrated circuit business on the San Francisco Peninsula. Before 1969, Fairchild and Signetics had dominated the Peninsula's IC industry. At the end of the downturn they were in a much weaker position, both financially and technologically. In contrast, National Semiconductor and some of the startups were in strong positions when the recession ended. They offered a wide array of innovative products, and they had enlarged their market share. These firms were also in a rather stable financial position. These results enabled the most successful startups to list their shares on the stock exchange. In 1971, Intel went public and raised $7 million in its initial public offering. Intersil did its own IPO the following year. In 1972, National Semiconductor also issued new stock, which brought $10 million to its treasury. As a result of this new influx of money and their solid product position, National Semiconductor, Intel, and Intersil were much better positioned than Fairchild and Signetics to take advantage of the booming demand for integrated circuits in the following years.

Explosion

The influx of capital coming from the initial public offerings, advanced product lines, and market expertise put National Semiconductor and the other startups in an excellent position to take advantage of the explosion in demand for integrated circuits in the early 1970s. The growth of the industrial market was phenomenal, particularly in testing and industrial automation applications. The automotive industry also incorporated integrated circuits in its products. In 1974 the federal government mandated that all car makers install a system that forced motorists to wear their seatbelts before switching on their engines. These ignition locks used integrated circuits. Demand for other consumer-oriented products also exploded. Japanese and American firms introduced digital calculators based on integrated circuits. The same applied to electronic watches. Until that time watch companies had produced a limited number of electronic watches. But as the price of integrated circuits declined, the price of digital watches followed suit. This opened up a large market for watch circuits. All the market building efforts of the specialized firms on the Peninsula bore fruit as the economy improved in 1972 and 1973.[57]

Changes in consumer electronics further stimulated demand for integrated circuits. Until then, manufacturers of radios, stereos, and televisions had used transistors rather than integrated circuits in their products. Starting in 1972, they increasingly switched to microcircuits. The market for semiconductor memories also boomed at that time. Mainframe computer makers, who had largely shunned semiconductor memories, started integrating them into their product lines in 1972. This opened up a large market for memory circuits. At the same time, the markets that had suffered most from the recession—the market for mainframes (for logic circuits) and the military market—recovered. Although they did not grow quite as fast as the consumer electronics and industrial markets, their growth was robust in 1972 and 1973 (Walker 1972).

National Semiconductor and the new spinoffs benefitted greatly from this expansion of the market. Although Fairchild and Signetics did well in their traditional markets (computer and military), they were unable to build a strong position in the industrial and consumer electronics markets. Fairchild introduced MOS RAMs and microprocessors. But it did not establish itself as a major player in these markets. In contrast, the new firms that had pioneered many of the MOS technologies and the building of industrial markets grew enormously. To meet the demands, these corporations aggressively funded rapid expansion. They recruited a large workforce both in engineering and production workers and they expanded their production capacity. In the process, they built strong manufacturing organizations. For example, Intersil moved to volume production at this time. Its sales went from $8.4 million in 1971 to $24.5 million in 1973. The firm, which had primarily produced watch circuits, began to manufacture memory circuits and analog precision products. In

1974, 15 percent of its sales were in low-power circuits (for watches and other consumer products), 25 percent in precision analog, and 60 percent in memories. To meet the growing demand for its products, Intersil built a new wafer-fabricating facility in 1973 and enlarged it the following year. It also opened a new assembly plant in Singapore. In addition, Hoerni started Eurosil, Intersil's European subsidiary. (Intersil owned 35 percent of Eurosil; the rest was held by German interests.) Eurosil, which was based in Munich, made CMOS circuits for the Bavarian watch and clock industry. It also sold timekeeping circuits to armament manufacturers, which used them in land mines and other military applications.[58]

Intel grew even faster. Its sales went from $9.4 million in 1971 to $134 million in 1974. This growth was predicated on the boom in the demand for semiconductor memories. Intel, which had established itself as the main producer of semiconductor memories with its 1103 DRAM chip, was the main beneficiary of the boom; it followed up the 1103 with denser chips. In parallel to the development of memory chips, Intel also developed the microprocessor business. Because microprocessors were particularly difficult to use, Intel developed design aids for the engineers who were interested in putting microprocessors in their own products. In the best Peninsula tradition, Intel made its microprocessor more user friendly with these design aids. It also developed new applications around its microprocessor product in order to show potential customers what could be done with this chip. The corporation also introduced new products. A team under Federico Faggin developed the 8008 and the 8080 microprocessors. These chips were substantially more powerful than the original circuit. The performance of Intel's microprocessors improved by approximately three orders of magnitude between 1971 and 1974 (Bylinsky 1973; Shima and Faggin 1974; Cole 1974).

To meet the growing demand for its products, Intel moved into high-volume production. It organized a massive training program and built new wafer-fabricating facilities in Mountain View and Livermore. In addition, it enlarged its manufacturing capacity by moving from 2-inch to 3-inch wafers. A 3-inch wafer has more than twice the surface area of a 2-inch wafer, which doubled the output of memory chips and microprocessors in the same manufacturing space. Converting to 3-inch wafers was a delicate move. The company had to develop a new manufacturing process and work on improving its yield of good chips. A manufacturing engineer later recalled: "We worked day and night to develop a three-inch process that gave us not only two times more output due to wafer size increase, but also another 50 percent increase in output due

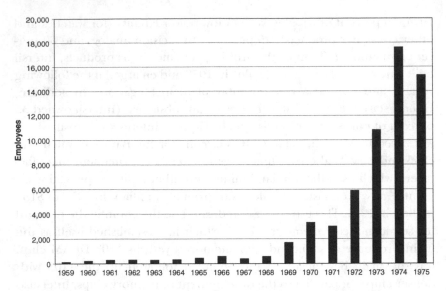

Figure 7.3
Employment at National Semiconductor, 1959–1975.

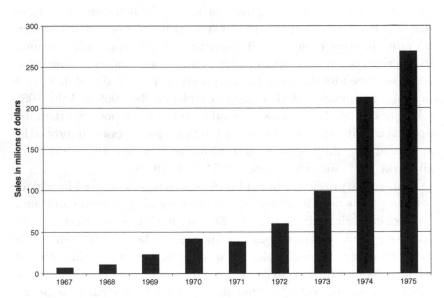

Figure 7.4
National Semiconductor's sales, 1967–1975.

to yield improvements. This resulted in a threefold increase in output and three times lower cost. That effort boosted our 1103 memory output dramatically and lowered the cost at the same time." (Yu 1998) The conversion to 3-inch wafers enabled Intel to meet fast-growing customer orders and to retain its 80 percent share of the 1103 market. Intel continued to increase production. By 1975 it was the third-largest semiconductor company on the San Francisco Peninsula, right behind Fairchild Semiconductor.[59]

While the specialized manufacturers did well, National Semiconductor did even better. In the first half of the 1970s, Sporck expanded his linear circuit business into consumer electronics. For example, National Semiconductor developed a series of amplifiers for audio equipment (Walker 1972). National also addressed the market for logic circuits among mainframe manufacturers and became a major supplier of circuits to IBM. In addition, National exploited the technologies that Intel, Intersil, and other specialized firms in the markets had pioneered. Unlike Fairchild and Signetics, it did so successfully—partly because of its expertise in MOS and partly because the management seems to have put enormous pressure on its engineering staff to move into other fields and develop new products. As a result, National Semiconductor became an important player in the memory business by introducing bipolar and MOS RAMs to the market. It also developed watch circuits as well as large-scale integration circuits for calculators and National Semiconductor went into the microprocessor business as well. By 1974, no other US manufacturer of integrated circuits was as diversified as National Semiconductor.[60]

National Semiconductor's strategy of diversification paid off handsomely. National's customer base grew from 1,300 in 1971 to more than 4,000 in 1973. Sales went from $38.5 million in 1971 to $213 million in 1974 and the workforce grew from 3,025 to 17,000 during the same period. National had 4,000 employees in Silicon Valley. During much of the early 1970s, National Semiconductor hired at a furious pace and used stock options heavily as a recruiting device. It also built new manufacturing facilities at a rapid rate. National substantially enlarged its main plant in Santa Clara. It also built a string of assembly plants in Southeast Asia. It constructed factories at Penang and Malacca in Malaysia and established new assembly facilities in Bangkok in Thailand and in Bandung in Indonesia. National Semiconductor became a super-mass-production manufacturer so that by 1976–77 it produced a million integrated circuits a day. By 1975 it was the largest semiconductor manufacturer on the

Peninsula and the second-largest semiconductor corporation in the United States, right after Texas Instruments.[61]

Conclusion

The growth of the integrated circuit business made the San Francisco Peninsula the microelectronics center of the United States. By 1975, five of the seven largest US semiconductor manufacturers were located in the area: National Semiconductor, Fairchild, Intel, Signetics, and American Micro-Systems. (The others were Texas Instruments and Motorola.) Silicon Valley also had innovative medium-size firms such as Intersil and many specialty-oriented enterprises. Critical to this explosive growth was a wave of firm formation. The startups focused on new process and design technologies, such as MOS. They stabilized MOS manufacturing processes and promoted a wide variety of MOS circuits. They also built large markets for these products. In the late 1960s, Sporck, Lamond, and their group at National Semiconductor developed the incipient industrial market by developing novel sales techniques and helping their customers design products around their micro-electronic components. Moore, Noyce, and Hoerni opened up the entirely new watch and memory markets by developing high-density and low-power integrated circuits. These extensive market-building efforts bore fruit in the first half of the 1970s. At that time, integrated circuits, especially MOS circuits, penetrated a wide variety of industrial sectors, including consumer electronics, instrumentation, machine tools, and automotive manufacturing. These circuits permitted the improvement of existing products as well as the development of entirely new ones. Integrated circuits had become, as Jay Last put it in 1969, "the vitamins of the entire industrial system" (quoted in Lindgren 1969).

The growth of integrated circuit manufacturing had a great effect on the electronics cluster in San Mateo and Santa Clara counties. Microcircuit firms pushed the manufacturing complex even more toward the commercial markets and transformed it into a region that was mostly oriented toward civilian pursuits. The rapid growth of the microcircuit business also reshaped the electronics manufacturing district and transformed it into the "Valley of Silicon." In the late 1960s and the first half of the 1970s, the semiconductor industry became by far the largest and most dynamic electronics sector on the Peninsula. The vacuum tube corporations that had dominated the industrial complex until the mid 1960s were relegated to a secondary position. These firms partially transformed themselves into suppliers of manufacturing equipment to the semiconductor industry. For example, Varian Associates, which had built a modest business selling evaporators and VacIon pumps to semiconductor firms in the early 1960s, made a major push into semiconductor manufacturing equipment at the end of the decade. In the late 1960s, Edward Ginzton, Varian's chairman and co-founder, identified the semiconductor machinery market as a growth opportunity: the semiconductor industry was expanding very fast, and it needed complex processing equipment in large quantities. To take advantage of these opportunities, Ginzton devoted significant resources to the engineering of semiconductor-making apparatus. These efforts soon bore fruit. By 1975, Varian had a large semiconductor equipment product line. In addition to evaporators and ultra-high-vacuum pumps, it made sputtering equipment and produced research instruments that determined the surface conditions, composition, and impurities of integrated circuits. Varian also produced crystal growers, silicon ingot-cutting machines, and detectors that helped monitor equipment operating under pressure or vacuum. In 1976, Varian expanded its product line by acquiring Extrion. Extrion, a Massachusetts-based company, manufactured ion implanters. These machines permitted the introduction of dopants into the silicon crystal in a much more precise and controlled fashion than diffusion furnaces. Ion implanters became essential for the manufacture of complex MOS circuits. Because of its product engineering efforts and its acquisition of Extrion, Varian Associates emerged as a major player in the semiconductor equipment industry. By 1976, it sold nearly $30 million worth of equipment to microelectronics firms and was the fourth-largest corporation in this business in the United States. Varian Associates, the maker of klystrons and scientific instruments, had become an important supplier of production equipment to Intersil, Signetics, and other local semiconductor corporations.[62]

The expansion of the semiconductor industry and the related growth of the venture capital business altered the cultural and institutional fabric of the Peninsula's manufacturing complex. They unleashed an extraordinary wave of entrepreneurship in the late 1960s and the 1970s. Inspired by the formation of new semiconductor companies and benefiting from greater access to venture capital, many local engineers and managers started new corporations. For example, former Varian Associates employees established more than 20 corporations in the second half of the 1960s. Most of these ventures produced semiconductor processing equipment. The other firms made precision instruments. The new semiconductor startups also provided a managerial model for local entrepreneurs. Following the pattern pioneered by the semiconductor ventures, these entrepreneurs had little interest in building long-lasting social institutions. Rather their goal was to use their firms as vectors for personal enrichment and technological innovation. They wanted to bring their company public and sell shares to outside investors. Many entrepreneurs adopted the new forms of management-employee relations, which had been developed at Intel, at National, and at other semiconductor startups. They gave stock options to their professional employees in order to align their interests with those of entrepreneurs and venture capitalists. These managerial techniques became increasingly central to the Silicon Valley way of doing business. By the early and mid 1970s, the manufacturing cluster on the San Francisco Peninsula was profoundly different from the one Charles Litton, Bill Eitel, and Jack McCullough had helped build 30 years earlier. It had crystallized into a new type of industrial district—one centered around semiconductor electronics and characterized by the culture of entrepreneurship, heavy reliance on inexpensive Asian labor, and close ties with financial markets.[63]

Conclusion

"It seems remarkable to me," William Eitel reflected in 1962, "that on the San Francisco Peninsula, off the beaten paths of commerce, grew so many independent new industries, all now of national and international importance."[1] In the 1960s, the Peninsula was indeed much different from the region Eitel had known as a young radio amateur 40 years earlier. In the late 1920s, the Peninsula was mostly a rural region and home to a handful of radio firms. These firms had a cumulative workforce of a few hundred engineers and operators. Four decades later, the region had become the main center for electronic component manufacturing in the United States, employing 58,000 engineers, technicians, and operators. Local firms dominated both the domestic and foreign markets for advanced tubes and semiconductors and had become major suppliers to the Department of Defense as well as a wide range of commercial industries. From an economic backwater, the San Francisco Peninsula had, in less than 40 years, become a major technological and commercial center.

The manufacturing district grew by bringing together a variety of technological groups. Three partially overlapping technical communities emerged and expanded on the San Francisco Peninsula: radio amateurs, microwave engineers, and semiconductor technologists. Radio amateurs were indigenous to the Peninsula. Starting in the 1900s, and partly because of its strong maritime tradition, the Bay Area was one of the nation's largest centers for amateur radio. The other two technological groups converged around a nucleus of western-born scientists and engineers who had worked at East Coast firms during and after World War II and then moved back to California after the war. For example, the Varian group, which had worked at Sperry Gyroscope on Long Island during the war, organized a microwave tube firm on the Peninsula in 1948. Similarly, William Shockley, a Palo Alto native who had made his career at the Bell

Telephone Laboratories, set up a laboratory specializing in the development of silicon transistors in the area in the mid 1950s. These entrepreneurs attracted many other microwave and semiconductor technologists from other regions. In turn, many of these transplants established their own firms on the San Francisco Peninsula.

Radio amateurs, microwave engineers, and semiconductor technologists brought with them new skills and technologies. Each entrepreneurial group was also characterized by its own distinct culture, values, as well as style of work and organization. Radio amateurs recreated and improved upon the power tube technologies developed by large East Coast firms in the 1920s and the 1930s. These men also developed a subculture characterized by its camaraderie, a strong democratic ideology, and genuine appreciation of ingenuity and innovation. The other two groups, microwave engineers and semiconductor technologists, introduced to the Peninsula advanced component technologies they had helped develop at East Coast firms during the war and in the immediate postwar period. These groups also brought their own professional ideology and political ideals. The microwave and silicon communities both valued egalitarianism and viewed engineers as independent professionals. However, the microwave and semiconductor communities differed in other ways: a substantial number in the microwave group had socialist leanings and utopian ideals and longed for a society where the distinction between capital and labor would be abolished. In contrast, the semiconductor community was meritocratic and resolutely capitalistic.

These groups established new firms on the San Francisco Peninsula, which addressed the needs of the armed services and various commercial industries for advanced electronic components. Like their competitors on the East Coast, these firms benefited enormously from the growing demand for vacuum tubes and semiconductors. The military provided a large market for their products during both World War II and the Korean War. This demand continued through the arms race of the late 1950s and early 1960s. In the case of microwave tube firms, the Department of Defense financed R&D efforts and supported the construction of new facilities. But when the military cut back its component expenditures and radically altered its procurement policies in the early 1960s, the component firms in Silicon Valley had to reorient themselves toward the commercial markets. The microwave tube corporations diversified into scientific instrumentation and semiconductor manufacturing equipment. The semiconductor firms opened up large markets for silicon transistors and integrated circuits in the civilian sector.

Silicon Valley firms fared well under these changing conditions. They grew faster than their East Coast counterparts and often surpassed them in the late 1950s and the 1960s. Several factors explain the "success" of Silicon Valley as a manufacturing district. The entrepreneurial groups and their firms developed very unusual manufacturing and product engineering competencies. Silicon Valley's innovator-entrepreneurs recognized manufacturing process knowledge that was critical and they pursued, captured, and leveraged this knowledge over the whole period. For example, Charles Litton, William Eitel, and Jack McCullough devoted themselves to the *making* of vacuum tubes. They engineered innovative manufacturing techniques, which enabled them to produce tubes with a very high vacuum. Varian Associates devised new ways of making reflex and high-power klystrons. But it was the group at Fairchild Semiconductor that was the most innovative in manufacturing. It developed the planar process—arguably the most important innovation in twentieth-century technology. The planar process made possible the manufacture of integrated circuits. In the 1960s and the early 1970s, the Fairchild group and the engineers they trained developed many variants of this process. In addition, these men gained unequaled expertise in the mass production of silicon devices. They maintained their leading edge well into the 1970s.

These manufacturing capabilities made Silicon Valley. They enabled local firms to engineer and produce highly reliable electronic components with very high performance. Eitel-McCullough, Litton Industries, Varian Associates, Fairchild Semiconductor, Intel, and National Semiconductor produced some of the highest-quality vacuum tubes and semiconductors in the country. Significantly, the district's bias toward quality matched the evolution in the demand for electronic components. In the mid and late 1950s, the military shifted its procurement of electronic components toward highly reliable tubes and semiconductors. Likewise, the makers of computers and consumer electronics were seeking faster and more reliable microcircuits in the following decades. While local firms greatly benefited from this shift in markets, East Coast corporations, which emphasized cost and volume production, lost market share and were often forced to leave the component business.

Manufacturing expertise also enabled Silicon Valley firms to achieve high production yields—the main economic variable in the electronic component industries. For example, during the Korean War Litton Industries obtained yields of 95 percent on some advanced magnetrons. This contrasted with the competitors' yields, which were in the 5–30 percent range. Similarly, semiconductor corporations on the Peninsula had higher yields than most of their competitors for bipolar and MOS integrated circuits. These yields gave Silicon Valley firms a tremendous competitive advantage over their East Coast rivals. Better yields enabled local firms to obtain higher profits (which was important in industries where the general level of profitability was low) and to lower their prices in order to increase market share.

Complementing their manufacturing expertise, component firms on the Peninsula developed strong marketing and system engineering capabilities. Eitel-McCullough, Litton, Varian, Fairchild, and Fairchild's spinoffs acquired deep knowledge of their customers and of the design of electronic systems. They knew nearly as much about their clients' products as the clients themselves knew. Because of its origins in ham radio and its employment of radio amateurs, Eitel-McCullough had an excellent expertise in radio system engineering. The founders of Varian Associates were experts both in klystrons and radar systems. Similarly, in the early 1960s, Fairchild Semiconductor acquired a competency in electronic system design, by hiring engineers from the computer, instrumentation, and consumer electronics industries. In the late 1960s and early 1970s, Fairchild's spinoffs followed suit by learning about the many electronic products that could use integrated circuits. This systems expertise enabled the vacuum tube and semiconductor corporations in Silicon Valley to anticipate the needs of their customers and engineer devices that met their requirements.

Silicon Valley firms also learned how to create new markets for their products. Because they made tubes and semiconductors that were complex and difficult to use, they had to help their customers learn to design them into their own products. In some cases, they actually engineered new products for their customers. Eitel-McCullough is a case in point. To create a market for its power klystrons in the 1950s, Eitel-McCullough designed a tropospheric scatter communication transmitter and gave this design to Radio Engineering Laboratories at no cost. Radio Engineering Laboratories produced this transmitter under its own name and bought Eitel-McCullough's klystrons to power it. Varian Associates followed similar tactics by educating academic chemists about the new technology and science of NMR spectroscopy. These educational efforts enabled Varian to create users for its NMR spectrometers.

The "art" of market building was pushed to new highs by Fairchild Semiconductor and its spinoffs. Fairchild's engineers wrote numerous applications notes to explain to customers what they could do with the transistors and integrated circuits. They also engineered new products such as transistorized television monitors around the firm's components and promoted these designs to their customers. Many Fairchild's designs were later put into production by consumer electronics firms such as Zenith and General Electric. In the late 1960s, National Semiconductor went one step further. It deployed field applications engineers who became actively involved in the engineering of their customers' products. Some field applications engineers went as far as directing the product engineering teams at user firms. These marketing and system engineering capabilities enabled Silicon Valley firms to open up new markets for their products, especially in the civilian sector, and to grow enormously in the late 1960s and the first half of the 1970s.

As these markets grew, the groups that built the power-grid tube, microwave tube, and semiconductor industries on the San Francisco Peninsula acquired a unique know-how in creating, financing, and managing startups. At these startups, Silicon Valley entrepreneurs pioneered new funding mechanisms and shaped novel management-employee relations. Charles Litton and the Varian brothers developed an uncanny ability to use military contracts as a means to finance the growth of their firms. The entrepreneurs and financiers who formed Fairchild Semiconductor went on to establish the venture capital industry on the San Francisco Peninsula.

Innovator-entrepreneurs in Silicon Valley also devised new ways of relating with employees. They were under the constant threat of unionization and they needed to secure the cooperation of a skilled workforce in order to build and control complex manufacturing processes. As a result, Silicon Valley firms developed a corporatist approach to management. They gave substantial autonomy to their engineering staffs and often organized engineering work around teams. They sought to involve their professional employees in the decision-making process. In addition, they developed unusual financial incentives for their work force: profit-sharing programs, stock ownership, and stock option plans.

Within this corporatist framework, one can distinguish three different approaches. Eitel-McCullough, Litton Industries, and many microwave tube firms adopted a participatory and paternalistic management style that emphasized profit sharing and generous employee benefits. Varian Associates had a socialist streak, developing a communal organizational and ownership structure. In contrast, Fairchild Semiconductor and most semiconductor firms pioneered an entrepreneurial form of corporatism organized around stock options. Starting in the early 1960s, Fairchild and its spinoffs granted stock options to engineers and middle managers. These options, which offered the right to buy stock at a predetermined price at a future date, had been reserved until then to upper management at East Coast firms. Instead, managers on the Peninsula offered these options to engineers and middle managers as well. As a result, employees were able to benefit from an increase in the valuation of their companies' stock.

Because of these new financial incentives and organizational forms, electronics firms on the Peninsula attracted some of the best design, processing, and manufacturing engineers in the United States from the 1940s to the 1970s. These innovations also permitted Silicon Valley firms to deploy their engineering workforce efficiently and to improve employee productivity. Flat organizational structures, profit sharing, and stock option programs gave them a substantial competitive advantage over their more traditional Eastern counterparts.

The constant circulation of design, production, and management skills within the district also explains the "success" of individual corporations and the manufacturing cluster as a whole. Manufacturing processes, design methodologies, and managerial techniques moved quickly from firm to firm and from industry to industry. Tube and semiconductor firms developed close alliances with their suppliers of manufacturing equipment. For instance, Eitel-McCullough used Litton Engineering's glass lathes and developed new assembly techniques. Skills and techniques flowed within industries as engineers moved from one corporation to the next. For example, many semiconductor entrepreneurs appropriated the manufacturing processes they had mastered (and often helped develop) at their previous employers and built new businesses on the basis of this know-how. Skills, practices, and techniques also moved across industries. Semiconductor manufacturers relied heavily on the competencies nurtured by the vacuum tube corporations. They hired technicians and process engineers from the vacuum tube firms. They borrowed new managerial techniques from Varian Associates and used them (with some modifications) to relate to their own employees.

But what sustained the circulation of practices and techniques throughout Silicon Valley? This circulation was partially predicated on the culture of cooperation that electronics hobbyists developed on the San Francisco Peninsula. Radio amateurs valued collaboration and the sharing of information—both in technology and business. Litton, a radio ham and microwave engineer, freely shared his knowledge of vacuum techniques and tube production methods. He also helped other entrepreneurs start electronics firms in the area. The flow of innovations was further facilitated by a sense of regional pride and the perception of common interests. The tube and semiconductor corporations on the Peninsula were late entrants in industries pioneered by large East Coast firms. Western pride and the imperative of surviving in industries dominated by the East led early Peninsula firms to band together against their larger competitors. Also critical were the dense networks of personal relationships that emerged in the region. In the early 1960s, Fairchild Semiconductor employed a large fraction of all semiconductor engineers on the San Francisco Peninsula. At Fairchild, these men built friendships and professional ties. After leaving the company, they kept in touch, helped each other, and traded information across firm boundaries. Stanford University also reinforced the regional circulation of skills by appropriating process and design technologies from Litton Engineering, Litton Industries, and Shockley Semiconductor. The university later codified and systematized this industrial knowledge and imparted it to engineering students, who in turn carried it to other corporations in Silicon Valley.

These flows of manufacturing techniques and managerial skills across firms and industries were critical to the rise of Silicon Valley. They enabled the rapid adoption of "best practices" across the district. They also led to the formation of region-specific bodies of knowledge that were crucial for the commercial success of local firms. Because they raided each other's employees and benefited from informal contacts among their engineers, MOS startups on the Peninsula developed a repertoire of process "tricks" which were known only in the area. These tricks enabled MOS companies to introduce new products rapidly to the market and to obtain good manufacturing yields. In contrast, MOS firms located outside of Northern California were not plugged into these networks and did not benefit from this shared knowledge. This put them at a distinct disadvantage since the MOS processes were very difficult to master and control.

The Peninsula's electronic component manufacturing complex provided the foundation for much of Silicon Valley's growth in the 1970s. Although local tube and silicon corporations retained and advanced their preeminent position in component manufacturing in the United States, much of the growth during this period came from the system sector. The late 1960s and the 1970s saw the emergence of new "industries" making calculators, computers, electronic watches, and video games on the Peninsula. Many semiconductor firms and entrepreneurial groups entered the digital watch and calculator businesses. There was a similar entrepreneurial flourish around video games—with the formation of firms such as Atari. Other groups made mainframes (Tenet) and fail-safe computers (Tandem). Hewlett-Packard and Varian Associates entered the minicomputer business. In the late 1970s, a new industry—personal computers—emerged with the formation of Apple Computer, Osborne Computer, and other startups.

Each of these industries made use of capital and technical and managerial competencies acquired by the tube and semiconductor communities. The electronic component industries were a source of funds for the new system ventures as fortunes made in components were reinvested in system businesses. For example, Robert Noyce and Gordon Moore financed computer and medical instrumentation startups in the 1970s. Component firms were a major source of managerial talent for the system sector and they supplied engineers knowledgeable about component and system design. Fairchild, Intel, and National supplied integrated circuits to local computer, instrumentation, and consumer electronics firms. These components made calculators, digital watches, and personal computers possible.

Apple Computer was emblematic of the deep ties that emerged between the system and component industries in Silicon Valley. Apple Computer was started by two young electronics hobbyists, Steve Wozniak and Steve Jobs, who were sophisticated users of microprocessors and other integrated circuits. Early on, they hooked up with the social networks in the semiconductor industry. In the first months of Apple's formation, Jobs approached Donald Valentine, a venture capitalist who had formerly directed sales and marketing at Fairchild and National Semiconductor, and asked him to invest in Apple. While Valentine declined at first to finance the startup, he advised Jobs to contact Mike Markkula, one of his former associates at Fairchild. Markkula, who had made a small fortune in the semiconductor business, had recently retired from Intel.[2]

Markkula understood the system possibilities opened up by the microprocessor and was quick to grasp the potential of Apple's first machine. Betting that Apple could become a large and successful firm, Markkula joined the company in 1977. He invested $92,000 of his own money in Apple Computer and convinced Jobs and Wozniak that they needed professional managers. To run Apple on a day-to-day basis, Markkula hired Michael Scott, a former manager at Fairchild and National. Markkula also raised the capital needed to finance Apple's growth. Among the investors were Valentine and Arthur Rock, the venture capitalist who had financed the formation of Intel and Intersil. Valentine and Rock joined Apple's board of directors. With this influx of managerial expertise, Apple rapidly adopted the "best practices" of Silicon Valley's semiconductor industry. The firm emphasized innovation, the development of quality products, and rapid sales growth. It also attracted and retained skilled engineers and managers by granting them attractive stock option packages (Gupta 2000).

The goal of Apple's managers was similar to that of most semiconductor entrepreneurs. They wanted to take the company public within 5 years and make a killing in the stock market. The design skills of Wozniak and other hobbyists in Apple's technical staff made this objective possible. In late 1976 and early 1977, Wozniak developed a successor machine to the Apple I. The revolutionary Apple II attracted a strong following among software programmers, who wrote application programs for the Apple II such as VisiCalc, a spreadsheet program. These programs "sold" the Apple II to the American public. As a result, the firm grew enormously. By 1980, Apple was the leading firm in the personal computer industry. It had a user base of 120,000 computers and was one of the 500 largest corporations in the United States. The company's rapid sales growth enabled its management to take Apple public and reap enormous capital gains. In turn, Apple's stellar success fueled the expansion of the computer, software, and disk drive industries on the San Francisco Peninsula, once again repeating the pattern of Silicon Valley. More than anything else, the making of Silicon Valley was the repetition of this pattern—as one generation of engineers mastered the mystery of design and manufacturing sophisticated electronics, so the next generation borrowed and learned from their predecessors. Manufacturing districts grow and thrive only so long as they remain communities of learning, practice, and collaboration.

Notes

Introduction

1. O'Mara (2004) makes a similar argument: that cities of knowledge such as Silicon Valley came into being because of large federal investments, the presence of a powerful university, and the availability of undeveloped land in attractive locations. See also Adams 2003, Lowen 1990 and 1997, and Gillmor 2005. Leslie (2001) later distanced himself from this university-centered argument.

2. Piore and Sabel draw on "la théorie de la régulation" and the work of Fréderic Le Play, especially his study of the Parisian luxury trades and other forms of "fabriques collectives" in the 1860s. On industrial clustering, see Le Play 1864, 302–315; Le Play 1878, 288–307.

3. On the importance of manufacturing in electronics, see Lécuyer 1999 and Leslie 2001.

Chapter 1

1. Charles Litton to Harold Buttner, December 31, 1946, Charles Litton Papers, 75/7c, box 11, folder Letters written by Litton, 1946, Bancroft Library, University of California, Berkeley; Frederick Terman to Harold Laun, December 17, 1953, in Frederick Terman Papers, SC 160, series V, box 6, folder 13, Archives and Special Collections, Stanford University; Jack McCullough, interview by Arthur Norberg, 1974, 90–92, Bancroft Library.

2. Little is known about Litton's background. The family evidently had some financial means; they owned a fairly large property in Redwood City and supported Litton through 5 years of college. On William Eitel's background, see William Eitel, interview by Arthur Norberg, 1974, Bancroft Library, 1–2 and 5–7. On McCullough's social background, see McCullough, interview by Arthur Norberg, 1974, 1–5. On Charles Litton, see Morgan 1967, 95–97.

3. Eitel, interview by Arthur Norberg, 1974, Bancroft Library.

4. Litton to Arthur Wynne, February 9, 1954, Charles Litton Papers, 75/7, box 13, folder letters written by Litton January-August 1954, Bancroft Library.

5. Litton to Wynne, February 9, 1954; Stanford University Register for 1924–1925 (Stanford Archives and Special Collections), 202–209; McCullough, interview by Arthur Norberg, 1974, 1–5.

6. Department of Commerce 1928; Morgan 1967, 22–30, 58–61; Eitel, interview by Norberg, 1974.

7. Eitel, interview by Norberg, 1974, 8–9, Bancroft Library; McCullough, interview by Norberg, 1974, 1–5.

8. Eitel, interview by Norberg, Bancroft Library, 8–18.

9. On the role of radio "experimenters" in the "exploration of the short waves," see the monthly column on experimenters in *QST* in the 1920s and the early 1930s; see also De Soto 1936; Merritt 1932. For a historical treatment of the development of short wave radio, see Headrick 1991, 1994.

10. Eitel, interview by Norberg, 10–11. For a description of Eitel's work on transmission at 10 meters, see Westman 1928; Hull 1929. On Litton's short wave radio activities, see Morgan 1967, 95–96; Eitel, interview by Norberg, 19.

11. Eitel, interview by Norberg, 11.

12. Ibid., 9–17.

13. Frederick Terman to Norberg, March 6, 1978, in Frederick Terman Papers, SC160, series XI, box 1, folder 4, Stanford Archives and Special Collections; Tyne 1977, 124–126, 167–180; Moorhead 1917, 1921.

14. See also Eitel, interview by Arthur Norberg, 19; Leonard Fuller, interview by Norberg, 1976, 133, Bancroft Library.

15. Register for 1924–1925 and Annual Report of the President of Stanford University for the 34th Academic Year Ending August 31, 1925, both in Stanford Archives and Special Collections.

16. Norberg, "Report on an Interview with Mr. Philip Scofield," September 25, 1973, 75/502 no. 13, Bancroft Library; Leonard Fuller, interview by Norberg, 1976, 127–128, Bancroft Library; Harold Buttner interview by Norberg; Roy Woenne and Norman Moore, interview by Norberg; Eitel, interview by Norberg, 13–18.

17. Lee de Forest developed the audion oscillator and amplifier at Federal Telegraph in the early 1910s. Although de Forest patented the audion oscillator under his own name, Federal Telegraph retained shop rights to this important invention. In turn, these shop rights helped Federal Telegraph secure the IT&T contract in 1927. On Federal Telegraph's history, see Pratt and Roosevelt 1944; Leib, Keyston, and Company 1928; Mann 1946; Aitken 1985. On the history of the Mackay Radio and Telegraph Company, IT&T's radio subsidiary, see "History of the Mackay Radio and Telegraph Company" and "A Brief Historical Outline and Description of the

Mackay Radio and Telegraph Company," both in Haraden Pratt Papers, 72/116, box 4, folder Radio History—Mackay Radio and Telegraph Company, Bancroft Library. For a copy of the agreement between Federal Telegraph and Mackay, see "Agreement between Radio Communication Corporation and Federal Telegraph Company," August 10, 1927, Bancroft Library. On Federal Telegraph's short wave radio program, see Leonard Fuller, "Some research experiences," notes for a talk given before the electrical and mechanical engineering faculties of the University of California at Berkeley, December 2, 1931, in Leonard Fuller Papers, 79/91c, box 1, folder Reference Material, misc., Bancroft Library; Pratt 1969; Deloraine 1976, 83–86; Leonard Fuller, interview by Norberg, 1976, 126–131, Bancroft Library.

18. On Heintz and Kaufman's contract with the Dollar Steamship Company, see "Dollaradio, Chronology" in carton 13, folder Dollaradio, Simpson Radio Corp., Pacific Radio Company; "Globe Wireless, Chronology," in box 14, folder Globe Wireless, Ltd & Dollaradio, History of Operations, 1928–1960; W. P. Boatwright, "Globe Wireless Ltd. Reestablished Communication Circuits," *Pacific Marine Review*, June 1946: 61–68, all in Dollar Papers, 69/113e, Bancroft Library; Ralph Heintz, interview by Norberg, 1977, 53–55, 73–74, 77–80, Bancroft Library.

19. Norberg, "Report on an Interview with Mr. Philip Scofield," September 25, 1973; Eitel, interview by Norberg, 13–18; Fuller, interview by Norberg, 127.

20. Heintz, interview by Norberg, 54 and 63; Fuller, interview by Norberg, 128.

21. On RCA's early history, see Aitken 1985, 302–513. On RCA's policies regarding competition in international radio communication and its control of vacuum tube patents, see Federal Trade Commission 1924.

22. McCullough, interview by Norberg, 11; Eitel, interview by Norberg, 20.

23. "Monthly Men in the Tube Lab," ca. January 1930, Charles Litton Papers, 75/7c, box 3, folder Federal Telegraph, 1929–1931; Litton, "Notebook," no date, in Charles Litton Papers, 75/7c, carton 5, folder Notebooks—Reference Notes ca. 1926–1929; Deloraine 1976, 83–86.

24. Heintz, interview by Norberg, 1977, 67, Bancroft Library.

25. Heintz, interview by Norberg, 68; Eitel, interview by Norberg, 27 and 38.

26. Fuller, interview by Norberg, 1976, 128–129, Bancroft Library.

27. Heintz, interview by Norberg, 1977, 86–87, 96–97; Eitel, interview by Norberg, 30; Fuller, interview by Norberg, 1976, 128–129.

28. Fuller to A. Clokey, February 6, 1930; Fuller to John Farrington, October 6 and October 13, 1930—all in Charles Litton Papers, 75/7c, box 3, folder Fuller, Leonard; Heintz, testimony to the commission on communications, United States Senate, January 23, 1930, in *Hearings before the Committee on Interstate Commerce, United States Senate*, volume 2 (Government Printing Office, 1930), 1917; Fuller, interview by Norberg, 128–130; Moore and Woenne, interview by Norberg, December 19, 1973, 1; Heintz, interview by Norberg, 95–98; Eitel, interview by Norberg, 25–29.

29. Litton's seal used a reinforcing ring, which counteracted the tendency of the metal to expand and contract. As a result, it relieved the seal's main weld and its glass element of excessive stresses, which resulted in breakage. In conjunction with the new copper to glass seal, Litton and his group developed a new process for making pure copper. Instead of casting copper in an open-air environment, Litton cast pellets of copper in a vacuum to avoid the introduction of oxygen into the copper elements. This process enabled the production of pure copper and as a result made possible the production of high-quality seals. Litton also developed new processes to bypass RCA's patents on oxide-coated cathodes. See Fuller, "Some research experiences," notes for a talk given before the electrical and mechanical engineering faculties of the University of California at Berkeley, December 2, 1931, in Leonard Fuller Papers, 79/91c, box 1, folder Reference Material, misc. On Litton's work on copper to glass seals, see Charles Litton, "Metal-to-Glass Seal," US Patent 1,940,870, filed September 15, 1933, granted December 26, 1933; Heintz, interview by Norberg, 43–44. On the work at Heintz and Kaufman on tantalum, see Heintz, interview by Thorn Mayes, 1974, 25, Bancroft Library; F. Hunter, "A Discussion of the Uses and Advantages of Tantalum in the Manufacture of Vacuum Tubes," July 13, 1928, Charles Litton Papers, 75/7c, carton 1, folder Tubes, miscellaneous data ca 1928–1930; Kohl 1951, 215–221.

30. Preist 1992, 3; McCullough, interview by Norberg, 12; Eitel, interview by Norberg, 20 and 28.

31. J. G. Copelin, "History of Litton Engineering Laboratories and Report for Year 1942," April 19, 1944, Charles Litton Papers, 75/7c, carton 1, folder renegotiation; Heintz, interview by Norberg, 37–38, 45, 52–53, and 96; McCullough, interview by Norberg, 11–12 and 54; Eitel, interview by Norberg, 51–52.

32. Woenne, interview by Norberg, December 19, 1973, 5, Bancroft Library.

33. Woenne, interview by Norberg, 4–5; Heintz, interview by Norberg, 63; Eitel, interview by Norberg; Harold Buttner, interview by Norberg, February 5, 1974, 41, Bancroft Library.

34. Pratt 1969, 46; Deloraine 1976; Fuller, interview by Norberg, 133–134. On the crisis at Heintz and Kaufman and the Dollar Steamship Company in the early and mid 1930s, see Niven 1987, 103–110; Heintz, interview by Norberg, 102–103; Eitel, interview by Norberg, 41; McCullough, interview by Norberg, 13–14.

35. Perrine 1932; Romander 1933; Eitel, interview by Norberg, 36–37; Fuller, interview by Norberg, 133–134.

36. Eitel and McCullough learned to make thoriated tungsten filaments, a complex and delicate process, by experimenting in Heintz and Kaufman's tube laboratory and by exchanging information with technicians working at the National Tube Company. The National Tube Company, a small San Francisco outfit, specialized in the repair of power tubes. Eitel and McCullough met employees of the National Tube Company at amateur radio clubs in San Francisco. On their use of thoriated tungsten filaments, see McCullough, interview by Norberg, 40. On Eitel's

and McCullough's development of a triode for amateur radio transmitters, see Eitel, interview by Norberg, 35–40; McCullough, interview by Norberg, 11, 16, and 40–41; communication from Jack Strother, July 21, 1996. On the tube's electrical specifications, see "Gammatron: Type 354," circa 1935, in Charles Litton Papers, 75/7c, carton 2, folder Tubes miscellaneous, gammatron, Bancroft Library.

37. Eitel, interview by Norberg, 40.

38. On the conflicts at Heintz and Kaufman and the establishment of Eitel-McCullough, see "First Directors' Meeting of Eitel-McCullough Inc." October 1, 1934, in Eitel-McCullough Records, 77/110c, box 7, folder Minutes Eitel-McCullough 10/34–10/45; McCullough, interview by Norberg, 14–15 and 35–36; McCullough, interview by Jack Strother and Robert Herdman, November 25, 1999, tapes 1 and 2, Stanford Archives and Special Collections; Eitel, interview by Norberg, 39–41 and 44–45; Heintz, interview by Norberg, 101–103; Norberg, "Report on an Interview with Mr. Philip Scofield," September 25, 1973, 75/502, no. 13, Bancroft Library.

39. The market for transmitting tubes shrank from $1,410,000 to $1,300,000 between 1931 and 1933. For statistics on power tube manufacture, see the US Department of Commerce's Census of Manufacture for 1931 and 1933. On the history of specialty manufacturing in the United States, see Scranton 1989, 1991, 1997.

40. Woenne, interview by Norberg, 4 and 6.

41. Litton to Albert Jason, June 30, 1953, Charles Litton Papers, 75/7c, box 13, folder Letters written by Litton, January-August 1953. On Litton's early work at Litton Engineering, see also Litton to A. Fuchs, December 17, 1953, Charles Litton Papers, 75/7c, box 13, folder Fuchs; Peter Benjaminson, "Homes Fill Historic Site," *Redwood City Tribune*, August 12, 1965, 13–15, in collection 74-423, San Mateo County History Museum, Redwood City; Packard 1995, 37. For a description of Litton's glass lathes, see "The Model E Glass Working Lathe," no date, Charles Litton Papers, 75/7c, carton 4, folder model E—glass working lathe.

42. J. G. Copelin, "History of Litton Engineering Laboratories and Report on Operations for Year 1942," April 19, 1944, Charles Litton Papers, 75/7c, carton 1, folder renegotiation.

43. Ibid.

44. On Litton's oil vapor pumps, see Litton, "High Vacuum pump," US Patent 2,289,845, filed January 12, 1939, granted July 14, 1942; Litton, "Vacuum Distillation Method," US patent 2,266,053, filed July 31, 1939, granted December 16, 1941; Litton to E. D. Phinney, January 5, 1939, Charles Litton Papers, 75/7c, box 11, folder 1931–39; Packard 1995, 37; Copelin, "History of Litton Engineering Laboratories and Report on Operations for Year 1942"; E. J. Walsh, "Litton Engineering Laboratories," October 17, 1942, Charles Litton Papers, 75/7c, carton 1, folder renegotiation. For a history of vacuum pump technology, see Hablanian 1984.

45. Eitel, "Electronics Considered Pace-Setter in Region's Development," *Redwood City Tribune*, December 27, 1962, 16A, in collection 64–94, San Mateo County History Museum.

46. Eitel, interview by Norberg, 45; McCullough, interview by Norberg, 50–55; De Soto 1936, 130–131. The radio amateur market offered attractive opportunities. The number of licensed radio amateurs more than doubled between 1929 and 1933, growing from 16,000 to 41,000. On the expansion of amateur radio during the Great Depression, see De Soto 1936.

47. McCullough, interview by Norberg, 54; Eitel, interview by Norberg, 51–53 and 61.

48. William Eitel and Jack McCullough, "Method and Apparatus for Exhausting Tubes," US patent 2,134,710, filed June 1, 1936, granted November 1, 1938; McCullough, interview by Jack Strothers and Robert Herdman, November 25, 1997, tape 2, Archives and Special Collections, Stanford University; McCullough, interview by Norberg, 56; Eitel, interview by Norberg, 53, 62, and 77.

49. To bring a new version of their ham radio tube to the market, Eitel and McCullough had to bypass the patents they had filed while working at Heintz and Kaufman. This task proved easy, however. They discovered both to their relish and their dismay that their patents were narrowly written. It was easy to bypass them. On Eitel-McCullough's legal wranglings with Heintz and Kaufman, see Eitel, interview by Norberg, 47–48; McCullough, interview by Norberg, 29.

50. "Tenth Anniversary Edition," *Eitel-McCullough News*, September 9, 1944, 6–11, collection 739, San Mateo County History Museum; McCullough, interview by Norberg, 18–20 and 26–27; McCullough, interview by Strother and Herdman, tape 2; Eitel, interview by Norberg, 51–55.

51. "Tenth Anniversary Edition," 34–41; Eitel, interview by Norberg, 73.

52. On the choice of Eitel-McCullough's tubes by the Naval Research Laboratory, see Allison 1981, 102–103; Eitel, interview by Norberg, 65.

53. Taylor 1961; Eitel, interview by Norberg, 64–65; McCullough, interview by Norberg, 76.

54. "Tenth Anniversary Edition," *Eitel-McCullough News*, 4–5; Eitel, interview by Norberg, 68–72; McCullough, interview by Norberg, 76; McCullough, interview by Strothers and Herdman, tape 2.

55. Minutes of board of Eitel-McCullough, December 26, 1939 and May 20, 1941, in Eitel-McCullough Records, 77/110c, carton 7, Bancroft Library.

56. "Tenth Anniversary Edition," *Eitel-McCullough News*, 4–5, 17–25, 44–47; Eitel, interview by Norberg, 68–72.

57. Minutes of board of Eitel-McCullough, December 26, 1939, in Eitel-McCullough Records, 77/110c, carton 7; McCullough, interview by Strothers and

Herdman, tape 2; McCullough, interview by Norberg, 76. On the history of welfare corporatism, see Brandes 1976 and Jacoby 1997.

58. Eitel, interview by Norberg, 74–76; McCullough, interview by Norberg, 46 and 76.

59. Eitel, interview by Norberg, 74–76; McCullough, interview by Norberg, 17, 44–46, 52.

60. Ralph Shermund to Philip Scofield, April 28, 1941; Shermund to Scofield, May 3, 1941; Shermund to Scofield, May 16, 1941—all in Charles Litton Papers, 75/7c, box 8, folder Ralph Shermund; Litton to Scofield, April 11, 1945, Charles Litton Papers, 75/7c, box 11, folder January-May 1945; Litton to Harrison Call, September 27, 1945, in Charles Litton Papers, 75/7c, box 11, folder June-December 1945; Norberg, "Report of an interview with Mr. Phil. Scofield," September 25, 1973, 75/502 no. 13, Bancroft Library.

61. Shermund to Scofield, April 28, 1941, May 3, 1941, and May 16, 1941—all in Charles Litton Papers, 75/7c, box 8, folder Ralph Shermund; Litton to Scofield, April 11, 1945, Charles Litton Papers, 75/7c, box 11, folder January-May 1945; Litton to Harrison Call, September 27, 1945, in Charles Litton Papers, 75/7c, box 11, folder June-December 1945. On Litton's profit-sharing programs, see Walsh, "Litton Engineering Laboratories," Charles Litton Papers, 75/7c, carton 1, Folder renegotiation; Litton to Melville Eastham, January 17, 1950, Charles Litton Papers, 75/7c, box 11, folder letters written by Litton.

62. On Litton's tenure at Federal Telegraph, see chapter 2.

63. Herbert Thielmayer, "Report," March 18, 1944 and Arthur Andersen, "Report for 1944," March 6, 1945, Charles Litton Papers, 75/7c, box 4, folder ICE; minutes of board of directors of Industrial and Commercial Electronics, September 27, 1943, Charles Litton Papers, 75/7c, box 11, folder 1940–1944; "Heintz and Kaufman: S. F. Electronic Inventors in War Service," *San Francisco Chronicle,* January 21, 1943, in Ralph Heintz Papers, 77/175c, carton 2, folder clippings, Bancroft Library.

64. Minutes of board of directors of Eitel-McCullough, March 17, 1942 and June 11, 1942; "Eitel-McCullough Uncompleted Contracts," January 16, 1943—all in Eitel-McCullough Records, carton 7, folder Eitel-McCullough Minutes 10/34–10/45; Tenth Anniversary Edition," *Eitel-McCullough News*, 4–5; Eitel, interview by Norberg, 68–72 and 85.

65. "Tenth Anniversary Edition," *Eitel-McCullough News*, 17–25.

66. "Marking Vertical Bar Grids by Machine," *Eitel-McCullough News*, March-April 1945, 3–7 in Eitel-McCullough Papers, M 1017, box 2, folder January-June 1945, Stanford Archives and Special Collections; "Tenth Anniversary Edition," *Eitel-McCullough News*, 18; Eitel, interview by Norberg, 68 and 81.

67. Eitel, interview by Norberg, 69.

68. "Rotary Exhaust Machine," *Eitel-McCullough News*, February 1945, 3–5 and 15; "The HV-I Oil-Diffusion Pump," *Eitel-McCullough News*, May-June 1945, 3–14—both in Eitel-McCullough Papers, M 1017, box 2, folder January-June 1945, Stanford Archives and Special Collections; "Tenth Anniversary Edition," *Eitel-McCullough News*, 17; Eitel, interview by Norberg, 69; McCullough, interview by Strothers and Herdman, tape 2.

69. Eitel, interview by Norberg, 72; McCullough, interview by Norberg, 78; McCullough, interview by Strothers and Herdman, tape 3.

70. Walsh, "Litton Engineering Laboratories."

71. On Litton Engineering during World War II, see Walsh, "Litton Engineering Laboratories"; Copelin, "Litton Engineering Laboratories and Report on Operations for Year 1942."

72. "Contract Cancellations," *Eitel-McCullough News*, March 11, 1944, in Eitel-McCullough Papers, M 1017, box 1, folder March-April 1944, Stanford Archives and Special Collections; McCullough, "Post-War Eitel-McCullough," *Eitel-McCullough News*, September 2, 1947, in Eitel-McCullough Papers, M 1017, box 2, volume 5; Litton to Scofield, April 11, 1945, Charles Litton Papers, 75/7c, box 11, folder January-May 1945; Litton to Harrison Call, September 27, 1945 and Litton to Scofield, December 18, 1945—both in Charles Litton Papers, 75/7c, box 11, folder June-December 1945.

73. "The 4-125A—A New Power Tetrode," *Eitel-McCullough News*, February 1945, in Eitel-McCullough Papers, M 1017, box 2, folder January-June 1945, Stanford Archives and Special Collections; "Eitel-McCullough Impresses at IRE Technical Meeting," *Eitel-McCullough News*, April 4, 1947, 3–14 in Eitel-McCullough Papers, M 1017, box 2, volume 5.

74. "Address by Paul Walker: The Future of Telecommunications as Affected by War Developments," December 5, 1945; Erickson 1973.

75. "REL Demonstrates FM Transmitter Using Eitel-McCullough Tubes," *Eitel-McCullough News*, November 12, 1946, 3–14 in Eitel-McCullough Papers, M 1017, box 2, volume 5; "Eitel-McCullough Impresses at IRE Technical Meeting," *Eitel-McCullough News*, April 4, 1947, 3–14 in Eitel-McCullough Papers, M 1017, box 2, volume 5; "Report on the Acquisition of Company X," Varian Associates Records, 73/65c, carton 3, folder Report on proposed acquisition of Company X, Bancroft Library.

76. Eitel, interview by Norberg, 77–78.

77. Preist 1992, 14–15, 23–28; Eitel, interview by Norberg, 96; McCullough, interview by Norberg, 95 and 101–105; McCullough, interview by Strother and Herdman, tapes 2 and 3.

78. Frederick Terman to Robert Swain, August 14, 1929, in Frederick Terman Papers, SC 160, series II, box 4, folder 4, Stanford Archives and Special Collections;

Terman, "Communication at Stanford," December 19, 1936, in Frederick Terman Papers, SC 160, series II, box 3, folder 9.

79. Terman, "Communication at Stanford," December 19, 1936, in Frederick Terman Papers, SC 160, series II, box 3, folder 9.

80. This grant involved a complex transaction by which Litton gave his rights on certain tube ideas that he had to Sperry Gyroscope. In return, Sperry gave $1,000 to Terman's radio engineering program.

81. On Litton's role in Stanford's tube programs, see Litton, "Varian Trips to Lab," no date, folder Tubes, technical notes 1926–1938, carton 1, Charles Litton Papers, 75/7c; Packard 1995, 34–43. Karl Spangenberg's textbook *Vacuum Tubes* was published by McGraw-Hill in 1948.

82. "Transcript of a telephone conversation among Litton, Jack Copelin, Robert Helm, Roy Woenne, and Joseph Gordon," April 25, 1945, folder telephone conversations about tubes, 1945, carton 1; Litton to Buttner, July 1, 1946, folder June–December 1946, box 11; Litton to E. Phinney, November 19, 1947, folder 1946, box 11—all in Charles Litton Papers, 75/7c; Moore and Woenne, interview by Norberg, 10–25. On Eitel-McCullough's venture into klystron engineering, see Eitel, interview by Norberg, 90–92; McCullough, interview by Norberg, 70–72; Preist 1992, 12–13, 16–19, 22.

Chapter 2

1. Russell Varian to John Mattil, April 18, 1948, folder 7, box 1, series: Russell Varian, Russell and Sigurd Varian Papers, SC 345, Archives and Special Collections, Stanford University; Hansen, "Statement of William Hansen," no date, folder statement of William Hansen, box 19, Edward Ginzton Papers, SC330, Archives and Special Collections, Stanford University; Sigurd Varian, hand-written notes on the development of the klystron, no date, in folder 23, box 1, series: Sigurd Varian, Russell and Sigurd Varian Papers, SC345, Archives and Special Collections, Stanford University; Frederick Terman, "Events Associated with the Invention of the Klystron Tube," October 20, 1958, folder 12, box 5, series: Russell Varian, Russell and Sigurd Varian Papers, SC345, Archives and Special Collections, Stanford University; Ginzton 1975; Varian 1983, 171–218. See also Galison et al. 1992; Hevly 1994.

2. The Varian brothers often called on Litton for technical advice. For example, in 1936 Sigurd Varian consulted him several times regarding his ruling engine project.

3. "Statement of William Hansen in the United States Patent Office Before the Examiner of Interferences," 1945; "Statement of Russell Varian in the United States Patent Office Before the Examiner of Interferences," 1945; "Statement of Sigurd Varian in the United States Patent Office Before the Examiner of Interferences," 1945, 68—all in Sperry Gyroscope collection, 1915, Hagley Museum and

Library; Charles Litton, "Varian Trips to Lab," no date, Charles Litton Papers, 75/7c, carton 1, folder Tubes, Technical Notes, 1926–1938, Bancroft Library; Everson 1974, 137–141.

4. Like many radio engineers on the San Francisco Peninsula, Litton was extremely interested in the new tube—all the more so since he had played with similar ideas in the early 1930s but had never patented or reduced them to practice.

5. There is evidence that the Stanford group asked Litton to make a sealed-off klystron at that time. But the group later decided not to go ahead with the project. See Ginzton and Cottrell 1995; Ginzton 1975; Hansen to Wilbur, July 20, 1939, folder 12, box 2, William Hansen Papers, SC 126, Stanford Archives and Special Collections; Litton to Hugh Jackson, May 22, 1939, in Frederick Terman Papers, SC 160, series II, box 4, folder 16; John Woodyard, interview by Arthur Norberg, 1974 and 1975, Bancroft Library.

5. Terman to Willis, February 23, 1939, in Frederick Terman Papers, SC 160, series II, box 3, folder 5; Harold Buttner, interview by Arthur Norberg, February 5, 1974, 3; Frederick Terman, interview by Norberg, Charles Susskind, and Roger Hahn, 1974–1978, 61, Stanford Archives and Special Collections; John Getting, "The Patent Position of Charles V. Litton," April 21, 1954, in Charles Litton Papers, 75/7c, carton 4, folder Patents—Litton. The Patent Position of Charles Litton, volume I.

7. Sperry Gyroscope established a vacuum tube shop in San Carlos to design and make sealed-off klystrons. The shop, which employed about fifteen engineers and mechanics, redesigned a klystron developed at Stanford for a blind landing system into a sealed-off device. Sperry's engineers also worked on a new sealed-off klystron that could operate at higher frequencies. They completed this design in the fall of 1940. Litton to E. Phinney, October 29, 1939, November 22, 1939, and December 15, 1939—all in Charles Litton Papers, 75/7c, box 11, folder 1912–1939; Litton to Phinney, January 3, 1940 and February 2, 1940, Litton to Buttner, January 17, 1940, and Report on Development Program for Month of January 1940—all in Charles Litton Papers, 75/7c, box 11, folder: 1940–1944; Terman to Willis, February 23, 1939, in Frederick Terman Papers, SC 160, series II, box 3, folder 5; Buttner, interview by Norberg, February 5, 1974, Bancroft Library; Frederick Terman, interview by Norberg, Charles Susskind, and Roger Hahn, 1974–1978, Stanford Archives and Special Collections. On the San Carlos shop, see R. Wathen, "The Sperry Company and Research," March 15, 1944, in Sperry Gyroscope Papers, 1915, box 37, folder 25, Hagley Museum and Library; Bryant 1990.

8. On Litton's work on radar triodes, see minutes of meeting held June 17 [1941], Charles Litton Papers, 75/7c, carton 2, folder Tubes—type L200, L400, L600; "Meeting on UHF Vacuum Tubes, November 25th [1941]," Charles Litton Papers, 75/7c, box 4, folder I miscellaneous; Glauber 1946.

9. John Getting, "The Patent Position of Charles V. Litton," April 21, 1954, in Charles Litton Papers, 75/7c, carton 4, folder Patents—Litton. The Patent Position of Charles Litton, volume I.

10. Federal Telegraph was now officially the Federal Telephone and Radio Corporation.

11. Litton, "Review of Vacuum Tube Division, November 15, 1942–November 15, 1943," Charles Litton Papers, 75/7c, box 3, folder Federal Telegraph, 1942–1943; Buttner, interview by Norberg; Jack McCullough, interview by Norberg, April 15 and 24, 1974, Bancroft Library.

12. Litton, "The Management Policy of Vacuum Tube Division, Federal Telephone and Radio Corporation" [circa 1942]; Litton, "Review of Vacuum Tube Division, November 15, 1942–November 15, 1943"—both in Charles Litton Papers, 75/7c, box 3, folder Federal Telegraph, 1942–1943; Litton to Gertrude Litton, November 28, 1943, Charles Litton Papers, 75/7c, box 11, folder Letters written by Litton, 1940–1944; J. Copelin, "Litton Engineering Laboratories," April 19, 1944, Charles Litton Papers, 75/7c, carton 1, folder Renegotiation.

13. Litton, "Review of Vacuum Tube Division, November 15, 1942–November 15, 1943," Charles Litton Papers, 75/7c, box 3, folder Federal Telegraph, 1942–1943; "Electrical Communication: 1940–1945, Part II, Federal Telephone and Radio Corporation, Newark, New Jersey," *Electrical Communication*, volume 23, 1946, 221–240.

14. Litton to Sosthenes Behn, August 20, 1945, in Charles Litton Papers, 75/7c, box 11, folder June-December 1945.

15. On the history of the magnetron and related radar systems, see Buderi 1996; Coupling 1948; Guerlac 1987; Boot and Randall 1976. On the design and production of magnetrons at Federal Telegraph during the war, see Vannevar Bush to C. Suits, July 1, 1943, July 1, 1943 and "OSRD Policy with Respect to International Telephone and Telegraph Corporation and Its Subsidiaries Including Federal Telephone and Radio Corporation for the Guidance of the Chiefs of Divisions 13, 14, and 15 of NDRC," July 1, 1943—both in Frederick Terman Papers, SC 160, series V, box 3, folder 11, Stanford Archives and Special Collections; "Electrical Communication: 1940–1945, Part II, Federal Telephone and Radio Corporation, Newark, New Jersey," *Electrical Communication*, volume 23, 1946, 221–240; William Eitel, "Electronics Considered Pace-Setter in Region's Development," *Redwood City Tribune*, December 27, 1962, collection 64–94, San Mateo County History Museum.

16. On Japanese radar during World War II, see Partner 1999. On radar counter-measures during World War II, see "Radio Research Laboratory and ABL-15, NDRC Review Meeting," February 21, 1945, Frederick Terman Papers, SC 160, series I, box 4, folder 3; Price 1984; "Administrative History of the Radio Research Laboratory," March 21, 1946, in Frederick Terman Papers, SC 160, series I, box 9, folder 1; Harris et al. 1978.

17. William Gray to Litton, May 22, 1944, in Charles Litton Papers, 75/7c, box 9, folder US Naval-Uz. On the development of continuous-wave magnetrons for electronic countermeasures at GE, see Isaac Rabi, "Meeting with General Electric, Schenectady, on March 20, 1944," March 29, 1944 in folder 14, box 4, series I, Terman Papers; J. Perkins and E. Dienst, "General Electric Company, Speed of Development of Electron Tubes," August 10, 1945, in folder 14, box 4, series I, Terman Papers.

18. William Gray to Litton, May 22, 1944, in Charles Litton Papers, 75/7c, box 9, folder US Naval-Uz; J. Copelin to J. Moore, December 27, 1944, in Charles Litton Papers, 75/7c, box 5, folder Litton Engineering 1944–January 1945; Litton to Moore, January 26, 1945, Charles Litton Papers, 75/7c, box 11, folder Letters written by Litton, January-May 1945; Litton to Captain Hutchins, February 6, 1945, Charles Litton Papers, 75/7c, box 111, folder January-May 1945.

19. Litton to Behn, August 20, 1945, in Charles Litton Papers, 75/7c, box 11, folder June-December 1945; Winfield Wagener, "Historical Report," September 26, 1945, in Charles Litton Papers, 75/7c, carton 2, folder Tubes Type 6J21 Historical Report; "Win Wagener Retires from WCEMA Leadership," *Varian Associates Magazine,* January 1957, Russell and Sigurd Varian Papers, series: Varian Associates, box 7, folder 2, Stanford Archives and Special Collections.

20. Litton to Isaac Rabi, June 26, 1943, Charles Litton Papers, 75c/7c, box 11, folder Letters written by Litton, 1940–1944; Wagener, "Historical Report," September 26, 1945, in Charles Litton Papers, 75/7c, carton 2, folder Tubes Type 6J21 Historical Report; Litton Engineering Laboratories, "Development of 1000 Watt Tunable Magnetron for S Band," September 30, 1945, in Charles Litton Papers, 75/7c, carton 2, folder Tubes—1000 watt tunable magnetron; Litton to A. Turner, November 8, 1945, Charles Litton Papers, 75/7c, box 11, folder June-December 1945.

21. Winfield Wagener, "Historical Report," September 26, 1945, in Charles Litton Papers, 75/7c, carton 2, folder Tubes Type 6J21 Historical Report; Litton Engineering Laboratories, "Development of 1000 Watt Tunable Magnetron for S Band," September 30, 1945, in Charles Litton Papers, 75/7c, carton 2, folder Tubes—1000 watt tunable magnetron; Litton to A. Turner, November 8, 1945, Charles Litton Papers, 75/7c, box 11, folder June-December 1945.

22. R. Morris to Phinney, December 18, 1944, Charles Litton Papers, 75/7c, box 3, folder Federal Telegraph; J. Moore to Litton, March 1, 1945, in Charles Litton Papers, 75/7c, box 9, folder OSRD/NDRC; Wagener, "Historical Report"; "Win Wagener Retires from WCEMA Leadership," *Varian Associates Magazine,* January 1957, Russell and Sigurd Varian Papers, series: Varian Associates, box 7, folder 2, Stanford Archives and Special Collections.

23. Wagener, "Development of 1000 Watt Tunable Magnetron for S Band," 1–9.

24. Copelin to Bureau of Ships, January 19, 1945, Charles Litton Papers, 75/7c, box 5, folder Litton Engineering 1944–January 1945; J. Gordon to US Navy, August

21, 1945 and September 10, 1945, in Charles Litton Papers, 75/7c, box 5, folder Litton Engineering, August-September 1945; "OSRD-VTDC Minutes," April 24 and May 25, 1945, Charles Litton Papers, 75/7c, box 10, folder OSRD VTDC minutes 1944–1945; Gordon to Army Air Force, November 13, 1945, Charles Litton Papers, 75/7c, box 11, folder June-December 1945; "Progress Report RRL," July 15, 1945, Frederick Terman Papers, SC 160, series I, box 14, folder 17.

25. Minutes, December 11, 1945, Charles Litton Papers, carton 1, folder Litton Engineering, miscellaneous; Litton to Albert Jason, June 30, 1953, Charles Litton Papers, box 12, folder Letters written by Litton, January-August 1953. On this wave of incorporations in California, see Paul Wendt, "The Availability of Capital to Small Business in California," unpublished mimeo, University of California, Berkeley, 1947.

26. Litton to A. Wing, January 11, 1946, in folder January-May 1945, box 11, Litton Papers.

27. Wagener, "Report No. 2," December 7, 1945, in Charles Litton Papers, 75/7c, box 5, folder Litton Engineering, October 45–46; memorandum, December 11, 1945, Charles Litton Papers, 75/7c, carton 1, folder Litton Engineering, miscellaneous; Litton to E. Showers, January 16, 1946, Charles Litton Papers, 75/7c, box 11, folder Letters written by Litton, January-May 1946; Wagener, "Report No. 3," February 4, 1946, in Charles Litton Papers, 75/7c, box 5, folder Litton Engineering, October 45–46.

28. Litton to Jason, June 30, 1953, Charles Litton Papers, 75/7c, box 12, folder Letters written by Litton, January-August 1953; Litton to R. Fuchs, December 17, 1953, Charles Litton Papers, 75/7c, box 12, folder Letters written by Litton, September-December 1953; Litton to Alphonse Dalton, September 27, 1954, Charles Litton Papers, 75/7c, box 12, folder Letters written by Litton, September-December 1954.

29. Chet Lob, interview by Lécuyer, April 9 and May 15, 1996; communication from Lob, January 22, 1997.

30. Litton to R. Framme, February 21, 1946, in Charles Litton Papers, 75/7c, box 11, folder January-May 1946; Litton to W. Floyd, June 4, 1946, Charles Litton Papers, 75/7c, box 11, folder June-December 1946; Litton to Buttner, July 1, 1946, Charles Litton Papers, box 11, folder Letters written by Litton, June-December 1946; Litton to L. Bedell, December 24, 1947, Charles Litton Papers, 75/7c, box 11, folder 1947; Litton to Grantham, April 28, 1948, Charles Litton Papers, 75/7c, box 11, folder 1948; Litton to W. Floyd, April 15, 1949, Charles Litton Papers, 75/7c, box 11, folder 1949; Litton to Buttner, January 10, 1950, in Charles Litton Papers, 75/7c, box 11, folder 1950.

31. Litton to Robert Schmidt, April 29, 1948, in Charles Litton Papers, 75/7c, box 11, folder 1948; Norman Moore and Roy Woenne, interview by Norberg, December 19, 1973, Bancroft Library.

32. On the history of the microwave tube programs at Stanford, see Leslie 1992, 44–75 and 159–187; Lowen 1997. See also Charles Süsskind, "Electron tube Research at Stanford University," June 1953, in Frederick Terman Papers, SC 160, series II, box 15, folder 11.

33. J. Berkley to Litton, April 11, 1947 and Litton to Berkley, April 23, 1947—both in Charles Litton Papers, 75/7c, box 11, folder 1948; Paul Crapuchettes, "Development of K-band local oscillator," January 1948, Charles Litton Papers, 75/7c, carton 2, folder Tubes—K-band local oscillator L-3006; Litton to Bureau of Ships, April 30, 1948 and R. Berry to Litton, June 18, 1948—both in Charles Litton Papers, 75/7c, box 11, folder Letters written by Litton, 1948. For treatments of the electrical engineering department in the postwar period, see Lowen 1997, 110–112; Leslie 1992, 55–60.

34. On Stanford's high-power klystrons, see Chodorow and Ginzton, "Development of High Powered Pulsed Klystrons," June 1949, and Chodorow, Ginzton, I. Nielsen, and Simon Sonkin, "Design and Performance of a High Power Pulsed Klystron," Microwave Laboratory Report No. 212, September 1953—both in Edward Ginzton Papers, SC 330, box 2, folder Publications I, Archives and Special Collections, Stanford University; Chodorow, interview by McMahon, 1984, IEEE History Center.

35. Litton to R. Larson, April 22, 1950, Charles Litton Papers, 75/7c, box 11, folder 1950; Litton to Jason, June 30, 1953, Charles Litton Papers, 75/7c, box 12, folder Letters written by Litton, January-August 1953; Moore and Woenne, interview by Arthur Norberg.

36. For statistical data on the microwave tube industry in the late 1940s and the 1950s, see the US Department of Commerce's Census of Manufacture for 1947 and 1954; also see Ginzton, "Speech at Shareholders' Meeting," February 20, 1964, Edward Ginzton Papers, SC 330, box 14A, folder Speeches 1960–65, Archives and Special Collections, Stanford University. On the evolution of military procurement during the Korean War, see Weston 1960.

37. On Raytheon in the late 1940s and during the Korean War, see "Short History of the Division," June 10, 1964 and Percy Spencer, "Raytheon, November 15, 1925–June 1, 1959," January 25, 1960, both in Raytheon Archives, courtesy of Norman Krim; Krim to Lécuyer, February 27, 1997, collection of C. Lécuyer; Wihtol, interview by Lécuyer, March 6 and March 8, 1996; Getting 1989, 247–342.

38. Moore, interview by Norberg, 19. On Hughes Aircraft. which has not yet found its historian, see Lay 1969 and Byrne 1993.

39. Litton to Daniel Reed, June 2, 1953, in Charles Litton Papers, 75/7c, box 12, folder Letters written by Litton, January-August 1953; Moore, interview by Norberg, 19–20.

40. Moore and Woenne, interview by Norberg. For a list of Litton's production and preproduction contracts in early 1952, see Litton to Office of Chief Signal Officer, January 28, 1952, in Charles Litton Papers, 75/7c, box 12, folder Letters

written by Litton; "Type JAN 4J50 Magnetron—Contract Summary," April 23, 1953, in Charles Litton Papers, 75/7c, carton 1, folder Litton Industries miscellaneous.

41. Litton, Moore, Woenne, and Crapuchettes thoroughly redesigned the 4J50 and 4J52 before putting these tubes into production. To do so, they used the design methodology they had developed for the continuous-wave magnetrons. As a result, their versions of the 4J50 and 4J52 differed from those developed by other firms.

42. Litton to Diggle, February 12, 1952, in Charles Litton Papers, 75/7c, box 12, folder Letters written by Litton, January-April 1952.

43. Litton to Grantham, January 25, 1951, in Charles Litton Papers, 75/7c, box 12, folder Letters written by Litton, January-April 1952.

44. A. Mortimer (Amperex), "Trip to Litton Industries, San Carlos, California," July 19, 1951, Charles Litton Papers, box 5, folder Litton Engineering 1947–1968.

45. Ibid.; Ted Taylor, interview by Lécuyer, January 24, 1996, February 12, 1996, March 6, 1996, March 29, 1996, and April 18, 1996.

46. Litton to L. Bedell, December 24, 1947, Charles Litton Papers, 75/7c, box 11, folder Letters written by Litton, 1947.

47. Ibid.; Mortimer, "Trip to Litton Industries, San Carlos, California," July 19, 1951, Charles Litton Papers, box 5, folder Litton Engineering 1947–1968.

48. Moore to Office of Air Regional Representative, December 23, 1952, Charles Litton Papers, 75/7c, carton 1, folder Litton Papers.

49. Litton Industries also serialized its tubes and recorded detailed information on each of them during the manufacturing process. Litton to Grantham, February 18, 1952, Charles Litton Papers, 75/7c, box 12, folder Letters written by Litton, January-April 1952; Moore to Office of Air Regional Representative, December 23, 1952, in Charles Litton Papers, 75/7c, carton 1, folder Litton Papers; Moore, "Quality Control in Power Vacuum Tube Manufacture," November 18, 1953, Charles Litton Papers, 75/7c, carton 1, folder Litton Industries; Mortimer, "Trip to Litton Industries, San Carlos, California," July 19, 1951, Charles Litton Papers, box 5, folder Litton Engineering 1947–1968.

50. Litton, "Notice" [circa 1950], in Charles Litton Papers, 75/7c, carton 1, folder Litton Engineering Bulletin Board Notices.

51. Litton to Grantham, August 7, 1952, Charles Litton Papers, 75/7c, box 12, folder Letters written by Litton, May-November 1952; Marian Goodman, "Litton's Home on a Hilltop," *Redwood City Tribune*, May 9, 1963, collection 74-423, San Mateo County History Museum.

52. Litton to Grantham, June 11, 1952 and Litton to Grantham, December 29, 1952—both in Charles Litton Papers, box 12, folder Letters written by Litton, May-December 1952; Litton to Grantham, February 25, 1953, Charles Litton Papers,

75/7c, box 12, folder Letters written by Litton, January-August 1953; "Type JAN 4J50 Magnetron—Contract Summary," April 23, 1953, Charles Litton Papers, 75/7c, carton 1, folder Litton Industries miscellaneous; Litton to Admiral Fox, June 10, 1953, Charles Litton Papers, 75/7c, box 12, folder Letters written by Litton, January-August 1953; Moore and Woenne, interview by Norberg.

53. Litton to National Security Industrial Association, March 3, 1953, Litton to E. Metzger, April 16, 1953, Litton to Reed, June 2, 1953, Litton to Grantham, June 8, 1953, Litton to Fox, June 10, 1953—all in Charles Litton Papers, 75/7c, box 12, folder Letters written by Litton, January-August 1953; C. Fox to Litton, June 2, 1953, in Charles Litton Papers, 75/7c, box 9, folder US Navy Department; Litton to David Combs, October 5, 1966, Charles Litton Papers, 75/7c, box 13, folder Letters written by Litton, 1966–1971; Moore and Woenne, interview by Norberg.

54. Litton to Grantham, August 7, 1952 and Litton to Joseph Rand, May 23, 1952—both in Charles Litton Papers, 75/7c, box 12, folder Letters written by Litton, May-November 1952; Litton to Albert Jason, June 23, 1953, Charles Litton Papers, 75/7c, box 12, folder Letters written by Litton, January-August 1953; Litton to R. Fuchs, December 17, 1953, Charles Litton Papers, 75/7c, box 12, folder Letters written by Litton September-December 1953; Litton to Alphonse Dalton, September 27, 1954, Charles Litton Papers, 75/7c, box 12, folder Letters written by Litton, September-December 1954.

55. Litton to Grantham, February 25, 1953, in Charles Litton Papers, 75/7c, box 12, folder Letters written by Litton January-August 1953; Frederick Terman to Harold Laun, November 5, 1953, in Frederick Terman Papers, SC 160, series V, box 6, folder 13, Archives and Special Collections, Stanford University; Moore and Woenne, interview by Norberg.

56. Litton to Combs, October 5, 1966, Combs, October 5, 1966, Charles Litton Papers, 75/7c, box 13, folder Letters written by Litton, 1966–1971.

57. Litton to Ines Mabon, May 2, 1957, Charles Litton Papers, 75/7c, carton 13, folder Letters written by Litton 1957; Lay 1969; Byrne 1993; Rodengen 2000.

58. Dennis Robinson, "Visit to Nathan Levin's Office, September 10, 1957," in Dennis Robinson Papers, MC481, box 5, folder July-December 1957, MIT Archives and Special Collections; Tex Thornton to Litton, March 26, 1954, Charles Litton Papers, 75/7c, box 9, folder Charles Thornton; Lay 1969; Byrne 1993; Rodengen 2000; Myrl Stearns, interview by Lécuyer, June 13, 1996.

59. Litton to Grantham, May 5, 1954, Charles Litton Papers, 75/7c, box 12, folder Letters written by Litton, January-August 1954, box 12; Litton to Dalton, September 27, 1954, Litton to Grantham, October 6, 1954, and Litton to R. Huggins, December 30, 1954—all in Charles Litton Papers, box 12, folder Letters written by Litton, September-December 1954; Marian Goodman, "Litton's Home on a Hilltop," *Redwood City Tribune*, May 9, 1963, in collection 74-423, San Mateo County History Museum.

60. Moore, interview by Norberg, 28. See also Litton to Grantham, October 6, 1954, Charles Litton Papers, 75/7c, box 12, folder Letters written by Litton September-December 1954; Terman to William Cooley, April 5, 1955 in Frederick Terman Papers, SC 160, series V, box 7, folder 5, Stanford Archives and Special Collections; Litton to Thornton, February 7, 1956, in Charles Litton Papers, 75/7c, box 13, folder Letters written by Litton, 1956; Litton to Ines Mabon, May 2, 1957, Charles Litton Papers, Charles Litton Papers, 75/7c, carton 13, folder Letters written by Litton, 1957; Terman to William Cooley, June 5, 1961 in Frederick Terman Papers, SC 160, series V, box 7, folder 8, Stanford Archives and Special Collections.

61. Litton to Herman Kuthe, March 5, 1954, in Charles Litton Papers, 75/7c, box 12, folder Letters written by Litton, January-August 1954; "Litton, Aircraft Radio Merger OK Pending Holder Approval," *Electronic News*, October 14, 1957, 3; "Litton, Monroe Calculating Agree to Merger Terms," *Electronic News*, October 21, 1957, 7; Rodengen 2000; Ted Taylor, interview by Lécuyer, February 12, March 6, and March 29, 1996.

62. Jay Last to parents, no date (probably early February 1961), courtesy of Jay Last; Rodengen 2000, 30–33.

63. Litton to Combs, October 5, 1966, Charles Litton Papers, 75/7c, box 13, folder Letters written by Litton, 1966–1971.

64. In the early 1960s, Litton was on the board of Huggins Laboratories. See "Watkins-Johnson Acquires Control of Santa Cruz Firm," *Palo Alto Times*, February 28, 1963, collection of C. Lécuyer.

Chapter 3

1. Speech by undersecretary of commerce Robert Williams reproduced in the November 1957 issue of *Varian Associates Magazine*.

2. On Edward Ginzton's family background, see Ginzton and Cottrell 1995; Ginzton, interview by Sharon Mercer, 1–9, Varian Associates Oral History Project, M0708, Archives and Special Collections, Stanford University. On Myrl Stearns' background, see Myrl Stearns, interview by Sharon Mercer, 1–10, Varian Associates Oral History Project, M0708, Archives and Special Collections, Stanford University; communication from Myrl Stearns, June 13, 1996.

3. Ernest Harrison, "The Mission of the Temple," *Temple Artisan*, July 1908, 22–25, 25, Bancroft Library.

4. "Temple Home Association Notes," *Temple Artisan*, July 1907, 36, Bancroft Library.

5. As the community ran into substantial financial difficulties, it gradually introduced a modicum of private enterprise in its economic activities in the 1910s. "The Temple Home Association. Report and Message of the President Read at Last Annual Meeting," *Temple Artisan*, August 1911, 198–201; "The Temple Home

Association," *Temple Artisan*, May 1918, 331–332, both at the Bancroft Library. See also Hine 1981, 1983. For an example of John Varian's socialist writings, see John Varian, "A Letter on Socialism," *Temple Artisan*, August 1908, 48–50.

6. On the group's political outlook, see Ginzton and Cottrell 1995, 71–72; Varian 1983, 65–78, 146–150, 234–235; communication from Stearns, June 13, 1996.

7. Varian 1983, 65–95 and 135–168; Dorothy Varian, "Excerpts from RHV letters while employed by Humble Oil," Russell and Sigurd Varian Papers, SC345, series: Russell Varian, box 2, folder 7, Archives and Special Collections, Stanford University. On Philo Farnsworth's Television Laboratory, see Everson 1949 and Farnsworth 1989. On Ginzton's and Stearns' studies at Stanford, see Ginzton and Cottrell 1995, 67–69; Ginzton, interview by Mercer, 9–11; Ginzton, interview by Michal McMahon, 1984, IEEE History Center; Stearns, interview by Mercer, 10–12. On Frederick Terman's electronics program in the 1930s, see Frederick Terman, interview by Arthur L. Norberg, Charles Susskind, and Roger Hahn, 1984, Archives and Special Collections, Stanford University; Leslie 1993, 46–51; Leslie and Hevly 1985; Gillmor 2005.

8. Varian 1983, 96–135; "Don Snow Retires," *Varian Associates Magazine*, April 1961, 11, SC345, series: Varian Associates, box 7, folder 6, Archives and Special Collections, Stanford University; William Hansen to Ray Wilbur, July 20, 1939, folder 12, box 2, and Hansen to L. Applegate, folder 13, box 2, both in William Hansen Papers, SC 126, Archives and Special Collections, Stanford University; Russell Varian to Richard Jenkins, April 9, 1958, Russell and Sigurd Varian Papers, SC345, series: Russell Varian, box 1, folder 9, Archives and Special Collections, Stanford University; Marvin Chodorow, interview by Sharon Mercer, 1989, 7–9, Varian Associates Oral History Project, M0708, Archives and Special Collections, Stanford University.

9. Varian and Varian 1939; Russell Varian to John Mattil, April 18, 1948, Russell and Sigurd Varian Papers, SC 345, series: Russell Varian, box 1, folder 7; Hansen, "Statement of William Hansen," no date, Edward Ginzton Papers, SC330, box 19, folder statement of William Hansen; Sigurd Varian, hand-written notes on the development of the klystron, no date, Russell and Sigurd Varian Papers, SC345, series: Sigurd Varian, box 1, folder 23; Frederick Terman, "Events Associated with the Invention of the Klystron Tube," October 20, 1958, Russell and Sigurd Varian Papers, SC345, series: Russell Varian, box 5, folder 12; Ginzton 1975; Varian 1983, 171–218. See also Galison et al. 1992; Hevly 1994.

10. Ginzton and Cottrell 1995; Hansen to Wilbur, July 20, 1939, William Hansen Papers, SC 126, box 2, folder 12, Stanford Archives and Special Collections; Bloch 1952.

11. The klystron department of the Sperry Gyroscope Company was small in the immediate postwar period. It sold $153,006 worth of klystrons in 1946. These sales grew to $294,819 in 1947 and $476,941 in 1948. Ginzton and Cottrell 1995, 81–86; Ginzton, interview by Mercer, 18–27; Ginzton, interview by Henry Lowood, Peter

Galison, and Bruce Hevly, February 3, 1988, 2–5; Stearns, interview by Lécuyer, November 25, 1996; Varian 1983, 221–233.

12. On microwave radar during the war, see Guerlac 1987 and Buderi 1996.

13. R. Wathen, "The Sperry Company and Research," March 15, 1944, Sperry Gyroscope collection, 1915, box 37, folder 25, Hagley Museum and Library; Sperry Gyroscope, "Klystrons and Accessories," 1946, Sperry Gyroscope collection, 1915, box 5, Hagley Museum and Library; Ginzton, interview by Henry Lowood, Peter Galison, and Bruce Hevly, February 3, 1988, 2–5; Chodorow, interview by Mercer, 8–9; Stearns, interview by Mercer, 20–25; Stearns, interview by Lécuyer, November 25, 1996.

14. R. Wathen, "The Sperry Company and Research," March 15, 1944, Sperry Gyroscope collection, 1915, box 37, folder 25, Hagley Museum and Library; Sperry Gyroscope, "Klystrons and Accessories," 1946, Sperry Gyroscope collection, 1915, box 5; Ginzton, interview by Mercer, 22–23; Ginzton, interview by Henry Lowood, Peter Galison, and Bruce Hevly, February 3, 1988, 2–5; Chodorow, interview by Mercer, 8–9; Stearns, interview by Mercer, 20–25; Stearns, interview by Lécuyer, November 25, 1996.

15. On the reverse migration of western engineers, see Arnold Wihtol, interview by Lécuyer, March 5 and March 8, 1996. On the group's interest in returning to California, see Varian 1983, 230–233 and 237–238; Ginzton and Cottrell 1995, 92–93; Stearns, interview by Mercer, 25; communication from Stearns, June 13, 1996.

16. Chodorow, interview by Mercer, 46. See also Russell Varian, "Ten Years Later . . . A New Industry Is Born," *Varian Associates Magazine*, September 1958, Russell and Sigurd Varian Papers, SC 345, series: Varian Associates, box 7, folder 4.

17. Russell Varian to James Luck, November 5, 1946, Russell and Sigurd Varian Papers, SC 345, series: Russell Varian, box 1, folder 6.

18. Ginzton, interview by Mercer, 27. On the group's "business plan," see Chodorow, interview by Mercer, 42–47; Stearns, interview by Mercer, 24–26.

19. Ginzton, interview by Mercer, 30.

20. Russell Varian, "Outline of Talk before Shareholders' Meeting," December 3, 1951, Russell and Sigurd Varian Papers, series: Russell Varian, box 41, folder 28, SC 345.

21. Ginzton, interview by Mercer, 85. Russell Varian, the most radical member of the group, went as far as promoting the formation of an industrial and agricultural commune centered around the new firm.

22. Russell Varian, "Company Philosophy, Company Objectives, Management Functions" [circa 1953], Edward Ginzton Papers, SC330, box 3, folder Varian Associates. On the General Radio Company, see Thiessen 1965 and Sinclair 1965.

23. Varian, "Ten Years Later," 10.

24. "Varians of Klystron Fame to Set up Lab in San Carlos," *Palo Alto Times,* July 2, 1948, newspaper clippings collection, folder Varian Associates, Palo Alto Historical Association.

25. On the Microwave Laboratory, see Hansen's accelerator project, see Galison et al. 1992; Bloch 1952; Leslie 1992. On Terman's support of the new venture, see Ginzton, interview by Michael McMahon, 1984, IEEE History Center.

26. Stearns, interview by Mercer, 39.

27. Stearns, Ginzton, and the Varians did not get founders' stock when they started Varian Associates. As a result, they owned a relatively small share of the company's stock.

28. Ginzton also acted as a go-between between Stearns and the Varian brothers who had a profound distrust of professional managers and worried that Stearns would take control of the company. On Varian's ownership policies, "Stock Option Agreement," November 30, 1948, Russell and Sigurd Varian Papers, SC 345, series: Varian Associates, box 5, folder 11; Russell Varian, "Report on Stock Policies," May 27, 1952, Russell and Sigurd Varian Papers, SC 345, series: Varian Associates, box 5, folder 11; Russell Varian, "Ideas Concerning Varian Associates' Policy in Sale of Stock," circa 1952, Edward Ginzton Papers, SC 330, box 8, folder 1948—organization of Varian Associates; minutes of executive committee, May 25, 1949; Russell Varian, "To the Shareholders of Varian Associates," April 6, 1950, both in Edward Ginzton Papers, SC 330/95–179, box 17, volume 2; Varian 1983, 255–256. On stock subscriptions, see "Varian Associates, our Growth," May 17, 1951, Russell and Sigurd Varian Papers, SC 345, series: Varian Associates, box 3, folder 17.

29. Stearns, interview by Mercer, 27. On Stearns' tour of government laboratories in the summer of 1948, see minutes of special meeting of board of directors of Varian Associates, September 27, 1948, Edward Ginzton Papers, SC 330/95–179, box 17, volume 2. For the brochure sent by Varian to military agencies, see "History, Personnel, and Plant Facilities of Varian Associates," 1948 and "Varian Associates, Microwave Engineering," 1948—both in Russell and Sigurd Varian Papers, SC 345, series: Varian Associates, box 3, folder 13.

30. The group also solicited contracts to develop Doppler radars and traveling-wave tubes from the Air Force and the Naval Research Laboratory. See Russell Varian to Airborne Instruments Laboratory, July 22, 1948 and Russell Varian to P. Hagan, September 30, 1948—both in Russell and Sigurd Varian Papers, SC 345, series: Varian Associates, box 2, folder 11.

31. Chodorow, interview by Mercer, 17.

32. Minutes of special meeting of board of directors of Varian Associates, September 27, 1948, Edward Ginzton Papers, SC 330, box 17, volume 2, Archives and Special Connections, Stanford University; Varian, "Ten Years Later"; Chodorow, interview by Mercer 1989, 16–18; Ginzton, interview by Mercer.

33. Ginzton, interview by Mercer, 32.

34. This contract seems to have attracted little interest from most large East Coast firms because of its technical difficulties and its low overhead (7%). In the late 1940s and the early 1950s, Varian received $1,492,638 from DOFL and the National Bureau of Standards for this contract. Minutes of Special meeting of board of directors of Varian Associates, September 27, 1948, Edward Ginzton Papers, SC 330/95 179, box 17, volume 2; Varian, "Ten Years Later," in Varian Associates, *Varian Associates Magazine*, September 1958, series: Varian Associates, Russell and Sigurd Varian Papers, SC345, box 7, folder 4; Chodorow, interview by Mercer, 17; Stearns, interview by Lécuyer, June 13, 1996; Ginzton, interview by Mercer, 32.

35. Sigurd Varian, Notes, July 1, 1958, in Russell and Sigurd Varian Papers, SC 345, series: Sigurd Varian, box 2, folder 3.

36. Sigurd Varian, Notes, July 1, 1958, in Russell and Sigurd Varian Papers, SC 345, series: Sigurd Varian, box 2, folder 3.

37. Chodorow, interview by Mercer, 17.

38. The R-1 was a fixed frequency reflex klystron and it had no tuning mechanism. It also operated in the K band. On the R-1, see Arnold Wihtol, interview by Lécuyer, March 8, 1996; Ted Moreno, interview by Lécuyer, July 19, 1996; Stearns, interview by Lécuyer, November 25, 1996; communication from Stearns, June 13, 1996. On Varian's early hirings, see Stearns, interview by Lécuyer, June 13, 1996; Cliff Gardner, interview by Sharon Mercer, 4–5, Varian Associates Oral History Collection, M 708.

39. On process innovations at Varian in the late 1940s, see Wihtol, interview by Lécuyer, March 8, 1996; Moreno, interview by Lécuyer, May 19, 1996. For an history of ceramic-to-metal sealing techniques, see Kohl 1960, 472–477

40. Wihtol, interview by Lécuyer, March 8, 1996; Moreno, interview by Lécuyer, May 19, 1996.

41. Withol, interview by Lécuyer, March 8, 1996.

42. Sigurd Varian, Notebook, entry for November 21, 1949, Russell and Sigurd Varian Papers, SC 345, series: Sigurd Varian, box 1, folder 14; "The Varian Honeycomb Grids," *Varian Associates Newsletter* (October–November 1951), 1, Russell and Sigurd Varian Papers, SC 345, series: Varian Associates, box 13, folder 7; Sigurd Varian and Russell Varian, "Method of Making a Grid Structure," Patent 2,619,438, filed April 16, 1945, and granted Nov. 25, 1952; Wihtol, interview by Lécuyer, March 8, 1996; Moreno, interview by Lécuyer, July 19, 1996.

43. On GE's concentration on television in the immediate postwar period, see Chet Lob, interview by Lécuyer, April 9, 1996. On the UHF tube project at Varian, see "Varian Associates-General Electric TV Transmitter," *Varian Associates Newsletter*, March 1951, 1–2, Russell and Sigurd Varian Papers, SC 345, series: Varian

Associates, box 13, folder 7; Norman Hiestand, "Development Program for G.E. Tubes," February 21, 1952, Edward Ginzton Papers, SC 330, box 3, folder Ginzton—not classified; "The History of High Power Tubes," *Varian Associates Magazine*, February 1961, 8–15, SC 345, series: Varian Associates, Russell and Sigurd Varian Papers, box 7, folder 6; Stearns, interview by Mercer, 37–39.

44. Richard Nelson, interview by Sharon Mercer, 1990, 4, Varian Oral History Collection, M 708.

45. In October 1951, GE asked Varian to develop a 15 KW klystron for UHF television transmitters. 300 of these tubes were made by Varian and GE between 1952 and 1954. In 1954, GE canceled the contract because of lower than anticipated demand for UHF transmitters. Sigurd Varian, Notebook, entries for June 27 1949, July 22, 1949, March 20, 1950, and April 17, 1950, Russell and Sigurd Varian Papers, SC 345, series: Sigurd Varian, box 1, folder 14; The Engineering Staff of Varian Associates, "High Power UHF-TV Klystron," *Electronics*, October 1951, 30–32; Chodorow to Ginzton, June 21, 1978, folder Marvin Chodorow, box 1, Edward Ginzton Papers, SC 330/95–179; Stearns, interview by Mercer, 37–39.

46. Varian Associates also received an important contract from Philco for the design of klystrons for microwave communication systems. Chodorow, interview by Mercer, 18.

47. On radar research in the postwar period, see Getting 1962 and Bryant 1994.

48. On the military interest in klystrons for Doppler radar, see Getting 1962, 86. On the Department of Defense's emphasis on tube reliability in the early 1950s, see Army Air Force Air Technical Service Command, "Electron Tube Branch Research and Development Program, 1946–1950," January 1946, folder US Department, Miscellaneous, box 9; F. R. Fuerth, "Talk given to the luncheon meeting of the Philadelphia Chapter of the Armed Forces Communication Association," February 10, 1953, folder US Navy Department, box 9; second issue of the *Electronics Applications Reliability Review*, December 2, 1953, folder Litton Industries, carton 1—all in Charles Litton Papers, 75/7c, Bancroft Library.

49. Richard Leonard, "Certain Salary Changes," no date, circa 1951, Varian Associates Records, MSS 73/65c, folder Financial Statements and Misc. (1949–September 1952), Bancroft Library. For Varian's decision to expand, see the introduction to "A Brief History of Varian Associates," circa 1951, Russell and Sigurd Varian Papers, SC 345, series: Varian Associates, box 3, folder 13.

50. "Myrl Explores Washington," *Varian Associates Newsletter*, September/October 1950, 3, Russell and Sigurd Varian Papers, SC 345, series: Varian Associates, box 13, folder 7; minutes of special meeting of board of directors, June 3, 1951, Edward Ginzton Papers, SC 330, box 17, volume 2; Defense Production Administration to Reconstruction Finance Corporation, August 8, 1951, Edward Ginzton Papers, SC 330/95–179, box 3, folder Varian Associates, not classified; Stearns, interview by Mercer, 29–30; "Test Equipment Sale to Hewlett-Packard," *Varian Associates Newsletter*, December 50/January 51, 5, Russell and Sigurd Varian Papers, SC 345,

series: Varian Associates, box 13, folder 7. On Varian's move to Stanford and the establishment of the Stanford Industrial Park, see Ginzton, interview by Mercer, 39–40; Lowood 1989.

51. Wihtol. "Progress Report," October 9, 1952, courtesy of Arnold Wihtol.

52. Wihtol, interview by Lécuyer, April 9, 1996.

53. "Organization, Varian Associates," January 11, 1952, Russell and Sigurd Varian Papers, SC 345, series: Varian Associates, box 4, folder 22; Stearns, interview by Lécuyer, November 25, 1996; Moreno, interview by Lécuyer, July 19, 1996.

54. Russell Varian, "Recent Developments in Klystrons," *Electronics*, April 1952, in Russell and Sigurd Varian Papers, SC 345, series: Russell Varian, box 5, folder 25; Cliff Rockwood, "Varian Associates and External Cavity Reflex Klystrons," *Varian Associates Newsletter*, March 1956, 1 and 4, Russell and Sigurd Varian Papers, SC 345, series: Varian Associates, box 13, green volume; minutes of regular meeting of board of directors, December 1, 1952, Russell and Sigurd Varian Papers, SC 345, series: Russell Varian, box 6, brown volume; Moreno, interview by Lécuyer, July 19, 1996.

55. On Stanford's high-power klystrons, see Chodorow and Ginzton, "Development of High Powered Pulsed Klystrons," June 1949, and Chodorow, Ginzton, I. Nielsen, and Simon Sonkin, "Design and Performance of a High Power Pulsed Klystron," Microwave Laboratory Report No. 212, September 1953—both in Edward Ginzton Papers, SC 330, box 2, folder Publications I; Ginzton and Cottrell 1995, 107–117; Ginzton, interview by Lowood, Galison, and Hevly, 27–30 and 47–52; Chodorow, interview by McMahon.

56. Chodorow, interview by Mercer, 23–24.

57. "The Merger of Varian Associates and Eitel-McCullough, Inc." July 9, 1965, in Edward Ginzton Papers, SC 330, box 12, folder Eimac merger/Department of Defense; Chodorow to Ginzton, June 21, 1978, Edward Ginzton Papers, SC 330/95–179, box 1, folder Marvin Chodorow; Preist 1992, 12–13; Chodorow, interview by Mercer, 21–24; Stearns, interview by Lécuyer, June 13, 1996.

58. Varian was unique among electronics firms on the Peninsula for its emphasis on the hiring and training of black workers and technicians. These men were then often hired by other electronics corporations in the area that wanted to have their "token blacks." For Varian's policies regarding minorities, see Stearns, interview by Lécuyer, June 13, 1996. On Varian's entry into klystron production, see Elliott Levinthal, "Memorandum for Expansion Planning Committee," May 23, 1951; Lloyd Sorg, "Plant Expansion Committee Meeting Agenda," May 23, 1951; "Expansion Planning Committee, Minutes of Second Meeting," May 24, 1951— all in Russell and Sigurd Varian Papers, SC 345, series: Russell Varian, box 3, folder 17; Emmet Cameron, interview by Sharon Mercer, 1989, Varian Associates Oral History Collection, M 708.

59. Minutes of executive committee, December 2, 1952, Russell and Sigurd Varian Papers, SC 345, series: Russell Varian, box 6, folder executive committee.

60. Wihtol, interview by Lécuyer, March 8, 1996.

61. The MAB was patterned after a similar committee at McCormick, a spice company in New York. The Management Advisory Board was later rescinded by the National Labor Relations Board because it violated labor laws. On Varian's team approach to manufacturing, see Wihtol, interview by Lécuyer, March 8, 1996. On Varian's Management Advisory Board, see minutes of Executive Committee, November 11 and December 2, 1952, February 17, April 14, May 12, May 26, June 16, August 25, September 29, October 27, and November 17, 1953—all in Russell and Sigurd Varian Papers, series: Russell Varian, box 6, folder executive committee; "What Is the MAB?" (minutes of management advisory board, April 1953); "Management Advisory Board Chairman's Report, Second Term, July to December, 1953"; "MAB Box Score," January 5, 1954"; "Management Advisory Board Chairman's Report, Fourth Term, July 1 to December 31, 1954"—all in Russell and Sigurd Varian Papers, series: Varian Associates, box 4, folder 12; "New Board Formed to Study VA Problems," *Varian Associates Newsletter,* January 1953, Russell and Sigurd Varian Papers, series: Varian Associates, box 8, folder 13; Moreno, interview by Lécuyer, July 19, 1996; Wihtol, interview by Lécuyer, March 8, 1996.

62. Wihtol, interview by Lécuyer, March 8, 1996.

63. Ibid.

64. On Varian's clean room, see "What's in the Goetz Building?" *Varian Associates Newsletter,* July 1951, Russell and Sigurd Varian Papers, series: Varian Associates, box 13, green volume; Wihtol, interview by Lécuyer, March 8, 1996. On tube production rates during the Korean War, see "Varian Associates, Our Growth," May 17, 1951, Russell and Sigurd Varian Papers, series: Varian Associates, box 3, folder 17; minutes of regular meeting of board of directors, December 1, 1952, Russell and Sigurd Varian Papers, series: Russell Varian, box 6, folder executive committee.

65. For data on Sperry's, RCA's, and Raytheon's sales in the early 1950s, see "Royalty from Domestic Licensees," January 1, 1955 and Ginzton, "Klystron Royalties," May 14, 1954—both in Frederick Terman Papers, SC 160, series III, box 34, folder 7; Wihtol, "Progress Report," October 9, 1952, courtesy of Arnold Wihtol.

66. The Department of Defense did not know that Stearns belonged to a family of IWW organizers. According to Stearns, he had "forgotten" to tell the military about it when he filled out his petition to obtain a security clearance at Sperry in 1941. On Varian's security problems in the summer and fall of 1953, see Stearns, "Facility Clearance Denial—INM Action," October 7, 1953, Varian Associates Records, 73/65c, carton 1, folder corporate records, 1953, Bancroft Library; Ginzton and Cottrell 1995, 98–99; Stearns, interview by Mercer, 41–45.

67. Stearns, interview by Mercer, 42.

68. "Trip Schedule for Francis Farquhar and Garfield Merner, September 2–5, 1953," Varian Associates Records, 73/65c, carton 1, folder corporate records 1953, Bancroft Library; Ginzton and Cottrell 1995, 98–99; Stearns, interview by Mercer, 41–45; Ginzton, interview by Mercer, 45–46.

69. On the evolution of military procurement in the second half of the 1950s and the early 1960s, see *Electronics Industry Association Yearbook*, 1964, courtesy of Jay Last.

70. Department of Defense procurement of magnetrons fell from $45 million in 1954 to $39 million in 1962. For statistical data on the microwave tube industry in the second half of the 1950s, see Ginzton, "Speech at Shareholders' Meeting," February 20, 1964, Edward Ginzton Papers, SC 330, box 14–A, folder speeches 1960–1965.

71. *Electronics Industry Association Yearbook*, 1964, courtesy of Jay Last. On radar engineering in the 1950s, see Buderi 1996, 354–420.

72. "Brief History of Eimac's Microwave Tube Business" [circa 1965], Eitel-McCullough Records, 77/110c, box 2, folder Varian, Bancroft Library; "Report on the Proposed Acquisition of Company X," April 9, 1965, Varian Associates Records, 73/65c, carton 3, folder Report on Proposed Acquisition of Company X.

73. Fred Speaks, "Mid-Year Review," July 16, 1959, Eitel-McCullough Records, 77/110c, box 6, folder Schwarz, Adolph litigation; "Brief History of Eimac's Microwave Tube Business," Eitel-McCullough Records, 77/110c, box 2, folder Varian; William Eitel, "Electronics Considered Pace-Setter in Region's Development," *Redwood City Tribune*, December 27, 1962, in collection 64–94, San Mateo County History Museum.

74. Frank Stoner, "A General Survey of the US Electronic Industry," August 19, 1954 and Stoner to Stearns, November 15, 1955—both in Varian Associates Records, 75/65c, box 1, folder corporate records.

75. Emmet Cameron, "V.A. Value Points," June 11, 1958, Russell and Sigurd Varian Papers, SC 345, series: Varian Associates, box 2, folder 14.

76. Stoner to Leslie Hoffman, July 16, 1958, Russell and Sigurd Varian Papers, SC 345, series: Russell Varian, box 2, folder 18.

77. "Varian Associates Catalog, 1960," business ephemera collection, Jackson library, Stanford University. On VacIon pump technology and its use in klystron design and manufacture, see "The VacIon: Varian Introduces Revolutionary New High Vacuum Pump," *Varian Associates Magazine*, March 1958, 4–6; "Varian VacIon Pumps and High Vacuum Accessories," circa 1960, Varian Associates Records, 73/65c, carton 3, folder catalogs; "Vacuum Development," *Varian Associates Magazine*, June 1978, 16–17, in Russell and Sigurd Varian Papers, series: Varian Associates, box 9, folder 4; Moreno, interview by Mercer; Taylor, interview by Lécuyer. For a discussion of VacIon pump technology and its commercial exploitation, see chapter 5.

78. "Brief History of Eimac's Microwave Tube Business," Eitel-McCullough Papers, 77/110c, box 2, folder Varian; Preist 1992; Ted Taylor, interview by Lécuyer, 1996.

79. "The Largest Klystron," *Varian Associates Magazine,* August 1960, 6, Russell and Sigurd Varian Papers, SC 345, series: Varian Associates, box 7, folder 75; "Varian Associates Catalog, 1960," business ephemera collection, Jackson library, Stanford University.

80. General Accounting Office, "Examination of catalog prices charged for kly-stron tubes under noncompetitive procurements negotiated by the military depart-ments and their prime contractors with Varian Associates, Palo Alto, California," January 15, 1963, Bancroft Library; Stearns, interview by Lécuyer; communication from Chester Lob, January 22, 1997.

81. Denis Robinson, notebook entry for April 16, 1954, in Denis Robinson Papers, MC 481, box 14, notebook 59, MIT Archives and Special Collections; Cameron, interview by Mercer, 4. For Varian's stock policies, see "Announcement on Stock Policy," April 14, 1955, Russell and Sigurd Varian Papers, series: Varian Associates, box 5, folder 12; Russell Varian to all members of the staff, June 13, 1955, Russell and Sigurd Varian Papers, SC 345, series: Varian Associates, box 4, folder 14; Stearns, "Some Dreams Come True," *Varian Associates Newsletter,* August 1957, 2–3, Russell and Sigurd Varian Papers, SC 345, series: Varian Associates, box 7, folder 2. On Varian's estimates of its capital needs, see minutes of regular meeting of board of directors, February 17, 1955, Edward Ginzton Papers, SC 330, box 17, volume 3. On its capital investments, see minutes of regular meeting of board of directors, May 11, 1956, Edward Ginzton Papers, SC 330, box 17, volume 4.

82. On the rise in the value of Varian's stock, see "Varian Associates Stock Price Compared to Dow Jones Average," Varian Associates Records, 73/65c, box 4, folder Minutes Board of Directors 1960.

83. The establishment of the profit-sharing program might have been accelerated by a failed organization drive in 1957. On Varian's management interest in main-taining the firm's cooperative character, see Russell Varian, "Announcement on Stock Policy," April 14, 1955, Russell and Sigurd Varian Papers, SC 345, series: Varian Associates, box 5, folder 12; Russell Varian to all members of the staff, June 13, 1955, Russell and Sigurd Varian Papers, SC 345, series: Varian Associates, box 4, folder 14. For Varian's stock purchase plan, see minutes of regular meeting of Board of Directors, September 16, 1955, volume 4; minutes of regular meeting of board of directors, October 14, 1955, volume 5—both in Edward Ginzton Papers, SC 330/95–179, box 17; "Basic Objectives and General Policies for Varian Associates," June 6. 1957, Russell and Sigurd Varian Papers, SC 345, series: Varian Associates, box 4, folder 13. For the firm's profit-sharing program, see Russell Varian to all employees, November 26, 1957, folder 15, box 4, series: Varian Associates, Russell and Sigurd Varian Papers. For the firm's benefits, see minutes of regular meeting of board of directors, August 29, 1957, Edward Ginzton Papers, SC 330 95–179, box 18, volume 6; Emmet Cameron, "Another Milestone: First Cash Profit Sharing Payments for Varian People," *Varian Associates Newsletter,* November 1958, 6–7, Russell and Sigurd Varian Papers, series: Varian Associates,

box 7, folder 3. On operators' and machinists' attitude toward stock purchase, see Moreno, interview by Lécuyer, July 19, 1996.

84. The Air Force granted $1,205,712 to Varian to buy new machinery for its Palo Alto plant in October 1956. Russell Varian, "Some Considerations on Organization," June 29, 1953 and "Effective Organization of Varian Associates," no date, both in Russell and Sigurd Varian Papers, SC 345, series: Varian Associates, box 2, folder 9; Ginzton, "Considerations Concerning the Development and Location of the Tube Manufacturing Facility," November 16, 1955, Varian Associates Records, 73/65c, box 1, folder corporate records 1955; "The Job Shop: Where Tubes Are Built from Ideas," *Varian Associates Newsletter,* July 1956, 9, Russell and Sigurd Varian Papers, SC 345, series: Varian Associates, box 7, folder 1; minutes of regular meeting of board of directors, October 11, 1956, Edward Ginzton Papers, SC 330/95–179, box 17, volume 5; Wihtol, interview by Lécuyer, April 3 and April 9, 1996.

85. Varian's tubes were used in every missile weapon system and constituted the backbone of SAGE, DEW, and BMEWS air defense systems.

86. On the evolution of Raytheon's sales, see Norman Krim to Lécuyer, February 20, 1997, collection of C. Lécuyer. On Varian's joint venture with Thomson-Houston, see Stearns 1963.

87. Terman, "Talk at Shareholders' Meeting," *Varian Associates Newsletter,* December 1951, 1, Russell and Sigurd Varian Papers, SC 345, series: Varian Associates, box 13, folder 7.

88. Richard Nelson, interview with Sharon Mercer, 5, Varian Oral History Collection, M 708.

89. Victor Grinich, interview by Lécuyer, February 7, 1996, April 23, 1996, and May 14, 1996.

90. Julius Blank, interview by Lécuyer, June 20, 1996; Taylor, interview by Lécuyer, March 6, 1996.

91. Taylor, interview by Lécuyer, March 6, 1996.

Chapter 4

1. An earlier version of this chapter, titled "Fairchild Semiconductor and Its Influence," was published in *The Silicon Valley Edge: A Habitat for Innovation and Entrepreneurship,* ed. C.-M. Lee et al. (Stanford University Press, 2000). For heroic accounts of Robert Noyce, see Wolfe 1983; Malone 1985; Berlin 2001a,b. On the role of the military in shaping semiconductor technology, see Braun and MacDonald 1982.

2. Gordon Moore, interview by Ross Bassett and Lécuyer, February 18, 1997; Jean Hoerni, interview by Lécuyer, February 4, 1996; Sheldon Roberts, interview by Lécuyer, July 6, 1996; Jay Last, interview by Lécuyer, April 1, 1996.

3. Eugene Kleiner, interview by Lécuyer, May 21, 1996; Julius Blank, interview by Lécuyer, June 20, 1996; Victor Grinich, interview by Lécuyer, February 7, 1996. Fairchild Semiconductor's founders came from a variety of social and national backgrounds. Many were from lower-middle-class or working-class families. Noyce was the son of a clergyman and Moore the son of a sheriff. Grinich had grown up in a working-class family in the Pacific Northwest. Last's parents were high school teachers, while Roberts' family had a small patent medicine business. The other two members of the group were of European origin. Kleiner, the son of a Jewish shoe manufacturer in Austria, had fled to the United States in the late 1930s. A French Swiss from a banking family, Hoerni had moved to the United States to do his postdoctoral work.

4. Roberts, interview by Lécuyer, July 6, 1996.

5. Moore—a native of Pescadero, a small farming town on the San Francisco Peninsula—was interested in going back to California. Blank and Last had visited California during the war and in the late 1940s and wanted to live there. Hoerni, who lived in Pasadena, wanted to stay in California. See Moore, interview by Bassett and Lécuyer, February 18, 1997; Last, interview by Lécuyer, April 1, 1996; Hoerni, interview by Lécuyer, February 4, 1996; Blank, interview by Lécuyer, June 20, 1996; communication from Jim Gibbons, March 10, 1999.

6. On Shockley Semiconductor, see Riordan and Hoddeson 1997. On semiconductor research at the Bell Telephone Laboratories, see Hoddeson 1981; Smits 1985.

7. Moore, interview by Allen Chen, July 9, 1992, Intel Museum.

8. Sah et al. 1957; L. N. Durya, "Fairchild Investigation," May 28, 1959, William Shockley Papers, SC222 95–163, box B4, folder Fairchild Info July 3, 1959, Stanford Archives and Special Collections; Grinich, interview by Lécuyer, February 7, 1996; Moore, interview by Bassett and Lécuyer, February 18, 1997; Roberts, interview by Lécuyer, July 6, 1996; Hoerni, interview by Lécuyer, February 4, 1996. On the Bell system's first double diffused mesa silicon transistor, see Miller 1958.

9. Hoerni, interview, February 4, 1996.

10. Last to Vic Jones, September 20, 1957, courtesy of Jay Last; Robert Noyce, interview by Herb Kleiman, November 18, 1965, Herb Kleiman collection, M827, Stanford Archives and Special Collections; Hoerni, interview by Lécuyer, February 4, 1996; Roberts, interview by Lécuyer, July 6, 1996; Moore, interview by Ross Bassett and Christophe Lécuyer, February 18, 1997; Kleiner, interview by Lécuyer, May 21, 1996; Last, interview by Lécuyer, April 1, 1996; Moore, interview by Allen Chen, July 9, 1992, Intel Museum.

11. Grinich, interview by Lécuyer, February 7, 1996; Hoerni, interview by Lécuyer, February 4, 1996; Moore, interview by Allen Chen, July 9, 1992, Intel Museum; communication from Jim Gibbons, March 16, 1999.

12. The group also criticized the fact that aside from Jim Gibbons Shockley Semiconductor had no marketing staff. Grinich, interview by Lécuyer, February 7, 1996; Hoerni, interview by Lécuyer, February 4, 1996; Roberts, interview by Lécuyer, July 6, 1996; William Shockley, notebook, entry for June 5, 1957, William Shockley Papers, SC222 95–153, box B4, Shockley, notebook, entry for June 3, 1957, William Shockley Papers, SC222 95–153, box B4.

13. Last to parents, July 7, 1957, courtesy of Jay Last; Kleiner, interview by Lécuyer, May 21, 1996; Moore, interview by Bassett and Lécuyer, February 18, 1997; Roberts, interview by Lécuyer, July 6, 1996; Moore, interview by Allen Chen, July 9, 1992, Intel Museum.

14. Eugene Kleiner to Hayden, Stone, and Company, June 14, 1957, courtesy of Jay Last. This letter was written by Eugene Kleiner and his wife Rose.

15. Kleiner, interview by Lécuyer, May 21, 1996; "Fairchild Camera and Instrument Corporation" (New York: Hayden, Stone & Co., 1958), 3, courtesy of Jay Last; Wilson 1985, 31–34; Anthony Perkins, interview with Arthur Rock, *The Red Herring* (March 1994), 54–59, 54. Wall Street firms and individual investors discovered the electronics industry in the mid 1950s, which led to a speculative boom in electronics stocks and to the emergence of a new literature on high-tech investing. See Grant Jefferey, *Science and Technology Stocks: A Guide for Investors* (Cleveland: Meridian Books, 1961); Arthur Merrill, *Investing in the Scientific Revolution: A Serious Search for Growth Stocks in Advanced Technology* (Garden City: Doubleday & Company, 1962).

16. Last to parents, July 7, 1957, courtesy of Jay Last; Last, interview by Lécuyer, April 1, 1996; Hoerni, interview by Lécuyer, February 4, 1996; Roberts, interview, July 6, 1996; Moore, interview by Allen Chen, July 9, 1992, Intel Museum; Anthony Perkins, interview with Arthur Rock, *The Red Herring* (March 1994), 54–59, 54.

17. Only one firm, possibly Lockheed, expressed an interest in Hayden Stone's proposal but backed down because of internal conflicts. The reasons for Eitel-McCullough's lack of interest in this proposal are not clear. One can speculate that Bill Eitel and Jack McCullough turned the deal down, because they were already investing heavily in the klystron business and the building of a new manufacturing plant on the San Francisco Peninsula. Noyce, interview by Herb Kleiman, November 18, 1965, Herb Kleiman collection, M827, Stanford Archives and Special Collections; Moore, interview by Allen Chen, July 9, 1992, Intel Museum; Wilson 1985, 31–34; Perkins, interview with Rock, 54; Hayden, Stone & Co, *Fairchild Camera and Instruments*, February 1958.

18. Prospectus: Fairchild Camera and Instruments, January 25, 1960, courtesy of Eugene Kleiner; Fairchild Camera and Instruments, Annual Report, 1957, collection of C. Lécuyer; Richard Hodgson, interview by Rob Walker, September 19, 1995, Silicon Genesis, M471, Archives and Special Collections, Stanford University; Nelson Stone, interview by Lécuyer, April 21, 1995; Thomas Bay, interview by Lécuyer, July 2, 1996.

19. Hayden, Stone & Co, *Fairchild Camera and Instruments*, February 1958, courtesy of Jay Last.

20. Fairchild Camera and Instruments, *Annual Report*, 1957. On Fairchild Camera and Instruments, see Hayden, Stone & Co, *Fairchild Camera and Instruments*, February 1958, courtesy of Jay Last; A. Coyle to New York Times, April 26, 1998, courtesy of Jay Last; Last to parents, July 7, 1957, courtesy of Jay Last; Bay, interview by Lécuyer, July 2, 1996; Stone, interview by Lécuyer, April 21, 1995; Hodgson, interview by Rob Walker, September 19, 1995, in Silicon Genesis, M471, Archives and Special Collections, Stanford University.

21. Noyce, who knew of the group's efforts to establish their own firm, was persuaded by Roberts and possibly Kleiner to join the group a few hours before the contract's signature. Roberts, interview with Lécuyer, July 6, 1996.

22. Fairchild Camera also controlled the voting trust in which all Fairchild Semiconductor's shares were deposited.

23. Contract among the California Group, Parkhurst (Hayden Stone), Fairchild Controls, and Fairchild Camera and Instruments, September 23, 1957, William Shockley Papers, SC222 95–153, box B4.

24. Lécuyer 2005; Terman to William Shockley, September 20, 1955, Frederick Terman Papers, SC160, series III, box 48, folder 8; Linvill to R. Wallace, September 20, 1955, Frederick Terman Papers SC160, series III, box 48, folder 8; Linvill, "Excerpts from Lecture for Stanford Alumni in Los Angeles: The New Electronics of Transistors," February 15, 1956, in historical files of the *Campus Report*, folder John Linvill; Linvill, "Application for Equipment Funds for Semiconductor Device Research," February 25, 1957, in historical files of the *Campus Report*, folder John Linvill; Terman to William Cooley, March 6, 1958, Frederick Terman Papers, SC160, series V, box 7, folder 7; "Solid-State Devices Lab Tackles Silicon Transistors in Partnership with Industry," June 1958, in historical files of the *Campus Report*, folder John Linvill; James Gibbons, "John Linvill—The Model for Academic Entrepreneurship," *The CIS Newsletter*, fall 1996, collection of C. Lécuyer; Linvill, interview by Lécuyer, April 25 and May 30, 2002; Gibbons, interview by Lécuyer, May 30, 2002.

25. For the founders' conception of the market at the time of the firm's founding, see "New Palo Alto Company Plans to Produce Transistors," *Palo Alto Times*, October 17, 1957, collection of C. Lécuyer; Grinich, interview by Lécuyer, February 7, 1996; Bay, interview, July 2, 1996; communications from Moore, May 21 and June 2, 1998; communication from Last, May 22, 1998; communication from Richard Hodgson, October 10, 1998. On the military's interest in silicon, see Misa 1985.

26. For Fairchild Semiconductor's policies regarding military research contracts, see Noyce, interview by Herb Kleiman, Herb Kleiman collection, M827, Stanford Archives and Special Collections; Last, interview by Lécuyer, April 1, 1996; Bay, interview by Lécuyer, July 2, 1996; Moore, interview by Bassett and Lécuyer, February 18, 1997; Roberts, interview by Lécuyer, July 6, 1996.

27. Fairchild Semiconductor's founders interviewed a manager from Varian Associates for the general manager position before settling on Edward Baldwin. Communication from Jay Last, February 26, 1999. For Fairchild's early hirings, see Last to Last, October 6, 1957, courtesy of Jay Last; Hester 1961; Grinich, interview by Lécuyer, February 7 and April 23, 1996; Kleiner, interview by Lécuyer, May 21 and June 5, 1996; Roberts, interview by Lécuyer, July 6, 1996; Thomas Bay, interview by Lécuyer, July 2, 1996; Richard Hodgson, interview by Rob Walker, September 19, 1995, Silicon Genesis, M471, Stanford Archives and Special Collections.

28. "Meeting Review: Eat Later," *IRE Grid*, May 1960, 32–33; Last, "Meeting Reports etc. . . 10/57–3/59," entries for January 2, February 7, February 24, and March 30, 1958; "Fairchild Devices in B-70 Bomber," *Leadwire*, December 1960, 3, courtesy of Jay Last; Bay, interview by George Rotsky, circa 1988, George Rotsky collection, M851, Archives and Special Collections, Stanford University; Bay, interview by Lécuyer, July 2, 1996; Grinich, interview by Lécuyer, April 23, 1996; communication from Bay, March 19, 1999; communication from Hodgson, October 10, 1998.

29. Bridges 1957; "USAF Reliability Emphasis Grows," *Aviation Week*, April 1, 1957, 28; Neufeld 1990, 215–218; interview with Robert Noyce, no date, Intel Museum. On the aerospace industry's shift from analog to digital techniques, see Ceruzzi 1989, 15–16, 20–30, 51–57, 80–111; Fishbein 1995, 111–121; Barling 1991.

30. Ceruzzi 1989, 15–16, 20–30, 51–57, 80–111; Fishbein 1995, 111–121; Bay, interview by Lécuyer, July 2, 1996.

31. Bay, interview by Lécuyer, July 2, 1996; communication from Last, February 25, 1999; communication from Bay, March 16, 1999.

32. Last, "Meeting Reports etc. . . 10/57–3/59," entries for January 2, 1957, courtesy of Jay Last.

33. Last, "Meeting Reports etc. . . 10/57–3/59," entries for January 2, February 7, February 24, and March 30, 1958; "Fairchild Devices in B-70 Bomber," *Leadwire*, December 1960, 3, courtesy of Jay Last; Bay, interview by George Rotsky, circa 1988, George Rotsky collection, M851; Bay, interview by Lécuyer, July 2, 1996; Grinich, interview by Lécuyer, February 7, 1996; communication from Bay, March 19, 1999; communication from Hodgson, October 10, 1998. On the history of core memories at IBM, see Pugh 1984, 63–159; Bashe et al. 1986, 231–272.

34. Communication from Richard Hodgson, October 10, 1998.

35. Bay, interview by Lécuyer, July 2, 1996; Bay, interview by George Rotsky, circa 1988, George Rotsky collection, M851; Last, "Meeting Reports, etc., 10/57–3/59," entries for January 2, February 3, February 10, February 17, and March 2, 1958, courtesy of Jay Last; "Fairchild Devices in B-70 Bomber," *Leadwire*, December 1960, 3, courtesy of Jay Last.

36. Hoerni, interview by Lécuyer, February 4, 1996; Bay, interview by Lécuyer, July 2, 1996; Grinich, interview by Lécuyer, February 6, 1996; Roberts, interview, July 6, 1996; Moore 1998, 55.

37. On Fairchild's process choice, see Moore 1998, 54; Last, interview by Lécuyer, April 1, 1996. On the Bell system's first double diffused silicon transistor, see Miller 1958.

38. Moore 1998, 56; Last, "Meeting Reports, etc., 10/57–3/59," Technical meeting, February 12, 1958, courtesy of Jay Last; Kleiner, interview with Lécuyer, May 21, 1996; Last, interview by Lécuyer, April 1, 1996; Moore, interview by Allen Chen, January 6 1993, Intel Museum; Roberts, interview by Lécuyer, July 6, 1996.

39. Hoerni, interview with Lécuyer, February 4, 1996.

40. Hoerni, interview with Lécuyer, February 4, 1996 and Blank, interview with Lécuyer, June 20, 1996.

41. Moore and Noyce, "Method for Fabricating Transistors," US Patent 3,108,359, filed June 30, 1959, granted October 29, 1963; Moore 1998, 56; Moore, interview by Bassett and Lécuyer, February 18, 1997; Moore, interview by Allen Chen, January 6, 1993, Intel Museum; Roberts, interview with Lécuyer, July 2, 1996.

42. Fairchild Semiconductor, "Fairchild Silicon Transistors," 1958, courtesy of Sheldon Roberts; Last, "Meeting Reports, etc. 10/57–3/59," entry for January 2, 1958, courtesy of Jay Last.

43. Hoerni ran into substantial contact and diffusion problems with the PNP version of IBM's core driver, which delayed the product's introduction until March 1959. On the engineering problems with the PNP, see Last, "Meeting Reports, etc. 10/57–3/59," entries for February 24, April 3, and September 5, 1958, courtesy of Jay Last; Hoerni, interview by Lécuyer, February 4, 1996.

44. Moore, interview by Bassett and Lécuyer, February 18, 1997; Kleiner, interview by Lécuyer, May 21, 1996, June 5, 1996, and June 16, 1998.

45. In 1957, Fairchild's founders considered renting Varian Associates' plant in San Carlos, which Varian was then vacating. They decided not to rent it, because they judged the plant not clean enough for silicon manufacture. "We concluded," recalled Kleiner, "that our requirements were much more stringent than for making microwave tubes. To modify that would have been more costly than creating from scratch." The tube plant was also too large for their immediate needs. Last to Jones, September 20, 1957, courtesy of Jay Last; Kleiner, interviews by Lécuyer, May 21, 1996, June 5, 1996, and June 16, 1998.

46. Kleiner, interview by Lécuyer, May 21, 1996.

47. Last, interview by Lécuyer, April 1, 1996; Kleiner, interview by Lécuyer, May 21, 1996; communication from Gordon Moore, June 2, 1998.

48. Kleiner, interview, May 21, 1996.

49. Fairchild's engineers also sought to hire operators with the right moral dispositions: "responsible, truthful, and handy" and willing to yield to the exigencies of silicon processing. It might also be noted that the promotion of the more diligent operators to "lead girls" and research technicians helped overcome the operators' resistance to the discipline of the process. Kleiner, interview by Lécuyer, May 21, 1996.

50. Communication by Moore, May 21, 1998; Last, "Meeting Reports etc 10/57–3/59," entry for February 17, 1958, courtesy of Jay Last.

51. Last, notes taken on Noyce's presentation at Fairchild Semiconductor's policy meeting on August 25, 1958, in Last, "Meetings Reports etc. 10/57–3/59," courtesy of Jay Last.

52. Last, "Meeting Reports etc," entry for August 25, 1958, courtesy of Jay Last; Grinich, interview by Lécuyer, February 7 and April 23, 1996. On the introduction of Fairchild's first product, see Last, "Meeting reports, etc. 10/57–3/59," entry for August 25, 1958, courtesy of Jay Last.

53. "Fairchild silicon transistors: Milli-micro-second switching speeds and high current too," *Solid-State Journal*, September-October 1958, 15; "From Fairchild: Mesa transistors in silicon," *Solid-State Journal*, January 1959, 14; "Transistor experts are betting that this is the winning combination: speed, power, reliability, availability, price," *Solid-State Journal*, March 1959, 6. For examples of Fairchild's early sales literature, see Fairchild Semiconductor, "Fairchild Silicon Transistors," 1958, courtesy of Sheldon Roberts. On Bay's sales and marketing innovations, see Bay, interview, by Lécuyer, July 2, 1996; communication from Bay, March 16, 1999; Bay, interview by George Rotsky, circa 1988, George Rotsky collection, M851; communication from Hodgson, October 10, 1998. For Fairchild's applications notes, see Isy Haas, "2mμsec Rise Time Pulses Obtainable from Fairchild 2N696–2N697 Silicon Mesa Transistors in Avalanche Mode," APP-1, no date (1958); Ray Kikoshima, "Blocking Oscillator Circuit," APP-2, no date (1958); G. Reddi, "Application of High Frequency Parameters," APP-3, December 28, 1958; Victor Grinich, "Voltage Ratings for Double Diffused Silicon Switching Transistors," no date (early 1959), all in Fairchild Semiconductor, "Product Catalog," 1962, courtesy of Jay Last.

54. An overwhelming share of Fairchild's sales in 1958 and 1959 was in the military sector: 92% of the firm's sales in the summer of 1958 were renegotiable, and this grew to 99% for the period September 1958–August 1959. Fairchild Camera and Instruments, "Prospectus," January 25, 1960, courtesy of Eugene Kleiner; Grinich, interview by Lécuyer, February 7, 1996 and April 23, 1996.

55. Communication from Moore, May 21, 1998; Moore, interview by Allen Chen, January 6, 1993, Intel Museum; Moore, interview by Bassett and Lécuyer, February 18, 1997; communication from Bay, March 16, 1999. On the Minuteman missile and its avionics equipment, see Reed 1986; MacKenzie 1990, 152–161; Wuerth 1976; Klass 1959; West 1962. On the Department of Defense's and especially the Air Force's increasing interest in reliability in the late 1950s and the early 1960s,

see Lambert 1958; Spiegel and Bennett 1961; Spiegel 1961; Ryerson 1962; Neufeld 1990, 215–218.

56. Bay, communication, March 16, 1999; Moore 1998. 57; Jay Farley, "Reliability Assurance at Fairchild Semiconductor Corporation," App-7, November 11, 1959, in Fairchild Semiconductor, *Product Catalog*, 1962, courtesy of Jay Last; Grinich, interview by Lécuyer, February 7, 1996 and April 23, 1996; Kleiner, interview by Lécuyer, June 16, 1998; Moore, interview by Bassett and Lécuyer, February 18, 1997; Hoerni, interview by Lécuyer, February 4, 1996; communication from Jay Last, June 1, 1998; communication from Gordon Moore, June 2, 1998.

57. Hoerni had toyed with the idea of the planar process in December 1957. He wrote his patent disclosure in January 1959. On the work of Hoerni, Noyce, and Moore on clean oxides at Shockley Semiconductor, see Roberts, interview by Lécuyer, July 6, 1996. On Hoerni's and the other founders' knowledge of the work at the Bell Telephone Laboratories, see Last, "Meeting Reports, etc. . . . ," entry for May 7, 1958, courtesy of Jay Last; Atalla et al. 1959. On Hoerni's idea of the planar process, see Hoerni, "Patent Disclosure: Method of Protecting Exposed P-N Junctions at the Surface of Silicon Transistors by Oxide Masking Techniques," January 14, 1959, and entries in his patent notebook for December 1, 1957 and March 2, 1959—all courtesy of Jay Last.

58. Hoerni, "Semiconductor Device," U.S. patent 3,064,167, filed May 1, 1959, granted November 13, 1962; Hoerni, "Method of Manufacturing Semiconductor Devices," U.S. patent 3,025,589, filed May 1, 1959, granted March 20, 1962; Hoerni, interview by Lécuyer, February 4, 1996; Last, interview by Lécuyer, April 1, 1996; Moore, interview by Bassett and Lécuyer, February 18, 1997.

59. Hoerni, "Semiconductor Device," U.S. patent 3,064,167, filed May 1, 1959, granted November 13, 1962; Hoerni, "Method of Manufacturing Semiconductor Devices," U.S. patent 3,025,589, filed May 1, 1959, granted March 20, 1962; Hoerni, "Planar Silicon Transistors and Diodes," Paper presented at the 1960 Electron Devices Meeting, Washington, D.C.—October 1960, Bruce Deal Papers, 88–033, Archives and Special Collections, Stanford University.

60. Hoerni, interview by Lécuyer, February 4, 1996.

61. Ibid.

62. Ibid.

63. Ibid.

64. Ibid.; Grinich, interview by Lécuyer, February 7, 1996 and April 23, 1996; communication from Last, February 24 and 25, 1999.

65. Communication from Moore, June 2, 1998.

66. Fairchild Semiconductor later sued Rheem for theft of trade secrets and settled out of court for $30,000. With this sum, Fairchild bought a one third interest in SGS, an Italian semiconductor company in 1960. On Baldwin's departure and

his establishment of Rheem Semiconductor, see Hester 1961; L. N. Duryea, "Fairchild Investigation," May 28, 1959, William Shockley Papers, SC 222 95–153, box B4, folder Fairchild Info 3 July 1959; Moore, interview by Bassett and Lécuyer, February 18, 1997; Hoerni, interview, February 4, 1996; Roberts, interview, July 6, 1996; Hodgson, interview by Rob Walker, September 19, 1995, Silicon Genesis, M471, Stanford Archives and Special Collections; communication from Hodgson, October 10, 1998; Harry Sello, interview by Rob Walker, April 8, 1995, Silicon Genesis, M741.

67. Progress Reports, Physics Section, February 1, 1960; March 1, 1960, box 5, folder 1; April 1, 1960, box 5, folder 1; October 1, 1960, box 5, folder 3, all in Technical Reports and Progress Reports, M 1055, Archives and Special Collections, Stanford University; Hoerni, interview by Lécuyer, February 4, 1996 and Grinich, interview by Lécuyer, February 7, 1996 and April 23, 1996.

68. Planar transistors had better noise levels and breakdown voltages. See Hoerni, "Planar Silicon Transistors and Diodes," paper presented at 1960 Electron Devices Meeting, Washington, Bruce Deal Papers, 88–033, Stanford University Archives; "Planar Transistor Offers Wide Beta," *Electronic Daily*, March 23, 1960, courtesy of Eugene Kleiner. On the development of gold-doped transistors, see Jean Hoerni, "Transistor Manufacturing Process," US Patent 3,108,914, filed June 30, 1959 and granted October 29, 1963; Hoerni, interview by Lécuyer, February 4, 1996; Moore 1998, 57.

69. Hoerni, "Planar Silicon Transistors and Diodes," paper presented at 1960 Electron Devices Meeting, Washington, Bruce Deal Papers, 88–033; Hoerni, interview by Lécuyer, February 4, 1996; Grinich, interview by Lécuyer, February 7, 1996.

70. Noyce, interview, no date, Intel Museum. On reliability concerns in the development of the integrated circuit, see Kleiman 1966.

71. Noyce, Patent notebook, January 23, 1959, courtesy of Jay Last; Noyce, interview, no date, Intel Museum; Grinich, interview by Lécuyer, May 14, 1996.

72. Last, "Development of the Integrated Circuit, August 1959–June 1960," 2000, courtesy of Jay Last; Kilby 1976.

73. Noyce, "Semiconductor Device-and-Lead Structure," US Patent 2,981,877, filed July 30, 1959. On the invention of the integrated circuit, see Reid 1984, 76–78; Wolff 1976; Moore 1998, 59–60.

74. Last, "Development of the Integrated Circuit."

75. Last, interview by Lécuyer, April 1, 1996.

76. Last, patent notebook, entry for February 23, 1960, courtesy of Jay Last; Last, "Solid-state Circuitry Having Discrete Regions of Semi-conductor Material Isolated by an Insulating Material," US patent 3,158,788, filed August 15, 1960 and granted November 24, 1964; Last, "Method of Making Solid-state Circuitry," US patent 3,313,013, filed August 15, 1960 and granted April 11, 1967; Lionel Kattner, patent

notebook, entry for May 31, 1961, courtesy of Lionel Kattner; Last, interview by Lécuyer, April 1, 1996; Last, communication, February 26, 1999; Kattner, interview by Lécuyer, October 14 and 15, 2000.

77. Isy Haas, patent notebook, entry for August 31, 1960, courtesy of Jay Last.

78. Progress Reports, Micrologic section, August 1, 1960, box 5, folder 3; October 1, 1960, box 5, folder 4; December 1, 1960, box 5, folder 7, all in Technical Reports and Progress Reports, M 1055; Progress reports, Device Development section, May 1, 1961, box 6, folder 4; June 1, 1961, box 6, folder 5, in Technical Reports and Progress Reports, M 1055; Fairchild Semiconductor, "The Inside Story on Fairchild Micrologic," 1961 or 1962, courtesy of Jay Last; Moore 1998, 59; Last, interview by Lécuyer, April 1, 1996.

79. Moore, progress report, April 1, 1961, May 1961, June 1961; Kattner, progress report, August 1, 1961—all in Technical Reports and Progress Reports, M 1055. On the marketing of the micrologic family, see Robert Graham, "Micrologic sales situation," December 1, 1961, courtesy of Jay Last.

80. On Autonetics' component reliability improvement program, see Scheffler 1981. On Autonetics' component reliability improvement program, see Klass 1959; Stranix 1960; "Minuteman Avionics Reliability Increases," *Aviation Week*, December 12, 1960, 99–105; Hilman 1961; Winterelt 1961; Culvers 1962; Hilman and Durand 1962; West 1962; Smith 1963; Wuerth 1976.

81. On the development and manufacture of reliable tubes for the transatlantic cable at the Bell Telephone Laboratories, see McNally et al. 1957; Smits 1985.

82. Autonetics' abridged "Statement of Work," in Stranix 1960, 95; Scheffler 1981.

83. Smullen, interview by Lécuyer, May 12, 1996.

84. On Autonetics' oversight of its vendors' manufacturing operations, see Stranix 1960; Scheffler 1981.

85. Kleiner, interview by Lécuyer, June 5, 1996; "FSC Signs Two Autonetics Contracts," *Leadwire*, June 1960, courtesy of Jay Last.

86. In the early 1960s Fairchild's operators collectively burnt their hairnets to protest against the manufacturing discipline. The workers also subverted the regulation imposing the wearing of smocks. "The minute smocks were put out," a manufacturing engineer recalled. "They all took them home and brought them up above their knees. They individualized the smocks, tailored them, and cut them up. 85% of the people wore them as they were. But 10 or 15% of the operators did not want that regimentation. So they did something to made their smock in some cases just a little more different, in other cases quite a bit different." (Roger Smullen, interview by Lécuyer, May 12, 1996) On the hairnet incident, see "Do You Remember. . ." *Leadwire* 9, no. 10, 1968, collection of C. Lécuyer.

87. Fairchild Semiconductor, *Micrologic Reliability Report*, March 1963, courtesy of Jay Last; Moncher and Johnstone 1963, 206. For a more detailed treatment of the reliability improvement program at Fairchild, see Lécuyer 1999a.

88. "Autonetics Contracts Now Total 8 Million," *Leadwire*, December 1960, courtesy of Jay Last. On the work of the Reliability Evaluation Division, see Progress Report, Reliability Evaluation Division, April 1, 1960, Technical Reports and Progress Reports, M 1055, box 5, folder 1; Hilman 1961; Hilman and Durand 1962.

89. Scheffler 1981; *Aviation Week*, December 12, 1960, 99–105. On the application of the new techniques and procedures developed under the Autonetics program to Fairchild's other product lines, see "FSC Signs Two Autonetics Contracts; Will Receive $1,511,210 for Reliability Program," *Leadwire*, June 1960; Kleiner, interview by Lécuyer, June 5, 1996; Smullen, interview by Lécuyer, May 12, 1996.

90. The Minuteman contract accounted for roughly one-fourth of Fairchild Semiconductor's revenues by 1960. Bay, interview by Lécuyer, July 2, 1996; John Hulme, interview by Lécuyer, March 18, 1997.

91. Bay, interview by Lécuyer, March 18, 1997.

92. *Leadwire*, October 1959–May 1960, courtesy of Jay Last.

93. Jack Yelverton, interview by Lécuyer, December 18, 2000; Yelverton, History of Silicon Valley seminar, Stanford University, October 23, 2001; Charles Sporck, interview by Lécuyer, May 15, 1995.

94. Engineers and other professional employees were the main beneficiaries of these programs. The working conditions of technicians and operators could be grim. On fabricating work, see Hayes 1989 and Siegel and Markoff 1985. On management-employee relations at Fairchild Semiconductor, see Yelverton, interview by Lécuyer, December 18, 2000; Yelverton, History of Silicon Valley seminar, Stanford University, October 23, 2001.

95. Yelverton, interview by Lécuyer, December 18, 2000; David James, interview by Lécuyer, December 14, 2000.

96. Fairchild had a major impact on firms making germanium transistors. Because its devices rapidly reached the speed of germanium transistors and were very reliable, Fairchild displaced germanium manufacturers in the military market in the late 1950s and the early 1960s. See Painter et al. 1960a,b.

97. Terman to William Cooley, June 3, 1963, Frederick Terman papers, SC 160, series V, box 7, folder 9; Terman to Cooley, July 31, 1967, Frederick Terman papers, SC 160, series V, box 7, folder 13; Chuck Gravelle, "Electroglas History," no date, courtesy of Electroglas; "Contact!" *Semiconductor Products*, August 1961, 17; "Electroglas DF-3," *Semiconductor Products*, March 1962, 19; "Semiconductor and Microcircuit Yields Go Up With Processing Systems from Electroglas," *Semiconductor Products*, October 1963, 7; Marshall 1965; "Hot Tip for Semiconductor Lead Bonding," *Semiconductor Products*, October 1963, 41; "New Tempress Auto-

matic Scribing Machine," *Semiconductor Products*, May 1965; "Precise People Make Precision Products," *Semiconductor Products*, February 1967.

98. Noyce, Moore, Blank, and Grinich, who were working at Fairchild at this time, requested Fairchild Camera's permission to invest in Davis' and Rock's fund. This permission was denied.

99. "Davis and Rock," July 13, 1961, courtesy of Jay Last; Davis 1964; "Arthur Rock," *Red Herring*, March 1994, 54–59; Rock 2000.

100. Kleiner, interview by Lécuyer, May 21, 1996 and June 5, 1996; Roberts, interview by Lécuyer, July 6, 1996; Hoerni, interview by Lécuyer, February 4, 1996; Perkins, interview with Rock, 54–59. For a journalistic treatment of the early history of the venture capital industry on the San Francisco Peninsula, see Wilson 1985.

Chapter 5

1. *Missiles and Rockets* 12, March 25, 1963 (special issue devoted to McNamara's policies); Davis 1964; Freund 1971, 44; Kaplan 1983.

2. *Missiles and Rockets* 12, March 25, 1963; Davis 1964.

3. On the impact of the Department of Defense's new procurement policies on component manufacturers, see Moreno 1963.

4. Edward Ginzton, "Speech at Shareholders' Meeting," February 20, 1964, Edward Ginzton Papers, SC330, box 14-A, folder Speeches 1960–65, Stanford Archives and Special Collections.

5. Saad 1964, 12; Orth 1964; "Microwave Tube Shakeout Nearing End," *Samson Trends*, November 1964, 6–8, in Eitel-McCullough Records, 73/65c, box 3, folder Samson Trends, Bancroft Library; Frederick Terman to John Hawkinson, February 2, 1965, in Frederick Terman Papers, SC 160, series V, box 7, folder 11, Stanford Archives and Special Collections; Ted Taylor, interview by Lécuyer, February 12, March 6, and March 29, 1996.

6. Nelson Stone, interview by Lécuyer, April 21, 1995; Bay 1961; Hoefler 1966a; Freund 1971, 52; Victor Grinich, "Progress Report, Engineering Department," April 13, 1960, box 5, folder 2, Technical Reports and Progress Reports, M 1055, Stanford Archives and Special Collections; "Transistor Entry Tightens," *Electronic News*, August 1, 1960, 1 and 4; Miller 1963; Stone, interview by Lécuyer, April 21, 1995; Pierre Lamond, interview by Lécuyer, November 17, 1995 and December 6, 1995; Grinich, interviews by Lécuyer, February 7, 1996 and April 30, 1996; communication from Nelson Stone, June 2, 1998.

7. Emmet Cameron, quoted in minutes of regular meeting of board of directors, June 23, 1960, Edward Ginzton Papers, SC 330 95–179, box 18, volume 9; Terman to William Cooley, March 3, 1960, June 1, 1960, June 5, 1961, and June 29, 1961— all in Frederick Terman papers, series V, box 7, folder 8; Terman to John

Hawkinson, February 2, 1965 and March 15, 1965, in Frederick Terman Papers, SC 160, series V, box 7, folder 11; "Report on the Proposed Acquisition of Company X," April 9, 1965, Varian Associates Records, 73/65c, carton 3, folder Report on Proposed Acquisition of Company X, Bancroft Library; "The Merger of Varian Associates and Eitel-McCullough, Inc." July 9, 1965, in Edward Ginzton Papers, SC 330, box 12, folder Eimac merger/Department of Defense; Jack McCullough, interview by Arthur Norberg, April 15 and 24, 1974, uncatalogued manuscript, Bancroft Library.

8. Minutes of regular meeting of board of directors, December 16, 1959, Edward Ginzton Papers, SC 330 95–179, box 18, volume 8; minutes of regular meeting of board of directors, September 1, 1960, Edward Ginzton Papers, SC 330 95–179, box 18, volume 9; minutes of regular meeting of board of directors, June 23, 1960, Edward Ginzton Papers, SC 330 95–179, box 18, volume 9. On Hewlett-Packard's policy, see Terman to Hawkinson, December 1, 1967, Frederick Terman Papers, SC 160, series 5, box 7, folder 13; "H-P vs. GAO, Continued," *Electronics*, February 5, 1968, 50–51; "H-P Loses Landmark Audit Case," *San Jose Mercury*, March 26, 1968; Murray Seagel, "Audit at Hewlett-Packard Raises a New Issue," *Los Angeles Times*, January 9, 1969, 1 and 7, in historical files of the *Campus Report*, folder David Packard, Stanford University.

9. Minutes of regular meeting of board of directors, December 16, 1959, Edward Ginzton Papers, SC 330 95–179/box 18, volume 8; Howard Cary to Ginzton, December 28, 1961, Edward Ginzton Papers, SC330, box 4, folder Board of Directors Correspondence, 1961–1967; Moreno to Ginzton, April 23, 1965, in SC 330, box 12, folder Eimac Merger/Department of Defense; McCullough, interview, 120–121; General Accounting Office, "Examination of catalog prices charged for klystron tubes under noncompetitive procurements negotiated by the military departments and their prime contractors with Varian Associates, Palo Alto, California," January 15, 1963; Ted Taylor, interviews by Lécuyer, March 6 and March 29, 1996.

10. The Supreme Court refused to examine the Hewlett-Packard case.

11. General Accounting Office, "Examination of catalog prices charged for klystron tubes under noncompetitive procurements negotiated by the military departments and their prime contractors with Varian Associates, Palo Alto, California," January 15, 1963; Joseph Campbell to Charles Gubser, January 17, 1963; Gubser to Ginzton, January 21, 1963; Emmet Cameron, "Draft of a proposed letter to express Varian's position re. PL 87–653 and ASPR revision #12," February 21, 1963; "Varian Associates' Position Statement Re GAO Report" [circa 1963]—all in Varian Associates Records, 73/65c, carton 5, folder Misc. reports, insurance plans, and pamphlets.

12. Cameron to Ted Moreno, May 13, 1963, Edward Ginzton Papers, SC 330, box 3, folder Personnel correspondence; Cameron to Frank Walker, August 13, 1963, Edward Ginzton Papers, SC 330, box 4, folder Board of directors correspondence, 1961–1967; minutes of regular meeting of board of directors, April 25, 1963, SC 330 95–179, box 19, volume 12; Terman to Hawkinson, February 2, 1965, Frederick

Terman Papers, SC 160, series V, box 7, folder 11; Moreno, interview by Lécuyer, July 19, 1996.

13. Minutes of regular meeting of board of directors, January 23, 1964, Edward Ginzton Papers, SC 330 95–179, box 19, volume 12; Ginzton and Cottrell 1995, 146–147; Moreno, interview by Lécuyer, July 19, 1996.

14. "Watkins-Johnson Acquires Control of Santa Cruz Firms," *Palo Alto Times*, February 28, 1963, collection of C. Lécuyer; "Microwave Tube Shakeout Nearing End," *Samson Trends*, November 1964, 6–8, Varian Associates Records, 73/65c, box 3, folder Samson Trends; "Report on the Proposed Purchase of Microwave Facility of Company Z," Varian Associates Records, 73/65c, carton 5, folder Acquisition and joint venture studies; Frederick Terman to John Hawkinson, February 2, 1965; Chet Lob, interview by Lécuyer, May 15, 1996; Ted Taylor, interview by Lécuyer, February 12, March 6, and March 29, 1996.

15. John Franklin to Terman, November 14, 1961; Terman to Franklin, November 22, 1961; Franklin to David Packard, November 24, 1961—all in Frederick Terman Papers, SC 160, series XIV, box 14, folder 19; Jack McCullough, interview by Norberg; William Eitel, interview by Arthur Norberg, no date, courtesy of Don Preist; Ginzton, interview by Mercer, March, April, June, October 1990, in Varian Associates Oral History Project, M708, Archives and Special Collections, Stanford University.

16. Ginzton, "Shareholders Meeting," February 20, 1964, Edward Ginzton Papers, SC 330, box 14–A, folder Speeches 1960–1965; "Report on the Proposed Acquisition of Company X," Varian Associates Records, MSS 73/65c, carton 3, folder Report on Proposed Acquisition of Company X; Varian Associates, Prospectus, May 28, 1965, in business ephemera collection, Jackson Library, Stanford University; minutes of meeting of board of directors of Eitel-McCullough, April 22, 1965, in Eitel-McCullough Records, 77/110c, carton 7, folder minutes of the board of directors 1963–1965, Bancroft Library; Jack McCullough, interview by Arthur Norberg, April 15 and 24, 1974, uncatalogued manuscript, Bancroft Library; William Eitel, interview by Arthur Norberg, no date, courtesy of Don Preist; Ginzton, interview by Mercer, March, April, June, October 1990, in Varian Associates Oral History Project, M708.

17. "Justice Department Stalls Eitel-McCullough Merger into Varian," *Electronic News*, July 5, 1965, 2; "The Merger of Varian Associates and Eitel-McCullough, Inc." July 9, 1965, in Edward Ginzton Papers, SC 330, box 12, folder Eimac merger/Department of Defense; Ginzton and Cottrell 1995, 145; McCullough, interview by Norberg, April 15 and 24, 1974; William Eitel, interview by Norberg, no date, courtesy of Don Preist; Ginzton, interview by Mercer, March, April, June, October 1990, in Varian Associates Oral History Project, M708.

18. Ginzton, "Visit to Washington, D.C., July 12 and 13," July 19, 1965, Edward Ginzton Papers, SC 330, box 12, folder Eimac merger/Department of Defense; Jack McCullough, interview by Norberg, April 15 and 24, 1974, uncatalogued manuscript, Bancroft Library; Ginzton, interview by Mercer, March, April, June, October 1990.

19. Ginzton, "Visit to Washington, D.C., July 12 and 13," July 19, 1965, Edward Ginzton Papers, SC 330, box 12, folder Eimac merger/Department of Defense.

20. "The Merger of Varian Associates and Eitel-McCullough, Inc," July 9, 1965, Edward Ginzton Papers, SC 330, box 12, folder Eimac merger/Department of Defense.

21. Preist 1992; McCullough, interview by Arthur Norberg, April 15 and 24, 1974; Moreno, interview by Mercer, October 1989, Varian Associates Oral History Project, M 708.

22. In 1963, Norman Moore, the head of Litton Industries' microwave tube division, branched out into high-end appliances. He set up a new division that made microwave ovens. His goals were to reduce Litton's reliance on military business and at the same time exploit the firm's low cost production expertise for commercial advantage. In the late 1950s and the early 1960s, engineers in Litton's microwave tube division developed innovative techniques for the manufacture of low cost magnetrons. These magnetrons cost $200 a piece in quantities of 10,000— a price which commercial users could bear. Moore thought that these magnetrons would give him a substantial cost advantage in the microwave oven business. Microwave ovens had been pioneered by Raytheon, the maker of magnetrons and radar sets, in the immediate postwar period. In 1947, a group of engineers at Raytheon developed the first microwave oven, an oven, which used a magnetron as its cooking source. In spite of its usefulness for the rapid cooking of foods, the oven found only a small market in the 1950s. The high cost of Raytheon's magnetrons limited its use. Under Moore's direction, Litton's engineers designed a cheaper microwave oven around the new Litton magnetron. But the firm, which had no knowledge of the appliance business ran into a distribution problem. At first, Litton marketed its ovens through the Tappan Stove Company, a manufacturer of appliances. It later introduced a microwave oven under its own brand name to the market. This product was geared toward restaurants and other commercial users. Litton's limited success in commercializing its microwave ovens along with the decline in microwave tube sales cost Moore his job. Moore left Litton in 1967.

23. Martin Packard, interview by Henry Lowood, November 22, 1988, Stanford Archives and Special Collections.

24. Minutes of special meeting of executive committee and finance committee of board of directors, July 25, 1966, in Ginzton Papers, SC330, box 19, volume 15; Myrl Stearns, interview by Mercer and Graig Nunnan, interview by Mercer, November 1989—all in Varian Oral History Collection, M 708.

25. "Vacuum Development," *Varian Associates Magazine*, June 1978, 16–17, in Russell and Sigurd Varian Papers, SC345, series: Varian Associates, box 9, folder 4.

26. "The VacIon: Varian Introduces Revolutionary New High Vacuum Pump," *Varian Associates Magazine*, March 1958, 4–6; "Vacuum Development," *Varian Associates Magazine*, June 1978, 16–17, in Russell and Sigurd Varian Papers, SC345, series: Varian Associates, box 9, folder 4.

27. Minutes of board of directors, September 25, 1958, Edward Ginzton Papers, SC330 95–179, box 18, volume 6; Wes Carnahan, "Varian and the Vacuum Business," December 17, 1958, Russell and Sigurd Varian Papers, SC345, series: Varian Associates, box 4, folder 26; Emmet Cameron to Malter, July 17, 1958; Malter to Stearns, September 12, 1958; Malter to Stearns, September 18, 1958 — all in Russell and Sigurd Varian Papers, SC345, series: Varian Associates, box 3, folder 1; Malter to Stearns January 7, 1959 and Robert Jepsen to Malter, May 21, 1959—both in Russell and Sigurd Varian Papers, SC345, series: Varian Associates, box 3, folder 2; Malter to Myrl Stearns, July 9, 1958, in Russell and Sigurd Varian Papers, SC345, series: Varian Associates, box 3, folder 1; "Dr. Malter to Head Vacuum Products Group," *Varian Associates Magazine,* January 1959, 8; "Management Profiles: Lou Malter," *Varian Associates Magazine,* July 1965, 10–11. On Ultek, see Paul Emerson, "Ultek Makes Big Strides in Field," *Palo Alto Times,* February 1, 1963, 4, collection of C. Lécuyer.

28. "Varian VacIon Pumps and High Vacuum Accessories," circa 1960, Varian Associates Records, 73/65c, carton 3, folder catalogs, Bancroft Library; "2 More High-Vacuum Firsts!" *Solid-State Journal,* March 1961, 49; "The Vacuum Products Division," *Varian Associates Magazine,* May 1961, 6–8; "Ever Lose a Vacuum? This New Black Box from Varian Finds Vacuum Leaks without Expensive Down Time," *Solid-State Journal,* February 1962; "New Vacuum Flange Developed," *Varian Associates Magazine,* May 1962, 7; "Vacuum Products Division," *Varian Associates Magazine,* September 1964, 7–9; "The Vacuum Products Division," *Varian Associates Magazine,* May 1961, 6–8; "Off the Shelf! Ultra-High Vacuum Systems," *Solid-State Journal,* May 1961, 47; "An Exceptional New Vacuum System," *Varian Associates Magazine,* December 1961, 13; "Vacuum Products Division," *Varian Associates Magazine,* September 1964, 7–9.

29. "The Vacuum Products Division," *Varian Associates Magazine,* May 1961, 6–8; "Off the Shelf! Ultra-High Vacuum Systems," *Solid-State Journal,* May 1961, 47; "An Exceptional New Vacuum System," *Varian Associates Magazine,* December 1961, 13; "Vacuum Products Division," *Varian Associates Magazine,* September 1964, 7–9.

30. Malter to Cameron, May 25, 1959, Russell and Sigurd Varian Papers, SC345, series: Varian Associates, box 3, folder 4; "The Vacuum Products Division," *Varian Associates Magazine,* May 1961, 6–8; "Off the Shelf! Ultra-High Vacuum Systems," *Solid-State Journal,* May 1961, 47; "An Exceptional New Vacuum System," *Varian Associates Magazine,* December 1961, 13; "Vacuum Products Division," *Varian Associates Magazine,* September 1964, 7–9; Victor Grinich, interview by Lécuyer, February 7, 1996, April 23, 1996, and May 14, 1996.

31. On the evolution of Varian's sales, see minutes of board of directors, November 17, 1960, in Ginzton Papers, SC330 95–179, box 18, volume 9; minutes of board of directors, July 25, 1966, Edward Ginzton Papers, SC330 95–179, box 19, volume 15. On Ultek, see Paul Emerson, "Ultek Makes Big Strides in Field," *Palo Alto Times,* February 1, 1963, 4; "Ultek Cuts it Close," *Palo Alto Times,* January 22, 1964; "Ultek Receives SLAC Contract," *Palo Alto Times,* January 27, 1964; "Ultek Gets Two Space Contracts," *Palo Alto Times,* June 24, 1966; "Increase in Sales

Reported by Ultek," *Palo Alto Times*, August 9, 1966—all in the collection of C. Lécuyer. On the use of vacuum technologies in the semiconductor industry, see Waits 1997, 105.

32. For a history of Stanford's medical linear accelerator, see Ueyama and Lécuyer (in press).

33. On Varian's NMR business, see Lenoir and Lécuyer; Lenoir 1997, 239–294; Bloch 1946; Bloch et al. 1946; Bloch, "History of the Invention of Nuclear Induction by Bloch and Hansen," June 1959, in 90–099, box 15, folder 1, Stanford Archives and Special Collections; Bloch and Hansen, "Method and Means for Chemical Analysis by Nuclear Induction," patent 2,561,489, filed December 23, 1946, issued July 24, 1951; Russell Varian, "On Use of Nuclear Induction for Chemical Analysis," July 23, 1946, and entry for August 10, 1946 in his "Research Notebook," Varian Papers, SC 345, series: Russell Varian, box 3, folder 8.

34. Arnold et al. 1951; Elliott Levinthal, "Memorandum," January 22, 1951, Edward Ginzton Papers, SC 330, box 3, folder Varian Associates; Shoolery 1993.

35. John Cullen, "ASP Pricing," January 17, 1961; Emery Rogers, "Report for Instrument Division," January 18, 1961—both in Varian Associates Records, 73/65c, box 4, folder Minutes Board of Directors 1961, Bancroft Library; "The Varian A-60 NMR Spectrometer as an Instrument and a Technical Achievement," February 27, 1961, Varian Associates Records, 73/65c, box 4, folder Promotional Materials; Shoolery 1993.

36. "Instrument Field Engineering Story: How Varian Sells Electronic Instruments," *Varian Associates Magazine*, February 1957, 4–5; "Varian Opens Applications Laboratory and Service Center: New Instrument Division Facility at Pittsburgh Airport," *Varian Associates Magazine*, August 1963, 12–14; Shoolery 1993.

37. "Instrument Field Engineering Story: How Varian Sells Electronic Instruments," *Varian Associates Magazine*, February 1957, 4–5; "Fourth NMR-EPR Workshop," *Varian Associates Magazine*, November 1960, 8–9; *NMR Spectroscopy: Papers Presented at Varian's Third Annual Workshop on Nuclear Magnetic Resonance and Electron Paramagnetic Resonance, held at Palo Alto, California* (Pergamon, 1960).

38. Rogers, "Report for Instrument Division," February 15, 1961: "Report for Instrument Division," April 20, 1961: "Report for Instrument Division," June 15, 1961: "Report for Instrument Division," July 27, 1961; "Report for Instrument Division," September 20, 1961—all in Varian Associates Records, 73/65c, box 4, folder Minutes Board of Directors 1961; "The HR-100," *Varian Associates Magazine*, July 1962, 4–6; "New NMR Spectrometers," *Varian Associates Magazine*, May 1964, 5; Meiboom 1963; Committee for the Survey of Chemistry, National Academy of Science National Research Council, *Chemistry: Opportunities and Needs* (1965).

39. Malter, "Vacuum Research Corporation," November 30, 1964, Edward Ginzton Papers, SC330/95–179, box 19, volume 13; "Dedication of Portland Plant," *Varian Associates Magazine*, June 1966, 8–9; minutes of regular meeting of board of directors, December 22, 1964, Edward Ginzton Papers, SC330 95–179, box 19, volume

13; minutes of regular meeting of board of directors, February 24, 1966, Edward Ginzton Papers, SC330 95–179, box 19, volume 15; Emmet Cameron, interview by Mercer, October 1989.

40. Shareholders' meeting, February 15, 1968, Edward Ginzton Papers, SC 330, box 14A, folder Shareholders' Speech 2–15–68; Harry Weaver to Ginzton, July 17, 1969, Edward Ginzton Papers, SC 330, box 3, folder personal correspondence; Ginzton, interview by Mercer; Stearns, interview by Mercer.

41. Fairchild Camera and Instruments, *Annual Report for 1963*, 18–20; Fairchild Camera and Instruments, *Annual Report for 1964*, 18–20; Moore and Davis 2001, 18–21, Stanford University; Sporck 2001; Sporck, interview by Rob Walker, February 21, 2000, Silicon Genesis, M 0741, box 2, Archives and Special Collections, Stanford University; communication from Sporck, July 8, 2002; communication from Thomas Bay, March 16, 1999; Pierre Lamond, interview by Lécuyer, August 12, 1997.

42. For a more detailed treatment of Fairchild Semiconductor's move to the commercial markets, see Lécuyer 1999a,b; Fairchild Semiconductor, *A Solid State of Progress*, ca. 1978, Technical Reports and Progress Reports, M1055, box 4, folder 9, Archives and Special Collections, Stanford University; Herbert Kleinman, interview with Robert Noyce, November 18, 1965, Herbert Kleinman collection, M827, Archives and Special Collections, Stanford University. Don Yost, interview by Lécuyer, August 12, 1997; Victor Grinich, interview by Lécuyer, February 7, 1996; Reed 1962; Martinez 1973.

43. By "commercial computers" I mean computers that were not part of larger weapon systems.

44. Marshall Cox, interview by Lécuyer, October 25, 1996.

45. Fairchild Camera and Instruments, *Annual Report for 1962*, 20; R. W. Allard, "Product Line History Control Data EDP Systems," July 21, 1980, Charles Babbage Institute; Thornton 1970; Thornton 1980; Flamm 1988; MacKenzie and Elzen 1996; Cox, interview by Lécuyer, October 25, 1996; Floyd Kvamme, interview by Lécuyer, September 25, 1996; Lamond, interviews by Lécuyer, November 17, 1995, December 6, 1995, and August 12, 1997. On computer packaging in the 1960s, see Harper 1969; Staller 1965.

46. Fairchild Camera and Instruments, *Annual Report for 1963*, 19; Bay, interview by Lécuyer, July 2, 1996; Lamond, interview by Lécuyer, November 7, 1995, December 6, 1995, and August 12, 1997.

47. Moore and Grinich, "R&D Progress Report," December 12, 1961, Technical Reports and Progress Reports, M 1055, box 6, folder 9, Archives and Special Collections, Stanford University; Moore, interview by Ross Bassett and Lécuyer, February 18, 1997; Grinich, interview by Lécuyer, February 7 and April 23, 1996; Jack Gifford, interview by Lécuyer, July 23, 1996; Bay, interview by Lécuyer, July 2, 1996.

48. Bay, interview, July 2, 1996.

49. Bay 1961; Gifford, interview, July 23, 1996; Cox, interview, October 25, 1996; Kvamme, interview by Lécuyer, September 25 and October 10, 1996; Bay, interview by Lécuyer, July 2, 1996.

50. John Hulme, interview by Lécuyer, March 18, 1997 and May 7, 1997; Moore, interview, February 18, 1997. "Application Engineering Personnel," 1964, courtesy of John Hulme.

51. Hulme, interview, March 18, 1997

52. Communications from Hulme, January 13, 1997 and April 8, 1999; Hulme, interview, March 18, 1997 and May 7, 1997.

53. Data sheets for the 2N709 and the 2N2368, Fairchild folder, computer ephemera collection, CBI 12, Charles Babbage Institute; Moore and Grinich, "R&D Progress Report," April 13, 1962, box 7, folder 3; Moore and Grinich, "R&D Progress Report," May 11, 1962, box 7, folder 4; Moore, "R&D Progress Report," July 6, 1962, box 7, folder 6; Moore and Grinich, "R&D Progress Report," April 13, 1962, box 7, folder 4; Moore, "R&D Progress Report," August 9, 1962, box 7, folder 5; Moore, "Research and Development Progress Report for 1962," January 24, 1963, box 7, folder 6; Progress report, Device development section, July 1, 1962, box 7, folder 5—all in Technical Reports and Progress Reports, M 1055; Fairchild Camera and Instruments, *Annual Report for 1964*, 20; "Dr. Noyce Presentation to FCI," December 1964, Photographic Files of Steve Allen, 91–167, box 1963–1964, Archives and Special Collections, Stanford University; Bay, interview, July 2, 1996; Hulme, interview, March 18 and May 7, 1997; communication from Hulme, April 8, 1999; Lamond, interview, November 17, 1995, December 6, 1995, and August 12, 1997; Roger Smullen, interview by Lécuyer, May 12, 1996.

54. "Fairchild Semiconductor," 1967, Ephemera collection, Jackson Library, Stanford University; Larry Blaser, "A Transistor Stereo FM Multiplex Adapter," APP-44, March 1962, in Fairchild Semiconductor, *Product Catalog*, 1962, courtesy of Jay Last; Kvamme, interview, September 25 and October 10, 1996; Hulme, interview, March 18 and May 7, 1997; Kvamme, interview, September 25 and October 10, 1996.

55. Hulme, interview, May 7, 1997

56. MacDougall 1967; Jay Last, "Visit to Continental Devices—Hong Kong," November 28, 1966, courtesy of Jay Last; Hulme, interview, March 18 and May 7, 1997; communication from Hulme, January 13, 1997; Kvamme, interview, September 25, 1996; Gifford, interview, July 23, 1996.

57. Kvamme, interview, October 10, 1997.

58. Fairchild Camera and Instruments, *Annual Report for 1963*, 19; Progress Reports, Application Engineering, August 1, 1961, box 6, folder 6; August 31, 1961, box 6, folder 7; September 30, 1961, box 6, folder 8; October 31, 1961, box 6, folder 8—all in Technical Reports and Progress Reports, M 1055; Richard Lane, "A Synchronously Tuned 60 MC I.F. Amplifier with A.G.C.," APP-26, circa 1961, in

Fairchild Semiconductor, *Product Catalog,* 1962, courtesy of Jay Last; "All Transistor TV," *Leadwire,* fall 1962, courtesy of Michael Brozda; Cox, interview, October 25, 1996; Hulme, interview, March 18, 1997; Kvamme, interview, October 10, 1997; Yost, interview, October 12, 1997; Bay, interview, July 2, 1996.

59. For later price cuts, see "Manufacturers Outlook," *EDN,* volume 11, July 1966, 56–59; Cox, interview, October 25, 1996.

60. Sporck 2001; Sporck, interview by Rob Walker, February 21, 2000, Silicon Genesis, M 0741, box 2, Stanford Archives and Special Collections.

61. Eugene Kleiner, interview by Christophe Lécuyer, May 21, 1996, June 5, 1996, and June 16, 1998.

62. Sporck 2001; Sporck, interview, May 15, 1995; Lamond, interview, November 15, 1995, and December 6, 1995.

63. Sporck, interview, May 15, 1995; Smullen, interview, May 12, 1996.

64. "Dr. Noyce talk to FCI," photographic files of Steve Allen, 91–167, box: 1963–1964; Reithard 1967; Kleiner, interviews, May 21, 1996 June 5, 1996, and June 16, 1998; Sporck, interview, May 15, 1995; Smullen, interview, May 12, 1996.

65. "Turning a Science into an Industry, an Interview with Dr. Robert Noyce, a Young Scientist-Executive of the Integrated Circuits Industry," *IEEE Spectrum,* January 1966: 99–102, 101; "Reliability '65," Fairchild Semiconductor folder, computer ephemera collection, CBI 12, Charles Babbage Institute; Yost, interview, August 12 and September 23, 1997; Smullen, interview, May 12, 1996. For a more detailed treatment of Fairchild's process-engineering efforts, see Lécuyer 1999a,b.

66. "Dr. Noyce talk to FCI," photographic files of Steve Allen, 91–167, box: 1963–1964; Yost, interview, August 12 and September 23, 1997.

67. Julius Blank to Noyce, January 15, 1963, Technical Reports and Progress Reports, M 1055, box 7, folder 6; Chen 1971; Sporck 2001; Sporck, interview, May 15, 1995; Blank, interview by Lécuyer, June 20, 1996.

68. Sporck, interview, May 15, 1995.

69. Last, "Visit to Continental Devices—Hong Kong," November 28, 1966, courtesy of Jay Last; Stone, interview, April 21, 1995; Blank, interview by Lécuyer, June 20, 1996; communication from Fred Bialek, August 26, 1997.

70. MacDougall 1967; Last, "Visit to Continental Devices—Hong Kong"; Bay, interview, July 2, 1996; communication from Bay, October 2, 1997; Yost, interview, August 12 and September 23, 1997. On early work at Fairchild on plastic packages, see Moore to Noyce, January 8, 1960, box 5, folder 3; Moore and Grinich, "R&D Progress Report," August 11, 1961, box 6, folder 7; Progress Reports, Chemistry Section, January 1, 1960, February 1, 1960, March 1, 1960, and April 1, 1960—all in Technical Reports and Progress Reports, M 1055, box 5, folder 1.

71. Chen 1971; "Fairchild Semiconductor," *Solid-State Design*, September 1964, 43; Fairchild Camera and Instruments, *Annual Report for 1964*; Stone, interview, April 21, 1995; Sporck, interview, May 15, 1995; Lamond, interviews by Lécuyer, November 9, 1995, November 17, 1995, December 6, 1995, and August 12, 1997.

Chapter 6

1. Bob Lindsey, "Minuscular Miracles Boom Electronics; Little Things Making Huge Payrolls" *San Jose Mercury*, no date [circa 1965], courtesy of Don Liddie.

2. On the marketing of the micrologic family, see Robert Graham, "Micrologic sales situation," December 1, 1961, courtesy of Jay Last.

3. Moore, "Approximate Distribution of Effort as Defined in 10/4/60 Personnel Forecast," January 18, 1961, courtesy of Jay Last; Moore and Davis 2001.

4. Hoerni might have been dissatisfied with the payout they had received from Fairchild. In accordance with the original agreement with Fairchild Camera, the founders had sold their Fairchild Semiconductor stock to Fairchild Camera and Instruments for the equivalent of $250,000 in Fairchild Camera shares in 1959. While this sum was substantial (they could have bought twenty houses in Palo Alto with it), it looked paltry in regard with their accomplishments. The firm they had established had been wildly successful and radically transformed the semiconductor industry. A quarter million dollars looked like an awfully good deal for Fairchild Camera.

5. "Note of Meeting: The Role of Microelectronics in Weapon Systems," April 24, 1961, courtesy of Jay Last; "Outline for Presentation to be Given in Washington, D.C. on April 21, 1961," courtesy of Jay Last; Pete Carey, "The Hero of Venture Capitalists," *San Jose Mercury*, February 19, 1978; Peter Rossmassler, luncheon address, in Proceedings of the Conference on "New Business: The Art of Joining Innovative Technology, Management, and Capital," May 22–23, 1969, Boston College; Davis and Rock to limited partners, no date, courtesy of Jay Last; "Infant with a Giant Appetite," *Business Week*, January 11, 1964; "Making Big Waves with Small Fish," *Business Week*, December 30, 1967.

6. Last to parents, no date (probably early February 1961), courtesy of Jay Last.

7. "Drs. Hoerni, Last Resign Posts at Fairchild to Join Teledyne," *Electronic News*, February 13, 1961, 15.

8. Kattner, "Signetics History," October 2000, courtesy of Lionel Kattner; David James, interview by Lécuyer, December 14, 2000; Lionel Kattner, interview by Lécuyer, October 13 and 14, 2000.

9. Signetics Corporation, "Facilities and Capabilities Report," no date (probably 1962), courtesy of David Allison; James, interview by Lécuyer, December 14, 2000; Kattner, interview by Lécuyer, October 13 and 14, 2000; Allison, interview by Lécuyer, June 23, 2000.

10. Kattner, "Signetics History"; James, interview by Lécuyer, December 14, 2000; Kattner, interview by Lécuyer, October 13 and 14, 2000. On the evolution of the "cheapie" project, see Gordon Moore to Robert Noyce, January 9, 1961, in Technical Reports and Program Reports, M 1055, box 5, folder 9; Moore to Noyce, March 8, 1968, April 11, 1961, May 10, 1961, June 14, 1961, July 12, 1961, and August 11, 1961—all in Technical Reports and Progress Reports, M 1055, box 6, folders 2–6.

11. "Can History Repeat? Scientists Strike Out in Pioneer Electronics Field," *San Jose Mercury*, November 10, 1961.

12. Ibid. See also "New Electronics Company Unveils Its First Product," *San Jose Mercury*, February 27, 1962.

13. On TI's integrated circuit program in the early 1960s, see Kilby 1976. On the group's negotiations with Texas Instruments, see Kattner, "Signetics History"; Kattner, interview by Lécuyer, October 13 and 14, 2000; James, interview by Lécuyer, December 14, 2000; Allison, interview by Lécuyer, June 23, 2000.

14. Kattner, "Signetics History"; James, interview by Lécuyer, December 14, 2000; Kattner, interview by Lécuyer, October 13 and 14, 2000; Allison, interview by Lécuyer, June 23, 2000.

15. Kattner, Proposal to Timex, circa 1966, courtesy of Lionel Kattner; Kattner, "Signetics History"; James, interview by Lécuyer, December 14, 2000; Kattner, interview by Lécuyer, October 13 and 14, 2000; Allison, interview by Lécuyer, June 23, 2000. On attitudes at Fairchild regarding Signetics, see Moore to Noyce, September 13, 1961, Technical Reports and Progress Reports, M 1055, box 6, folder 7.

16. Last, "Amelco Formation Notes," 1961; Last to Peter Avakian, May 10, 1961; Last, "A Summary of the Microelectronics Program," August 1, 1961; Last to James Meindl, August 9, 1961; Amelco, Inc, "Capabilities and Qualifications," no date; Amelco Semiconductor employment figures, 1961–1965—all courtesy of Jay Last.

17. Amelco Electron device division, October 1961, courtesy of Jay Last.

18. Roberts worked on solar cells and transducers. Last, "A Summary of the Microelectronics Program," August 1, 1961, courtesy of Jay Last; Jim Battey, "The Transistor Business," December 1966, courtesy of Jay Last. On the development of the FADAC computer, see Teledyne, annual report for 1962. Last, communication to the author, April 9, 2002. On the development of planar transistors, see Hoerni, interview by Lécuyer, February 4, 1996. On Amelco's relations with military agencies, see Leon Steinman to Lowell Anderson, March 28, 1961; Last to Richard Alberts, June 1, 1961; Hoerni and Last to George Kozmetsky, June 30, 1961; Last to Meindl, August 9, 1961; "A proposal for integrated circuit electrical connection techniques," no date (probably February or March 1962) submitted to the Aeronautical Systems Division, Wright-Patterson Air Force base; "A Proposal for High Efficiency Transistor Structures for the US Army Signal Supply Agency, Fort Monmouth," March 5, 1962 —all courtesy of Jay Last.

19. Davis and Rock to limited partners, September 6, 1962; Last to Henry Singleton, May 7, 1962, courtesy of Jay Last.

20. Last to Singleton, May 7, 1962, courtesy of Jay Last.

21. Davis and Rock to limited partners, September 6, 1962; proposal for "thin film techniques for silicon integrated circuits," March 12, 1963, courtesy of Jay Last.

22. Jim Battey, "The Transistor Business," December 1966, courtesy of Jay Last; Last to Max Palevsky, May 10, 1962, courtesy of Jay Last.

23. Ever the contrarian, Hoerni developed N-channel field-effect transistors. Siliconix, the only other manufacturer of field-effect transistors, was selling P-channel devices. On Hoerni's work on field transistors, see "Minuteman Integrated Circuit Questionnaire," September 26, 1962, courtesy of Jay Last; Don Ganson to Hoerni, May 4, 1962, courtesy of Jay Last; Robert Robson to Robert Freund, June 28, 1962, courtesy of Jay Last; TSMC research department, "Teledyne," March 26, 1963, Frederick Terman Papers, SC160, series V, box 7, folder 1, Archives and Special Collections, Stanford University. Hoerni, interview by Lécuyer, February 4, 1996; Roberts, interview by Lécuyer, July 6, 1996; Teledyne, annual report for 1962, courtesy of Jay Last.

24. On the development of special assemblies at Amelco, see Hoerni, interview by Lécuyer, February 4, 1996. On the first Amelco integrated circuits, see Amelco, "Price List," May 1962, courtesy of Jay Last. On the development of DCTL integrated circuits, see Last "Research Meeting," February 1962; Proposal for US Army signal supply agency, Fort Monmouth "High Efficiency transistor structures," March 5, 1962; Last, "Microcircuitry," October 1962; Last to Michael Wolff, December 7, 1962; Last, "Integrated Circuits in Production," April 18, 1963; "Management Meeting," April 28, 1963; Last, "Logic Comparison," April 28, 1964; "OMIC Sales to Date," June 1965; Jim Battey, "The Transistor Business," December 1966—all courtesy of Jay Last; "OMIC," *Solid-state Design*, July 1964, 66; Teledyne, annual report for 1962; Teledyne, annual report for 1963.

25. Last to C. Craighead, April 26, 1963, courtesy of Jay Last.

26. On the cardiac monitor, see Last to Richard Allen, August 9, 1962 and June 10, 1963. On other hybrid circuits, see Last to C. Craighead, April 26, 1963; Last to Kozmetsky and Battey, August 26, 1963; Last, "R&D Income," December 27, 1963—all courtesy of Jay Last. Communication from Jay Last, April 9, 2002.

27. Last, "R&D Income," December 27, 1963; communication from Jay Last, April 9, 2002.

28. A. Alvarez, "Program Plan and Schedule for HHX-CPU Mechanization Study," August 15, 1963; Teck Wilson to George Kozmetsky, September 18, 1963; Kozmetsky to Wilson, September 19, 1963; Kozmetsky to Wilson, September 27, 1963; Wilson, "H-H(X) Microelectronics Summary," September 30, 1963; Kozmetsky, "Mountain View Visit by Meggs Brearley, Bu Weaps," October 16, 1963; Teledyne, Annual Report for 1963; Last to Battey, December 13, 1963; "Making Big Waves

with Small Fish," *Business Week*, December 30, 1967; Miller 1964a, 79; Last, "Teledyne Operational Amplifier Companies," August 3, 1967, courtesy of Jay Last.

29. Last, "Packaging of Integrated Circuits," February 12, 1964; Isy Haas, Memorandum, October 5, 1964; Last to Battey, June 2, 1964; Davis and Rock to limited partners, August 3, 1964; Miller 1964b, 79 and 81; "Avionics Order Nears for Teledyne," *Electronics*, November 16, 1964, 37–38; Teledyne, annual report for 1965; Last to Battey, "IHAS Circuits," February 9, 1965; Battey, "IHAS Program," April 2, 1965; Last, "Long Term Goals for Amelco Semiconductor," April 27, 1965; Mr. Madland," LSI without LSD," *Electronic Products*, August 1967, 10 and 113; "Making Big Waves with Small Fish," *Business Week*, December 30, 1967. On the state of Amelco's processing capabilities in early 1964, see Haas, "Microcircuit Design Rules and Production Techniques," circa May 1964, courtesy of Jay Last.

30. Davis and Rock to limited partners, September 6, 1962; Kozmetsky to Hoerni, April 1, 1963; courtesy of Jay Last; Minutes, management meeting, April 17, 1963; Minutes, management meeting, April 25, 1963; Minutes, sales-production meeting, April 29, 1963—courtesy of Jay Last.

31. On sales of Amelco products, see Teledyne, annual report for 1962; Teledyne, annual report for 1963; "OMIC sales," April 1963–June 1965, courtesy of Jay Last; Amelco, "Device Analysis, 1964," courtesy of Jay Last.

32. Allison, interview by Lécuyer, June 23, 2000; James, interview by Lécuyer, December 14, 2000.

33. James, cited in "The New Shape of Electronics," *Business Week*, April 14, 1962.

34. Signetics leased semiconductor processing equipment in order to reduce its expenses. On processing equipment, see Allison, interview by Lécuyer, June 23, 2000; James, interview by Lécuyer, December 14, 2000; Kattner, interview by Lécuyer, October 13 and 14, 2000. For Yelverton's and Baker's backgrounds, see Signetics, "Facilities and Capabilities Report" [circa 1962], courtesy of David Allison; Yelverton, interview by Lécuyer, December 18, 2000; Baker, interview by Lécuyer, December 15, 2000.

35. T. Pitts, "Production Report," May 8, 1962, courtesy of Orville Baker; Signetics, "Facilities and Capabilities Report," probably September 1962, courtesy of David Allison; James 1962; Kattner, interview by Lécuyer, October 13 and 14, 2000; Allison, interview by Lécuyer, June 23, 2000; James, interview by Lécuyer, December 14, 2000; Baker, interview by Lécuyer, December 15, 2000.

36. On Signetics' early orientation toward custom circuits, see Signetics Corporation, "Concept," 1961, courtesy of Don Liddie; press release, no date, courtesy of Lionel Kattner; "The New Shape of Electronics," Signetics advertisement in *Proceedings of the IRE, Electronic Daily,* and *Electronic News*, March 1962, courtesy of Orville Baker; Cunningham & Walsh; "Signetics Integrated Circuits Seminar," spring 1963, 59 and 74; Kattner, "Signetics History"; James, interview by Lécuyer, December 14, 2000; Baker, interview by Lécuyer, December 15, 2000; Allison, interview by Lécuyer, June 23, 2000.

37. Signetics, Facilities and Capabilities Report, probably September 1962, courtesy of David Allison.

38. Texas Instruments had introduced a family of RTL integrated circuits a few months earlier. RTL and DCTL circuits are closely related.

39. Signetics, "Facilities and Capabilities Report," no date, probably September 1962, courtesy of Lionel Kattner; Baker, autobiographical note, December 14, 2000; Kattner, "Signetics History"; Baker, interview by Lécuyer, December 15, 2000; Allison, interview by Lécuyer, June 23, 2000.

40. On the development of the first DTL family and related processes, see "New Electronics Company Unveils Its First Product," *San Jose Mercury*, February 27, 1962, 24; Signetics, "Facilities and Capabilities Report," probably September 1962, courtesy of Lionel Kattner; Signetics, Press release, December 19, 1963, courtesy of Lionel Kattner; Signetics, "Process Specifications," February 5, 1964, courtesy of Lionel Kattner; "A Big Industry That Thinks Small," *Corning Gaffer*, March 21, 1967, courtesy of Don Liddie; Kattner, "Signetics History"; Baker, interview by Lécuyer, December 15, 2000; Allison, interview by Lécuyer, June 23, 2000.

41. Graham and Shuldiner 2001, 271–274; Lay 1969, 146; James, interview by Lécuyer, December 14, 2000; Allison, interview by Lécuyer, June 23, 2000; Yelverton, interview by Lécuyer, December 18, 2000.

42. "Agreement between Corning Glass, Lehman Brothers, and the other Stockholders," November 19, 1962, courtesy of Lionel Kattner; Graham and Shuldiner 2001, 271–274; Baker, autobiographical note, December 14, 2000, courtesy of Orville Baker; Yelverton, interview by Lécuyer, December 18, 2000; Kattner, interview by Lécuyer, October 13 and 14, 2000; James, interview by Lécuyer, December 14, 2000; Baker, interview by Lécuyer, December 15, 2000.

43. For Signetics' sales literature, see Mark Weissenstern and Robert Beeson, "Constraints in Designing Integrated Circuits," *Electronic Design*, February 15, 1963; Signetics, "Spring Seminar," 1963, courtesy of Lionel Kattner. On preFEBs and variFEBs, see "Signetics variFEBs," January 1963; "Signetics preFEBs," January 1963; "Signetics Integrated Circuits: Condensed Catalog," March 1963—all courtesy of Orville Baker; Joseph Van Poppelen, interview by Lécuyer, December 6, 2000.

44. Van Poppelen, interview by Lécuyer, December 6, 2000.

45. James, interview by Lécuyer, December 14, 2000.

46. "Signetics Receives Army Contract for Integrated Circuits," *Electronic News*, July 15, 1963. On early attitudes regarding integrated circuits, see Cunningham & Walsh, "Signetics Integrated Circuits Seminar," spring 1963, courtesy of Lionel Kattner; Baker, interview by Lécuyer, December 15, 2000; Yelverton, interview by Lécuyer, December 18, 2000; James, interview by Lécuyer, December 14, 2000. On the use of Fairchild's DCTL circuits by the Instrumentation Laboratory, see Hall 1996, 2000.

47. Cited in Bridges 1963, 31–32

48. Kleiman 1966; Tucker 1968, 11–18; Signetics, "Spring Seminar," 1963, courtesy of Lionel Kattner.

49. Signetics might have also benefited from the fact that TI and Fairchild concentrated their effort on the large contracts they had received from Autonetics and the Instrumentation Laboratory respectively.

50. Allison, interview by Lécuyer, June 23, 2000; Van Poppelen, interview by Lécuyer, December 6, 2000; Liddie, interview by Lécuyer, December 6, 2001.

51. Graham and Shuldiner 2001, 271–274; Allison, interview by Lécuyer, June 23, 2000; Van Poppelen, interview by Lécuyer, December 6, 2001; Liddie, interview by Lécuyer, December 6, 2001; James, interview by Lécuyer, December 14, 2000; Kattner, interview by Lécuyer, October 13 and 14, 2000.

52. "Signetics Starts Expansion," *Electronic News,* November 14, 1963, 36; "Signetics to Build $5 Million Complex in Sunnyvale, Calif." *Electronic News,* January 20, 1964, 48. Van Poppelen, interview by Lécuyer, December 6, 2001. Allison, interview by Lécuyer, June 23, 2000; Kattner, interview by Lécuyer, October 13 and 14, 2000; Liddie, interview by Lécuyer, December 6, 2001.

53. "Turning a Science into an Industry, an Interview with Dr. Robert Noyce, a Young Scientist-Executive of the Integrated Circuits Industry," *IEEE Spectrum,* January 1966, 99–102.

54. By the early 1960s, Fairchild had turned away from its policy of no government development contracts. It accepted contracts for integrated circuits, mainly as a way of accommodating important customers that wanted the firm to take these contracts.

55. On this controversy, see Sporck 2001; Lamond, interview, October 23, 2001. According to Lamond, another source of tension between Norman and the marketing staff was the fact that Norman was uncomfortable about Fairchild's increasing commitment to the commercial markets.

56. Sporck 2001; "1964 IEEE Convention," *Solid-state Design,* May 1964, 46–49.

57. Pierre Lamond, "Progress Report-Device Development Department," September 3, 1963, 4–5, Technical Reports and Progress Reports, M 1055, box 8, folder May-August 1963; Lamond, "Progress Report-Device Development Department," November 1, 1963, 13, Technical Reports and Progress Reports, M 1055, box 9, folder September-December 1963; Lamond, "Progress Report-Device Development Department," December 2, 1963, 3 and 15, Technical Reports and Progress Reports, M 1055, box 9, bound volume; Lamond, "Progress Report-Device Development Department," December 30, 1963, 5–6, Technical Reports and Progress Reports, M 1055, box 9, bound volume; Lamond, "Progress Report-Device Development Department," April 1, 1964, 4, Technical Reports and Progress Reports, M 1055, box 9, folder 3; Lamond "Progress Report-Device Development

Department," May 1, 1964, 3, Technical Reports and Progress Reports, M 1055, box 10, folder 1.

58. On these price cuts, see Frederick Terman to John Hawkinson, June 1, 1964, SC 160, series V, box 7, folder 10. On this decision at Fairchild, see Sporck 2001; Sporck, interview by Walker, February 21, 2000. On the previous attempt to cripple Signetics, see Allison, interview by Lécuyer, June 23, 2001.

59. "Signetics Shows How in Semiconductors," *Business Week*, July 26, 1969, 41–42; communication from Don Liddie, March 26, 2002. On Riley's professional career, see Hoefler 1966b.

60. Kattner, "Timex proposal," circa 1966, courtesy of Lionel Kattner. This is confirmed by Allison, interview by Lécuyer, June 23, 2000; James, interview by Lécuyer, December 14, 2000; Kattner, interview by Lécuyer, October 13 and 14, 2000; Van Poppelen, interview by Lécuyer, December 6, 2001; Baker, interview by Lécuyer, December 15, 2000.

61. Of the original team, only Baker and Allison remained at Signetics. Sources: Moore and Davis 2001; Kattner, "Timex proposal," circa 1966, courtesy of Lionel Kattner; James, interview by Lécuyer, December 14, 2000; Kattner, interview by Lécuyer, October 13 and 14, 2000; Van Poppelen, interview by Lécuyer, December 6, 2001.

62. Sporck 2001, 134; Roger Smullen, interview by Lécuyer, May 12, 1996; Kvamme, interview, September 25, 1996 and October 10, 1996. For a list of Fairchild "design wins," see Miller 1964a.

63. "System Uses Integrated Circuits," *Electronics*, October 5, 1964, 75–77. RCA also introduced an integrated-circuit-based computer, the Spectra 70, in 1964.

64. "Turning a Science into an Industry, an Interview with Dr. Robert Noyce, a Young Scientist-Executive of the Integrated Circuits Industry," *IEEE Spectrum*, January 1966, 99–102; communications from Hulme, January 13, 1997 and April 8, 1999; Hulme, interview, March 18, 1997 and May 7, 1997.

65. See also Mathews 1964; Grinich 1964.

66. Edmond Pelta, cited in "System Uses Integrated Circuits," *Electronics*, October 5, 1964, 75–77, 76.

67. Moore, "Research and Development Progress Report for 1962," January 24, 1963, Technical Reports and Progress Reports, M 1055, box 7, folder 6; Koeper 1966. Automatic insertion machines and wave soldering were developed in the 1950s and introduced in the commercial computer industry in the early 1960s, see Flamm 1988; Batcher 1962; Danko 1962.

68. Rex Rice, "Progress Report, Digital Systems Research Department," November 3, 1963, Technical Reports and Progress Reports, M 1055, box 11, folder 1.

69. For early work in R&D on the in-line package, see Rice 1964; Progress Reports, Digital Systems Research Department, October 1, 1963, box 9, brown folder; November 1, 1963, box 11, brown folder; December 3, 1963, box 9, brown folder; January 6, 1964, box 9, brown folder; March 3, 1964, box 9, folder 2; April 1, 1964, box 9, folder 3; May 5, 1964, box 10, folder 1; June 3, 1964, box 10, folder 2; June 30, box 10, folder 3; November 3, 1964, box 11, folder 1; August 7, 1964, box 10, folder 4; September 2, 1964, box 10, folder 6; October 1, 1964, box 10, folder 6—all in Technical Reports and Progress Reports, M 1055. See also Progress Reports, Device Development Section, July 1, 1964, box 10, folder 3; August 1, 1964, box 10, folder 4; September 1, 1964, box 10, folder 5 in Fairchild Semiconductor Papers, 88–095. On the redesign of the package from a single to a dual in-line configuration, see Progress Reports, Digital Systems Research Department, November 3, 1964, box 11, folder 1; December 2, 1964, box 11, folder 1; January 4, 1965, box 11, folder 2; Progress Reports, Device Development Section, November 1, 1964, box 10, folder 7; December 1, 1964, box 11, folder 1; January 1, 1965, box 11, folder 2—all in Technical Reports and Progress Reports, M 1055. For a more detailed treatment of the development of the dual in-line package, see Lécuyer 1999a,b.

70. On the adoption of the dual in-line, see Kvamme and Bieler 1966; Rogers 1968; "Microelectronics Goes Commercial," *EDN*, vol. 11, July 1966.

71. Fairchild Camera and Instruments, *Annual Report for 1964*, 20; Kvamme, interview by Lécuyer, September 25, 1996 and October 10, 1996; Lamond, interview by Lécuyer, October 23, 2001.

72. Communication from John Hulme, June 25, 2003.

73. Robert Widlar, "Biasing Scheme Especially Suited for Integrated Circuits," US patent 3,364,434 filed April 19, 1965, granted January 16, 1968; Widlar, "Low-Value Current Source for Integrated Circuits," US patent 3,320,439 filed May 26, 1965, granted May 16, 1967; communication from John Hulme, June 25, 2003; Jack Gifford, interview by Lécuyer, July 23, 1996.

74. The 709 had 14 transistors and 14 resistors.

75. US sales of linear circuits grew from $6 million in 1964 to $31 million in 1966 (communication from John Hulme, June 25, 2003; Jack Gifford, interview by Lécuyer, July 23, 1996). For a contemporary discussion of the linear circuit business, see Leeds 1967.

76. "What Made a High Flier Take Off at Top Speed," *Business Week*, October 30, 1965, 118–122. Jim Riley, Signetics' new general manager, re-oriented Signetics toward the commercial business. To penetrate the commercial market, Allison and Baker developed a new family of DTL circuits, the 600 series. Signetics also adopted Fairchild's dual in-line package. In order to lower its production costs and meet the price requirements of commercial users, Signetics set up a plant in South Korea in 1966.

77. "H-P Activity in ICs Stirs Speculation," *Electronic News,* July 5, 1965; Barney 1967; "ICs: New Generation of Mighty Midgets," *Measure,* July 1967, 2–5; Band et al. 1967; O'Brien and McMains 1967; Rotsky 1988.

78. Linvill et al. 1964; Pritchard 1968; Linvill, "Notes for Conversation with FET," September 25, 1972, in Frederick Terman Papers, SC 160, series XIV, box 11, folder 1; Linvill and Hogan 1977; Linvill, interview by Lécuyer, April 25 and May 30, 2002; communication from Jacques Beaudoin, December 7, 1995; communication from Linvill, August 21, 2002. On the solid-state electronics program at Stanford, see Lécuyer 2005.

Chapter 7

1. For accounts of Intel, see Bassett 2002; Berlin 2001b; Moore 1994; Moore 1996a.

2. This entrepreneurial flowering was not exclusive to the Peninsula. New integrated circuit firms sprung up in other parts of the country. For example, a group of Texas Instruments engineers formed Mostek. IBM spawned a few integrated circuit firms such as SEMI and Cogar Corporation. General Instruments spun off MOS Technology. But most of the new corporations entering the semiconductor industry located in the Santa Clara Valley.

3. Jean Hoerni, "Proposal for Custom Integrated Circuit Organization," no date [1966], courtesy of Jack Yelverton.

4. Thomas Bay and other sales and marketing managers at Fairchild invested in Data General, a spinoff of Digital Equipment Corporation. A sales manager at Fairchild also became Data General's first marketing manager.

5. Kenney and Florida 2000; Eugene Kleiner, interview by Lécuyer, May 21, 1996, June 5, 1996, and June 16, 1998; Sheldon Roberts, interview by Lécuyer, July 6, 1996; Marshall Cox, interview by Lécuyer, October 25, 1996.

6. John Hulme, telephone conversation, June 25, 2003. On Molectro, see "Molectro Opens New Plant, Lab in Santa Clara," *Palo Alto Times,* August 16, 1963, collection of C. Lécuyer; "Santa Clara Firm Acquired by Company on East Coast," *Palo Alto Times,* July 26, 1965, collection of C. Lécuyer; "NSC Assumes Molectro Assets," *Palo Alto Times,* August 8, 1966, collection of C. Lécuyer; "New Digital Logic Circuits Only from Molectro," *Solid-state Design,* July 1964, 59; "Molectro Corp." *Solid-state Design,* July 1964, 74; "Molectro Announces RTL Integrated Circuits," *Solid-state Design,* September 1964, 39.

7. Lamond had an option for 1,000 shares. Roger Smullen had an option for 50 shares. "What Made a High Flier Take off at Top Speed," *Business Week,* October 30, 1965, 118 and 120; Prospectus "Fairchild Camera and Instruments," June 7, 1966, ephemera collection, Jackson Library, Stanford University; Hoefler 1968a; Sporck 2001; Sporck, interview by Lécuyer, May 15, 1995; Lamond, interview by Lécuyer, November 9, 1995, November 17, 1995, December 6, 1995, and August 12, 1997, October 23, 2001; Roger Smullen, interview by Lécuyer, May 12, 1996;

Floyd Kvamme, interview by Lécuyer, September 25, 1996 and October10, 1996; Roger Borovoy, interview by Lécuyer, June 14, 1996 and October 23, 1996.

8. Sporck 2001; Sporck, interview by Lécuyer; Lamond, interview by Lécuyer.

9. Sporck 2001; Kvamme 2000; National Semiconductor Corporation, "Proxy Statement for the Annual Meeting of Stockholders to be Held on September 24, 1971," August 20, 1971, ephemera collection, Jackson Library; Sporck, interview by Lécuyer; Lamond, interview by Lécuyer; Regis McKenna, interview by Lécuyer, Ross Bassett, and Henry Lowood, January 29, 1996.

10. On the impact of Sporck's departure at Fairchild, see Borovoy, interview by Lécuyer; Joseph Van Poppelen, interview by Lécuyer, December 6, 2001. On Fairchild Semiconductor's growing difficulties in 1967 and 1968, see "Reshaping Fairchild," *Electronics,* July 24, 1967, 44; "Going Down," *Electronics,* October 30, 1967, 46; "FC&I Resigns; Earnings Plummet," *Electronic News,* October 23, 1967, 3; "Fast Footwork in an Industry Talent Hunt," *Business Week,* March 11, 1967; Victor McElheny, "Dissatisfaction as a Spur to Career," *New York Times,* December 15, 1976; Thomas Bay, interview by Lécuyer, July 2, 1996.

11. There were also rumors in the business press that Fairchild Camera had "cooked its books" in 1965 and 1966. See articles cited in previous note.

12. "Going Down," *Electronics,* October 30, 1967, 46; Moore 1996a; Noyce, interview, no date, Intel Museum; Gordon Moore, interview by Lécuyer and Ross Bassett, February 18, 1997; Nelson Stone, interview by Lécuyer, April 21, 1995.

13. Moore 1996a; Noyce, interview, no date, Intel Museum; Moore, interview by Lécuyer and Ross Bassett.

14. On the massive hirings of Motorola engineers and managers, see Hoefler 1968b; "The Fight That Fairchild Won," *Business Week,* October 5, 1968, 106–108, 112–115.

15. On the formation of Intersil, see Hoerni "Proposal for Custom Integrated Circuit Organization," no date (1966), courtesy of Jack Yelverton; "Dr. Jean Hoerni Starting Own Semiconductor Company," *Palo Alto Times,* August 3, 1967, collection of C. Lécuyer; "Hoerni: From the Alps to the Mesas," *Electronic Engineering Times,* May 21, 1973, 1 and 8; Hoerni, interview by Lécuyer, February 4, 1996; John Hall, interview by Lécuyer, September 5, 2003; Luc Bauer, interview by Lécuyer, September 16, 2003. On the first quartz watch, see "IC Time," *Electronics,* March 6, 1967, 357; Forrer 1969.

16. Fields 1969a, 129–130, 132, and 137; Murray 1972; Sporck 2001; Sporck, interview by Lécuyer, May 15, 1995.

17. On transfer problems at Fairchild, see Bassett 2002; Berlin 2001a; Moore 1996a,b.

18. Kvamme 2000; communication from Floyd Kvamme, September 3, 2003.

19. Fairchild defended itself rather poorly against National's raids—for several reasons. Sporck had played a major part in Fairchild's growth. Many Fairchild managers also owned National shares. They had bought these shares after Widlar's and Talbert's move to National and did not want to jeopardize their investments. As a result, the company did not go aggressively after Sporck and his acolytes. In 1967, Fairchild's managers forced Sporck to sign an agreement, whereby National would not hire Fairchild's employees in the next 6 months. But Sporck and his troops found creative ways to circumvent this agreement. For example, they brought in Fairchild people as consultants rather than as regular employees (some remained consultants for years). They also convinced some of their Fairchild recruits to work at another semiconductor firm for a few months before moving to National. After Hogan's takeover, Fairchild was even less successful in slowing down the brain drain. Hogan granted more stock options than Carter and Hodgson ever did. But these options proved ineffectual—as the prospects of Fairchild Camera stock did not seem promising. The company also stopped defending itself legally against these raids. During this period, the legal department focused on fighting a lawsuit from Motorola, which alleged that Lester Hogan, Fairchild's new CEO, had stolen trade secrets from Motorola. Sporck 2001; "War Drums Start to Beat in Phoenix," *Electronic News*, August 19, 1968, 12; Roger Borovoy, interview by Lécuyer and Ross Bassett, June 14, 1996 and October 23, 1996; Sporck, interview by Lécuyer, May 15, 1995; Lamond, interview by Lécuyer, November 9, 1995, November 17, 1995, and December 6, 1995; Smullen, interview by Lécuyer, May 12, 1996; Kvamme, interview by Lécuyer, September 25, 1996 and October 10, 1996; Stone, interview by Lécuyer, April 25, 1995.

20. Sporck 2001; Sporck, interview by Lécuyer, May 15, 1995; Sporck, interview by Rob Walker, February 11, 2000, Silicon Genesis, M0741, box 2, Stanford Archives and Special Collections; Lamond, interview by Lécuyer, November 9, 1995, November 17, 1995, and December 6, 1995; Fred Bialek, interview by George Rotsky, no date, George Rotsky collection, M851 (96 -074), Archives and Special Collections, Stanford University. For National's accounting scheme, see Regis McKenna, interview by Lowood, Bassett, and Lécuyer, January 29, 1996 and communication from Kvamme, September 3, 2003.

21. National Semiconductor, *1968 Annual Report*, ephemera collection, Jackson library, Stanford University; Sporck, interview by Lécuyer, May 15, 1995; Lamond, interview by Lécuyer, November 9, 1995, November 17, 1995, and December 6, 1995; Lamond, talk given to the History of Silicon Valley seminar, Stanford University, October 28, 2001, collection of C. Lécuyer.

22. "Clever Move," *Electronics*, April 1, 1968, 34–35; Fields 1969a.

23. On Widlar's clout at National Semiconductor, see Sporck 2001; Smullen, interview by Lécuyer, May 12, 1996; Sporck, interview by Lécuyer, May 15, 1995; Sporck, interview by Walker, February 21, 2001.

24. Leeds 1967; "Design Ingenuity Is the Key to Success," *Electronics*, August 7, 1967, 100–106; National Semiconductor Corporation, *Annual Report*, 1968; "Clever Move," *Electronics*, April 1, 1968, 34–35; Fields 1969a,b; Gold 1991; McKenna, inter-

view by Lécuyer, Bassett, and Lowood, January 29, 1996; Smullen, interview by Lécuyer, May 12, 1996.

25. On MOS circuits at National, see National Semiconductor Corporation, *Annual Report*, 1968; Fields 1969a. On TTL at National, see "They're at it again," *Electronics*, October 16, 1967, 41–42; Fields 1969a; LeVieux, "The Historical Context of National Semiconductor," February 26, 1991, courtesy of Michael Brozda; Sporck 2001; Lamond, interview by Lécuyer, November 9, 1995, November 17, 1995, and December 6, 1995; Smullen, interview by Lécuyer, May 12, 1996..

26. Fields 1969a; LeVieux, "The Historical Context of National Semiconductor"; Sporck 2001; "Equal Opportunity: National/TTL," *Electronic Design*, September 27, 1969, 12–13; Lamond, interview by Lécuyer, November 9, 1995, November 17, 1995, and December 6, 1995; Smullen, interview by Lécuyer, May 12, 1996; Orville Baker, interview by Lécuyer, December 15, 2000; Jerry Sanders, interview by Rob Walker, October 18, 2002, Silicon Genesis, M0741; Lamond, interview by Lécuyer, November 9, 1995, November 17, 1995, and December 6, 1995.

27. Sporck 2001; Murray 1972; Lamond, interview by Lécuyer, November 9, 1995, November 17, 1995, and December 6, 1995; McKenna, interview by Henry Lowood, Ross Bassett, and Lécuyer, January 29, 1996.

28. Sporck 2001; Lamond, interview by Lécuyer, November 9 and 17 and December 6, 1995.

29. Communication from Kvamme, September 3, 2003.

30. Ibid. For a typical applications note written by Widlar, see Widlar, "Monolithic OP Amp—The Universal Linear Component," National Semiconductor Applications note 4, April 1968, collection of C. Lécuyer. On National's applications engineering effort, see Larry LeVieux, "The Historical Context of National Semiconductor: How We Got Here and Where We Are Going," February 26, 1991, courtesy of Michael Brozda.

31. Sporck 2001; Sporck, interview by Lécuyer; Sporck, interview by Rob Walker; communication from Kvamme, September 3, 2003. On the Singapore government's industrial policies, see Yew 2000, 49–69; Mathews and Cho 2000; Mathews 1999.

32. National also sold microcircuits to military system contractors. In 1970, roughly 10% of its integrated circuit output went to avionics companies and other manufacturers of military hardware. National Semiconductor, *Prospectus*, August 28, 1970, amended October 14, 1970, in ephemera collection, Jackson Library, Stanford University; communication from Kvamme, September 3, 2003.

33. There were a few exceptions such as AMD. AMD focused exclusively on the second sourcing of linear and logic circuits. It produced circuits designed by other firms—companies such as Fairchild, Texas Instruments, and National Semiconductor. In the late 1960s, AMD built a large business producing Fairchild's linear circuits. It also copied Fairchild's DTL line. To differentiate itself from other

second-source companies, AMD produced these devices to military specifications. This enabled AMD to address the reliability requirements of the military market and at the same time create a reputation for quality. AMD further innovated by naming its products with the same numerals as the original circuits. This helped customers to identify easily which circuits AMD was second sourcing. Fairchild and other companies sued AMD for using the same numerals, but the courts sided with AMD. This enabled AMD to grow rapidly in the early 1970s. By 1972, AMD had $11 million in sales.

34. There is a large literature on Intel, produced mostly by its founders and its executives. For examples of this literature, see Moore 1994, 1996a. For a journalistic account of Intel, see Jackson 1997. See also Bassett 2002; Berlin 2001b. For IBM's decision to use monolithic memories, see Santo 1988; Pugh et al. 1991, 458–476; Bassett 2002, 100–106. For forecasts of the growth of the memory market, see Sideris 1973.

35. This phenomenon can be explained in the following way: Engineers at user firms would invest substantial time and efforts in understanding the first circuit and learning how to use it in their designs. They had little incentive to make similar intellectual investments for later circuits.

36. Lydon 1966; Lindgren 1969; communication from Luc Bauer, September 16, 2003.

37. "Watch out," *Electronics*, February 5, 1968, 209; Hoerni, interview by Lécuyer, February 4, 1996; communication from Roger Wellinger, October 30, 2003. See also Forrer et al. 2002.

38. Hoerni, interview by Lécuyer, February 4, 1996; Hall, interview by Lécuyer, September 5, 2003.

39. It was a frequency divider circuit.

40. Hoerni, interview by Lécuyer, February 4, 1996; Hall, interview by Lécuyer, September 5, 2003. On RCA's MOS program, see Herzog 1975–76.

41. Hoerni, interview by Lécuyer, February 4, 1996; Hall, interview by Lécuyer, September 5, 2003.

42. Hoerni, interview by Lécuyer, February 4, 1996.

43. "Watch out," *Electronics*, February 5, 1968, 209; Hoerni, interview by Lécuyer, February 4, 1996.

44. Hoefler 1971a; Hall, interview by Lécuyer, September 5, 2003; Hoerni, interview by Lécuyer, February 4, 1996.

45. "Silicon Gates Opening Up for MOS," *Electronics*, September 15, 1969, 67–68; Sideris 1973; Moore 1994, 1996.

46. Gene Flath, interview with Noyce, Bill Davidow, and Gene Flath, August 13, 1983, Intel archives.

47. See also Santo 1988.

48. Hoff 1970; Sideris 1973, 108–113; Santo 1988; interview with Noyce and Davidow, August 13, 1983, Intel archives.

49. "MOS, Linears Bloom in a Flat Land," *Electronics*, August 17, 1970, 71–75; Lundell 1971.

50. "Intel Sees New Products Cutting Into Profit Margins," *Electronic News*, March 4, 1974, 59.

51. "FC&I Sustains $1,614,000 1st Qtr. Loss," *Electronic News*, May 10, 1971; "2d-Quarter FC&I Report Less Gloomy Than Feared," *Electronic News*, August 2, 1971; "Fairchild Camera Had Loss in Third Quarter," *Wall Street Journal*, October 28, 1971; "FC&I Reports $3 million loss in 3d quarter, $5.9 million in 9 Mos," *Electronic News*, November 11, 1971; Rhea 1971; "Fairchild Profit in Quarter Signals Semicon Rebound," *Electronic News*, May 1, 1972; "Signetics headcounts," courtesy of Don Liddie.

52. Amelco left the TTL business in 1970. Sporck 2001; "54/74 or FIGHT: IC PRICE WAR," *Electronic News*, January 5, 1970, 1 and 36.

53. "54/74 or Fight: IC Price War," *Electronic News*, January 5, 1970, 1 and 36; "MOS, Linears Bloom in a Flat Land," *Electronics*, August 17, 1970, 71–75; Sporck 2001; National Semiconductor, *Prospectus*, September 28, 1972, ephemera collection, Jackson Library, Stanford University; Lamond, talk at the DBF, July 10, 1995, collection of C. Lécuyer; Kvamme, interview by Lécuyer, September 3, 2003.

54. Sheets 1970; "3-State TTL Line Expanded," *Electronics*, October 12, 1970, 157–158; "National Presents the Tri-State-of-the-Art," *Electronics*, May 24, 1971, 9–12; Kvamme, interview by Lécuyer, September 3, 2003.

55. Weber 1967, 124; "Interview with Robert Noyce," *Business Week*, April 14, 1980, 98–99; Noyce and Hoff 1981; interview with Noyce and Davidow, August 13, 1983, Intel archives; Moore 1996, 60.

56. "New Demands on Semicons," *Electronic News*, May 18, 1970, 1 and 18; Hoefler 1970a,b; "Court Stays Intersil on Talent Raids," *Electronic News*, April 20, 1970, 1 and 22; Hoerni, interview by Lécuyer, February 4, 1996.

57. "Solid-state Booms in First 5 Months," *Electronics*, August 28, 1972, 30–31; Altman 1972; "MOS, Bipolar ICs Whirl Semi Sales to New Heights," *Electronics*, January 4, 1973, 76–80; Loehwing 1973; Riley 1973; "Boom Times Again for Semiconductors," *Business Week*, April 20, 1974, 65–78.

58. "Marriage of Intersils Confirmed," *Electronic News*, April 26, 1971; George Sideris, "Custom LSI Fades into Background," *Electronics*, January 10, 1974, 74 and 76; Hoerni, interview by Lécuyer, February 4, 1996; communication from Luc Bauer, September 16, 2003.

59. "Production 'Finesse' Permits the 'Barely Possible' at Intel," *Peninsula Electronics News*, December 31, 1973, 1 and 16; Moore 1996a, 87–92; Yu 1998, 83–99.

60. "IC Aimed at Consumer Job," *Electronics*, January 17, 1972, 111; "Charlie Sporck: Mandate for Profit," *Electronic Engineering Times*, August 13, 1973, 1 and 16; Altman 1972, 69–70; "Notoriety: NatSem Vows Windy City Confab," *Electronics*, September 11, 1972, 7.

61. "National Semiconductor, *Prospectus*, September 28, 1972, ephemera collection, Jackson Library of Business, Stanford University; "Charlie Sporck: Mandate for Profit," *Electronic Engineering Times*, August 13, 1973, 1 and 16; Cole 1974; Sporck, interview by Lécuyer, May 15, 1995; communication from Kvamme, September 3, 2003.

62. Thomas Sege, "Report for Equipment Group, Fiscal Month of May 1970," June 12, 1970, Varian Associates Records, 73/65c, carton 5, folder Varian Associates miscellaneous, Bancroft Library; "Introducing Rod Palmborg Vacuum Division Salesman," *Varian Associates Magazine*, July 1971, 6–9, Russell and Sigurd Varian Papers, SC 345, series: Varian Associates, box 8, folder 8, Archives and Special Collections, Stanford University; "Semiconductor Firm Likes 'Porta-Test' Portability, Sensitivity," *Varian Associates Magazine*, January 1974, 6–7, Russell and Sigurd Varian, SC 345, series: Varian Associates, box 8, folder 11; "Remarks by Edward Barlow," February 20, 1975, SC 345, Russell and Sigurd Varian Papers, SC 345, series: Varian Associates, box 5, folder 8; "Remarks by Larry Hansen," *Varian Associates Magazine*, March 1977, 8–12, Russell and Sigurd Varian Papers, SC 345, series: Varian Associates, box 9, folder 3; "Remarks by Norman Parker," May 11 and 12, 1981, Russell and Sigurd Varian Papers, SC 345, series: Varian Associates, box 5, folder 3.

63. James Hait to Edward Ginzton, June 3, 1970 and Ginzton to Hait, June 11, 1970 in Edward Ginzton Papers, SC 330, box 4, folder Board of directors correspondence, 1970–71, Archives and Special Collections, Stanford University.

Conclusion

1. William Eitel, "Electronics Considered Pace-Setter in Region's Development," *Redwood City Tribune*, December 27, 1962, collection 64–94, San Mateo County History Museum, Redwood City.

2. This discussion of Apple Computer is based on the following: Steve Wozniak, talk at Silicon Valley History Seminar, Stanford University, November 12, 2001; Bill Coleman, talk at Silicon Valley History seminar, November 26, 2001; Freiberger and Swaine 2000; Linzmayer 2004.

Bibliography

Adams, Stephen. 2003. "Regionalism in Stanford's Contribution to the Rise of Silicon Valley." *Enterprise and Society* 4: 521–543.

Aitken, Hugh. 1985. *The Continuous Wave: Technology and American Radio 1900–1932.* Princeton University Press.

Allison, David. 1981. *New Eye for the Navy: The Origin of Radar at the Naval Research Laboratory.* Naval Research Laboratory.

Altman, Lawrence. 1972. "Semiconductor RAMs and Computer Mainframe Jobs." *Electronics,* August 28: 63–77.

Arnold, James, S. S. Dharmatti, and Martin Packard. 1951. "Chemical Effects on Nuclear Induction Signals from Organic Compounds." *Journal of Chemical Physics* 19: 507.

Asher, Norman, and Leland Strom. 1977. *The Role of the Department of Defense in the Development of Integrated Circuits.* Institute for Defense Analysis.

Atalla, M., E. Tannenbaum, and E. Scheibner. 1959. "Stabilization of Silicon Surfaces by Thermally Grown Oxides." *Bell System Technical Journal* 38: 749–783.

Band, Ian, Ed Hilton, and Max Schuller. 1967. "Implementing Integrated Circuits in HP Instrumentation." *Hewlett-Packard Journal,* August: 2–4.

Barling, G. 1991. "The Changing World of Avionics." *Measurement and Control* 24: 78–95.

Barney, Walter. 1967. "Hewlett-Packard's Do-It-Yourself ICs." *Electronics,* June 26.

Bashe, Charles, Lyle Johnson, John Palmer, and Emerson Pugh. 1986. *IBM's Early Computers.* MIT Press.

Bassett, Ross. 2002. *To the Digital Age: Research Labs, Start-Up Companies, and the Rise of MOS Technology.* Johns Hopkins University Press.

Batcher, R. 1962. "Product Engineering and Production." *Proceedings of the IRE* 50, May: 1289–1305,

Bay, Thomas. 1961. "New Problems in Marketing Semiconductors." *Solid State Journal,* August: 55–56.

Berlin, Leslie. 2001a. "Robert Noyce and Fairchild Semiconductor, 1957–1968." *Business History Review* 75: 63–102.

Berlin, Leslie. 2001b. Entrepreneurship and the Rise of Silicon Valley: The Career of Robert Noyce, 1956–1990. Ph.D. dissertation, Stanford University.

Bloch, Felix. 1946. "Nuclear Induction." *Physical Review* 70: 460–474.

Bloch, Felix. 1952. "William Webster Hansen, 1909–1949." *National Academy of Sciences Biographical Memoirs* 26: 121–137.

Bloch, Felix, William Hansen, and Martin Packard. 1946. "The Nuclear Induction Experiment." *Physical Review* 70: 474–485.

Boot, Henry and John Randall. 1976. "Historical Notes on the Cavity Magnetron." *IEEE Transactions on Electron Devices* 23, no. 7: 724–729.

Brandes, Stuart. 1976. *American Welfare Capitalism, 1880–1940.* University of Chicago Press.

Braun, Ernest, and Stuart MacDonald. 1982. *Revolution in Miniature.* Cambridge University Press.

Bresnahan, Timothy, and Manuel Trajtenberg. 1995. "General Purpose Technologies, 'Engines of Growth'?" *Journal of Econometrics* 65 83–108.

Bridges, James. 1957. "Progress in Reliability of Military Electronics Equipment during 1956." *IRE Transactions for Reliability and Quality Control* 11: 1–7.

Bridges, James. 1963. "Government Needs and Policies in the Age of Microelectronics." In *The Impact of Microelectronics,* ed. G. Jacobi and S. Weber. McGraw-Hill.

Bryant, John. 1990. "Microwave Technology and Careers in Transition: The Interests and Activities of Visitors to the Sperry Gyroscope Company's Klystron Plant in 1939–1940." *IEEE Transactions on Microwave Theory and Techniques* 38, no. 11: 1545–1558.

Bryant, John. 1994. "Generations of Radar." In *Tracking the History of Radar,* ed. O. Blumtritt et al. IEEE.

Buderi, Robert. 1996. *The Invention That Changed the World.* Simon and Schuster.

Bylinsky, Gene. 1973. "How Intel Won Its Bet on Memory Chips." *Fortune,* November.

Byrne, John. 1993. *The Whiz Kids: The Founding Fathers of American Business and the Legacy They Left Us.* Doubleday.

Calkins, Robert, and Walter Hoadley. 1941. *An Economic and Industrial Survey of the San Francisco Bay Area*. California State Planning Board.

Ceruzzi, Paul. 1989. *Beyond the Limits: Flight Enters the Computer Age*. MIT Press.

Ceruzzi, Paul. 1998. *A History of Modern Computing*. MIT Press.

Chen, Edward K. Y. 1971. The Electronics Industry of Hong Kong: An Analysis of its Growth. M. Soc. Sc. thesis, University of Hong Kong.

Christaller, Walter. 1966. *Central Places in Southern Germany*. Prentice-Hall.

Christensen, Clayton. 1997. *The Innovator's Dilemma: When New Technologies Cause Great Firms to Fail*. Harvard Business School Press.

Cole, Bernard. 1974. "So Different—But So Much the Same." *Electronics*, October 3: 86–87.

Committee for the Survey of Chemistry. 1965. *Chemistry: Opportunities and Needs*. National Research Council.

Coupling, J. J. 1948. "Maggie." *Astounding Science Fiction*, February: 77–94.

Culvers, G. 1962. "Derating Philosophy on Minuteman Transistors." In *Semiconductor Reliability*, volume 2, ed. W. Von Alven. Engineering Publishers.

Danko, S. F. 1962. "Printed Circuits and Microelectronics." *Proceedings of the IRE* 50: 937–945.

Davis, Thomas 1964. "Electronics and the Investor," *Western Electronics News*, March.

DeGrasse, Robert. 1984. "The Military and Semiconductors." In *The Militarization of High Technology*, ed. J. Tirman. Ballinger.

Deloraine, Maurice. 1976. *When Telecom and ITT Were Young*. Lehigh Books.

Department of Commerce, US. 1928. *Amateur Radio Stations of the United States*. Government Printing Office.

De Soto, Clinton. 1936. *Two Hundred Meters and Down: The Story of Amateur Radio*. American Radio Relay League.

Douglas, Susan. 1987. *Inventing American Broadcasting, 1899–1922*. Johns Hopkins University Press.

Erickson, Don. 1973. *Armstrong's Fight for FM Broadcasting*. University of Alabama Press.

Everson, George. 1949. *The Story of Television: The Life of Philo T. Farnsworth*. Norton.

Fagen, M. D., ed. 1975. *A History of Engineering and Science in the Bell System: The Early Years (1875–1925)*. Bell Telephone Laboratories.

Farnsworth, Elma. 1989. *Distant Vision: Romance and Discovery on an Invisible Frontier.* PemberlyKent.

Federal Trade Commision, US. 1924. *Report of the Federal Trade Commission in Response to House Resolution 548, Sixty Seventh Congress, Fourth Session.* Government Printing Office.

Fields, Stephen. 1969a. "Charting a Course to High Profits." *Electronics*, October 13.

Fields, Stephen. 1969b. "Monolithic Regulator Stands Alone." *Electronics*, September 29.

Fishbein, Samuel. 1995. *Flight Management Systems: The Evolution of Avionics and Navigation Technology.* Praeger.

Flamm, Kenneth. 1988. *Creating the Computer: Government, Industry, and High Technology.* Brookings Institution.

Forrer, Max. 1969. "A Flexure-Mode Quartz for an Electronic Wrist-Watch." In *Proceedings of the 33rd Annual Frequency Symposium*, Atlantic City.

Forrer, Max, René Le Coultre, André Beyner, and Henri Oguey. 2002. *L'Aventure de la Montre à Quartz: Mutation Technologique Initiée par le Centre Electronique Horloger, Neuchâtel.* Centredoc.

Freiberger, Paul, and Michael Swaine. 2000. *Fire in the Valley: The Making of the Personal Computer.* McGraw-Hill.

Freund, Robert. 1971. Competition and Innovation in the Transistor Industry. Ph.D. dissertation, Duke University.

Galison, Peter, Bruce Hevly, and Rebecca Lowen. 1992. "Controlling the Monster: Stanford and the Growth of Physics Research, 1935–1962." In *Big Science*, ed. P. Galison and B. Hevly. Stanford University Press.

Gebhard, Louis. 1979. *Evolution of Naval Radio-Electronics and Contributions of the Naval Research Laboratory.* Naval Research Laboratory.

Getting, Ivan. 1962. "Radar." In *The Age of Electronics*, ed. C. Overhage. McGraw-Hill.

Getting, Ivan. 1989. *All in a Lifetime: Science in the Defense of Democracy.* Vantage.

Gillmor, C. Stewart. 2005. *Fred Terman at Stanford: Building a Discipline, a University, and Silicon Valley.* Stanford University Press.

Ginzton, Edward. 1975. "The $100 Idea." *IEEE Spectrum* 2: 30–39.

Ginzton, Anne Cottrell, and Leonard Cottrell, eds. 1995. *Times to Remember: The Life of Edward L. Ginzton.* Blackberry Creek.

Glauber, J. 1946. "Radar Vacuum-Tube Developments." *Electrical Communication* 23: 306–319.

Gold, Martin. 1991. "Linear Industry Remembers Widlar." *Electronic Enginering Times*, March 11.

Golding, Anthony. 1971. The Semiconductor Industry in Britain and the United States: A Case Study in Innovation, Growth, and the Diffusion of Technology. D. Phil. thesis, University of Sussex.

Goldstein, Andrew. 1994. "Finding the Right Material: Gordon Teal as Inventor and Manager." in Frederick Nebeker ed. *Sparks of Genius: Portraits of Engineering Excellence.* IEEE.

Graham, Margaret, and Alec Shuldiner. 2001. *Corning and the Craft of Innovation.* Oxford University Press.

Grinich, Victor. 1964. "Semiconductors for Commercial Applications." *Solid State Design*, September: 12–13.

Grinich, Victor, and Horace Jackson. 1975. *Introduction to Integrated Circuits.* McGraw-Hill.

Guerlac, Henry. 1987. *Radar in World War II.* American Institute of Physics.

Gupta, Udayan, ed. 2000. *Done Deals: Venture Capitalists Tell Their Stories.* Harvard Business School Press.

Hablanian, M. 1984. "Comments on the History of Vacuum Pumps." *Journal of Vacuum Science and Technology A* 2: 118–125.

Hall, Eldon. 1996. *Journey to the Moon: The History of the Apollo Guidance Computer.* American Institute of Aeronautics and Astronautics.

Hall, Eldon. 2000. "From the Farm to Pioneering with Digital Control Computers: An Autobiography." *IEEE Annals of the History of Computing* April-June: 22–31.

Hansen, William 1938. "A Type of Electrical Resonator," *Journal of Applied Physics* 9: 654–663.

Harper, Charles. 1969. *Handbook of Electronic Packaging.* McGraw-Hill.

Harris, D., H. Lorenzen, and S. Stiber. 1978. "History of Electronic Counter-measures." In *Electronic Countermeasures*, ed. J. Boyd et al. Peninsula Publishing.

Hayes, Dennis. 1989. *Behind the Silicon Curtain: The Seduction of Work in a Lonely Era.* South End Press.

Headrick, Daniel. 1991. *The Invisible Weapon: Telecommunications and International Politics, 1851–1945.* Oxford University Press.

Headrick, Daniel. 1994. "Shortwave radio and Its Impact on International Telecommunications Between the Wars." *History and Technology* 11: 21–32.

Heinrich, Thomas. 2002. "Cold War Armory: Military Contracting in Silicon Valley." *Enterprise and Society* 3: 247–284.

Herzog, G. B. 1975–76. "COS/MOS from Concept to Manufactured Product." *RCA Engineer*, December/January: 36–38.

Hester, William. 1961. "Dr. E. M. Baldwin." *Solid State Journal*, March: 18–19.

Hevly, Bruce. 1994. "Stanford's Supervoltage X-Ray Tube." *Osiris* 9: 85–100.

Hilman, Julian. 1961. "Transistor Evaluation Program," In *Semiconductor Reliability*, volume 1, ed. J. Schwop and H. Sullivan. Engineering Publishers.

Hilman, Julian, and Frank Durand. 1962. "Prediction, Screening, and Specifications." In *Semiconductor Reliability*, volume 2, ed. W. Von Alven. Engineering Publishers.

Hine, Robert. 1981. *California Utopianism: Contemplations of Eden*. Boyd & Fraser.

Hine, Robert. 1983. *California's Utopian Colonies, 1850–1950*. University of California Press.

Hoddeson, Lillian. 1981. "The Discovery of the Point-Contact Transistor." *Historical Studies in the Physical Sciences* 12, no. 1: 41–76.

Hoefler, Don. 1966a. "I Didn't Raise My Boy to Be a Manager." *Electronics News*, October 17.

Hoefler, Don. 1966b. "From Cookware to Quad Gates." *Electronic News*, December 12: 20.

Hoefler, Don. 1968a. "Dr. Noyce Happy Doing His Thing." *Electronic News*, October 28.

Hoefler, Don. 1968b. "Underground Express to California." *Electronic News*, September 9: 2.

Hoefler, Don. 1970a. "Man with the Emerald Arm." *Electronic News*, April 17.

Hoefler, Don. 1970b. "Hear Riley Leaving Signetics for Intersil." *Electronic News*, August 31.

Hoefler, Don. 1971a. "Intersil Gets Seiko Pact." *Electronic News*, January 25.

Hoefler, Don. 1971b. "Fairchild Semicon, MOD Divvy Up Research Unit," *Electronic News*, November 1, 1971

Hoerni, Jean. 1960. "Planar Silicon Transistors and Diodes." Paper presented at Electron Devices Meeting, Washington.

Hoff, Ted. 1970. "Silicon-Gate Dynamic MOS Crams 1,024 Bits on a Chip." *Electronics*, August 3: 68–73.

Holbrook, Daniel. 1995. "Government Support of the Semiconductor Industry: Diverse Approaches and Information Flows." *Business and Economic History* 24, no. 2: 133–165.

Holbrook, Daniel. 1999. Technical Diversity and Technological Change in the American Semiconductor Industry, 1952–1965. PhD dissertation, Carnegie Mellon University.

Hull, Ross. 1929. "The Status of 28,000-kc. Communication." *QST*, January: 9–16.

Jackson, Tim. 1997. *Inside Intel: Andy Grove and the Rise of the World's Most Successful Chip Company.* Dutton.

Jacoby, Sanford. 1997. *Modern Manors: Welfare Capitalism Since the New Deal.* Princeton University Press.

James, David. 1962. "Automation: The Key to Integrated Circuit Manufacture?" *Solid State Design*, August: 63–64.

Kaplan, Fred. 1983. *The Wizards of Armageddon.* Simon and Schuster.

Kargon, Robert, and Stuart Leslie. 1994. "Imagined Geographies: Princeton, Stanford and the Boundaries of Useful Knowledge in Postwar America," *Minerva* 32: 121–143

Kargon, Robert, Stuart Leslie, and Erica Schoenberger. 1992. "Far Beyond Big Science: Science Regions and the Organization of Research and Development." In *Big Science*, ed. P. Galison and B. Hevly. Stanford University Press.

Kenney, Martin, and Richard Florida. 2000. "Venture Capital in Silicon Valley: Fueling New Firm Formation." In *Understanding Silicon Valley*, ed. M. Kenney. Stanford University Press.

Kilby, Jack. 1976. "Invention of the Integrated Circuit." *IEEE Transactions on Electron Devices* 23, July: 648–654.

Klass, Philip. 1959. "Reliability Is Essential Minuteman Goal." *Aviation Week*, October 19: 69–79.

Klass, Philip. 1962. "Defense to Speed Development, Cut Cost." *Aviation Week and Space Technology*, February 5: 26.

Kleiman, Herbert. 1966. The Integrated Circuit: A Case Study of Production Innovations in the Electronics Industry. Ph.D. dissertation, George Washington University.

Koeper, R. E. 1966. "Microcircuit Packaging Evolution." *EDN* 11, September 28: 52–65.

Kohl, Walter H. 1951. *Materials Technology for Electron Tubes.* Reinhold.

Kohl, Walter H. 1960. *Materials and Techniques for Electron Tubes.* Reinhold.

Kovit, Bernard. 1961. "Getting Reliable Components for Minuteman Guidance." *Space Aeronautics*, January: 143–160.

Kvamme, Floyd. 2000. "Life in Silicon Valley: A First-Hand View of the Region's Growth." In *The Silicon Valley Edge*, ed. C.-M. Lee et al. Stanford University Press.

Kvamme, Floyd and L. H. Bieler. 1966. *Fairchild Semiconductor Integrated Circuits*. Fairchild Semiconductor.

Lambert, J. S. 1958. "Air Force Reliability Program." *IRE Transactions PGRQC* 14, 17–21.

Lay, Beirne. 1969. *Someone Has to Make It Happen: The Inside Story of Tex Thornton, the Man Who Built Litton Industries*. Prentice-Hall.

Layton, Edwin. 1976. "Scientists and Engineers: The Evolution of the IRE." *Proceedings of the IEEE* 64: 1390–1392.

Lécuyer, Christophe. 1999a. Making Silicon Valley: Engineering Culture, Innovation, and Industrial Growth, 1930–1970. Ph.D. dissertation, Stanford University.

Lécuyer, Christophe. 1999b. "Silicon for Industry: Component Design, Mass Production, and the Move to Commercial Markets at Fairchild Semiconductor, 1960–1967." *History and Technology* 16: 179–216.

Lécuyer, Christophe. 2000. "Fairchild Semiconductor and Its Influence." In *The Silicon Valley Edge*, ed. C.-M. Lee et al. Stanford University Press.

Lécuyer, Christophe. 2003. "High Tech Corporatism: Management-Employee Relations in U.S. Electronics Firms, 1920s–1960s." *Enterprise and Society* 4: 502–520.

Lécuyer, Christophe. 2005. "What Do Universities Really Owe Industry? The Case of Solid State Electronics at Stanford." *Minerva* 43: 51–71.

Leeds, Mark. 1967. "Scrambling for Linear IC Business." *Electronics*, August 7: 89–99.

Leib, Keyston, and Company. 1928. *Kolster Radio Corporation: The Strategic Exploitation of a New Science*. Leib, Keyston & Company.

Lenoir, Timothy. 1997. *Instituting Science: The Cultural Production of Scientific Disciplines*. Stanford University Press.

Lenoir, Timothy, and Christophe Lécuyer. 1995. "Instrument Makers and Discipline Builders: The Case of NMR." *Perspectives on Science* 4: 97–165.

Le Play, Fréderic. 1864. *La Réforme Sociale en France*, volume 1. Henri Plon.

Le Play, Fréderic. 1878. *La Réforme Sociale en France*, volume 2. Alfred Manne.

Leslie, Stuart. 1992. *The Cold War and American Science: The Military-Industrial-Academic Complex at MIT and Stanford*. Columbia University Press.

Leslie, Stuart. 1993. "How the West Was Won: The Military and the Making of Silicon Valley." In *Technological Competitiveness*, ed. W. Aspray. IEEE.

Leslie, Stuart. 2001a. "Regional Disadvantage: Replicating Silicon Valley in New York's Capital Region." *Technology and Culture* 42: 236–264.

Leslie, Stuart. 2001b. "Blue Collar Science: Bringing the Transistor to Life in the Lehigh Valley." *Historical Studies in the Physical and Biological Sciences*, 42: 71–113.

Leslie, Stuart, and Robert Kargon. 1994. "Electronics and the Geography of Innovation in Post-War America." *History and Technology* 11: 217–231.

Leslie, Stuart and Robert Kargon. 1996. "Selling Silicon Valley: Frederick Terman's Model for Regional Advantage." *Business History Review* 70: 435–472.

Levin, Richard. 1982. "The Semiconductor Industry." In *Government and Technical Progress*, ed. R. Nelson. Pergamon.

Lindgren, Nilo. 1969. "The Splintering of Solid State Electronics." *Innovation* 8: 2–15.

Linvill, John, and Lester Hogan. 1977. "Intellectual and Economic Fuel for the Electronics Revolution." *Science* 195, March 18: 1107–1113.

Linvill, John, James Angell, and Robert Pritchard. 1964. "Integrated Electronics vs Electrical Engineering Education." *Proceedings of the IEEE* 52: 1425–1429.

Linzmayer, Owen. 2004. *Apple Confidential 2.0: The Definitive History of the World's Most Colorful Company.* No Starch Press.

Litton, Charles, and Phillip Scofield. 1925. The Development and Construction of the Electron Jet Recorder. Thesis for the Engineering Degree, Stanford University. Archives and Special Collections, Stanford University.

Loehwing, David. 1973. "Semiconductor Explosion: For the Versatile Little Chips, Markets and Uses Proliferate." *Barron's*, September 10.

Lösch, August. 1967. *The Economics of Location.* Wiley.

Lowen, Rebecca. 1990. Exploiting a Wonderful Opportunity: Stanford University, Industry, and the Federal Government, 1937–1965. Ph.D. dissertation, Stanford University.

Lowen, Rebecca. 1997. *Creating the Cold War University: The Transformation of Stanford.* University of California Press.

Lowood, Henry. 1989. *From Steeples of Excellence to Silicon Valley.* Varian Associates.

Lundell, E. Drake. 1971. "Memory Shops Fill Northern California." *Computerworld*, March 29.

Lydon, James. 1966. "Recruiting Fever Mounts in Microelectronics Field." *Electronic News*, July 25: 1, 40.

MacDougall, John. 1967. "Low Cost Semiconductors for the Consumer Market." *Electronics World* 78, no. 3: 37–39.

MacKenzie, Donald. 1990. *Inventing Accuracy.* MIT Press.

MacKenzie, Donald, and Boelie Elzen. 1996. "The Charismatic Engineer." In *Knowing Machines,* ed. D. MacKenzie. MIT Press.

Maclaurin, W. Rupert. 1949. *Invention and Innovation in the Radio Industry.* Macmillan.

Malone, Michael. 1985. *The Big Score: The Billion Dollar Story of Silicon Valley.* Doubleday.

Mann, F. J. 1946. "Federal Telephone and Radio Corporation: A Historical Overview, 1909–1946." *Electrical Communication* 23: 377–405.

Marshall, Alfred. 1890. *Principles of Economics.* Macmillan.

Marshall, Samuel. 1965. "The Men Who Make Our Machinery," *Semiconductor Products,*" October: 15.

Martinez, A. 1973. "Transistor Circuits in Television: Some Evolutionary Aspects." *Radio and Electronic Engineer* 43, no. 1/2: 103–114.

Mathews, John. 1999. "A Silicon Island of the East: Creating a Semiconductor Industry in Singapore." *California Management Review* 41, no. 2: 55–78.

Mathews, John, and Dong Sung Cho. 2000. *Tiger Technology: The Creation of a Semiconductor Industry in East Asia.* Cambridge University Press.

Mathews, Walter. 1964. "Promised IC Benefits Here, Dr. Moore Tells IEEE Section." *Electronic News,* November 30: 32.

McNally, J. O., G. H. Metson, E. A. Veazie, and M. F. Holmes. 1957. "Electron Tubes for the Transatlantic Cable System." *Bell System Technical Journal* 36: 163–188.

Meiboom, Saul. 1963. "Nuclear Magnetic Resonance." *Annual Review of Physical Chemistry* 14: 58.

Merritt, Ernest. 1932. "The Optics of Radio Transmission." *Proceedings of the Institute of Radio Engineers* 20: 29–39.

Miller, Barry. 1963. "Competition to Tighten in Semiconductors." *Aviation Week & Space Technology,* September 23: 56–68.

Miller, Barry. 1964a. "Microcircuitry Production Growth Outpaces Applications." *Aviation Week and Space Technology,* November 16: 76–89.

Miller, Barry. 1964b. "Navy May Buy Minimum Helicopter Navaid." *Aviation Space and Technology,* October 5: 79–81.

Miller, L. E. 1958. "The Design and Characteristics of a Diffused Silicon Logic Amplifier Transistor." *Proceedings of Wescon* 2: 132–140.

Millman, S., ed. 1984. *A History of Engineering and Science in the Bell System: Communications Sciences (1925–1980).* AT&T Bell Laboratories.

Misa, Thomas. 1985. "Military Needs, Commercial Realities, and the Development of the Transistor, 1948–1958." In *Military Enterprise and Technological Change*, ed. M. Smith. MIT Press.

Moncher, F. L., and D. T. Johnstone. 1963. "Production Controls for Reliability and Reproducibility." In *Aerospace Reliability and Maintainability Conference*. AIAA.

Moore, Gordon. 1965. "Cramming More Components onto Integrated Circuits." *Electronics*, April 19: 114–117.

Moore, Gordon. 1994. "The Accidental Entrepreneur." *Engineering & Science*, summer.

Moore, Gordon. 1996a. "Intel—Memories and the Microprocessor." *Daedalus* 125, no. 2: 55–80.

Moore, Gordon. 1996b. "Some Personal Perspectives on Research in the Semiconductor Industry." In *Engines of Innovation*, ed. R. Rosenbloom and W. Spencer. Harvard Business School Press.

Moore, Gordon. 1998. "The Role of Fairchild in Silicon Technology in the Early Days of 'Silicon Valley.'" *Proceedings of the IEEE* 86, no. 1: 53–62.

Moore, Gordon, and Kevin Davis. 2001. Learning the Silicon Valley Way. Discussion Paper No. 00–45, Stanford Institute for Economic Policy Research.

Moorhead, Otis. 1917. "The Manufacture of Vacuum Detectors." *IRE Proceedings* 5: 427–432.

Moorhead, Otis. 1921. "The Specifications and Characteristics of Moorhead Vacuum Valves." *IRE Proceedings* 9: 95–129.

Moreno, Theodore. 1963. "Government Pricing and Procurement Policies." *Microwave Journal*, January: 165–166.

Morgan, Jane. 1967. *Electronics in the West: The First Fifty Years*. National Press Books.

Morris, Peter. 1990. *A History of the World Semiconductor Industry*. Peregrinus.

Murray, Thomas. 1972. "Live Wire at National Semi." *Dun's*, August: 51–54.

Neufeld, Jacob. 1990. *Ballistic Missiles in the United States Air Force, 1945–1960*. Office of Air Force History.

Niven, John. 1987. *The American President Lines and Its Forebears, 1848–1984: From Paddlewheelers to Containerships*. University of Delaware Press.

Norberg, Arthur. 1976. "The Origins of the Electronics Industry on the Pacific Coast." *Proceedings of the IEEE* 64: 1314–1322.

Noyce, Robert, and Marcian Hoff. 1981. "A History of Microprocessor Development at Intel." *IEEE Micro* 1, no. 1: 8–21.

O'Brien, Thomas, and John McMains. 1967. "Integrated-Circuit Counters." *Hewlett-Packard Journal*, August: 9–13.

Olson, Hank and Al Jones. 1996. "Defiance in the West: The Heintz and Kaufman Story." *Antique Wireless Association Review* 10: 188–221.

O'Mara, Margaret Pugh. 2004. *Cities of Knowledge: Cold War Science and the Search for the Next Silicon Valley*. Princeton University Press.

Orth, Richard. 1964. "The Microwave Business: Problems or Opportunities?" *Microwave Journal*, September: 104–105

Packard, David. 1995. *The HP Way: How Bill Hewlett and I Built Our Company*. HarperCollins.

Page, Robert. 1988. "Early History of Radar in the US Navy." In *Radar Development to 1945*, ed. R. Burns. Peter Peregrinus.

Painter, R., et al. 1960a. "Across-Board Competency New Transistor Field Need," *Electronic News*, July 25.

Painter, R., et al. 1960b. "Transistor Entry Tightens," *Electronic News*, August 1.

Partner, Simon. 1999. *Assembled in Japan: Electrical Goods and the Making of the Japanese Consumer*. University of California Press.

Pease, Bob. 1991. "What's All This Widlar Stuff, Anyhow?" *Electronic Design*, July 25: 146–149.

Perrine, Charles. 1932. "Thirty-Three Watts Per Dollar from a Type 52." *QST*, September: 17–20.

Piore, Michael, and Charles Sabel. 1984. *The Second Industrial Divide: Possibilities for Prosperity*. Basic Books.

Porter, Michael. 1990. *The Competitive Advantage of Nations*. Free Press.

Porter, Michael. 1998a. "Clusters and the New Economics of Competition." *Harvard Business Review*, November-December: 77–90.

Porter, Michael. 1998b. *On Competition*. Harvard Business School Press.

Pratt, Haraden. 1969. "Sixty Years of Wireless and Radio: Autobiographical Reminiscences." *IEEE Spectrum* 6: 41–49.

Pratt, Haraden, and J. K. Roosevelt. 1944. Developments in the Field of Cable and Radio Telegraph Communication. AIEE technical paper 44–126. Haraden Pratt papers, 72/116, box 3, folder: articles, Bancroft Library, University of California, Berkeley.

Preist, Don. 1992. The Story of Eimac and the Remarkable Men Who Made It. Unpublished manuscript, Stanford Archives and Special Collections.

Price, Alfred. 1984. *The History of US Electronic Warfare.* Association of Old Crows.

Pritchard, Robert. 1968. "Training New Engineers." In *Conference Proceedings on the Impact of Microelectronics II.*

Pugh, Emerson. 1984. *Memories That Shaped an Industry: Decisions Leading to IBM System/360.* MIT Press.

Pugh, Emerson, Lyle Johnson, and John Palmer. 1991. *IBM's 360 and Early 370 Systems.* MIT Press.

Pursell, Carroll. 1976. "The Technical Society of the Pacific Coast, 1884–1914." *Technology and Culture* 17: 702–717.

Real, Mimi. 1984. *A Revolution in Progress: A History of Intel to Date.* Intel.

Reed, George. 1986. U.S. Defense Policy, U.S. Air Force Doctrine and Strategic Nuclear Weapon Systems, 1958–1964: The Case of the Minuteman ICBM. Ph.D. dissertation, Duke University.

Reed, Oscar. 1962. "Broadcasting Development Now Taking Place." *Proceedings of the IRE* 50, May: 837–847.

Reid, T. R. 1984. *The Chip: How Two Americans Invented the Microchip and Launched a Revolution.* Simon and Schuster.

Reithard, E. M. 1967. "Growing Two-Inch Silicon Crystals." *SCP and Solid State Technology,* May: 46–47.

Rhea, John. 1971. "FC&I Completes Shuffle." *Electronic News,* December 6.

Rhea, John. 1972. "Fairchild Cuts Back MOS Division." *Electronic News,* September 11.

Rice, Rex. 1964. "Systematic Procedures for Digital System Realization from Logic Design to Production." *Proceedings of the IEEE,* December: 1691–1702.

Riley, Wallace. 1973. "Semiconductor Memories Are Taking Over Data-Storage Applications." *Electronics,* August 2: 75–90.

Riordan, Michael, and Lillian Hoddeson. 1997. *Crystal Fire: The Birth of the Information Age.* Norton.

Rock, Arthur. 2000. "Arthur Rock & Co." In *Done Deals,* ed. U. Gupta. Harvard Business School Press.

Rodengen, Jeffrey. 2000. *The Legend of Litton Industries.* Write Stuff Enterprises.

Rogers, Bryant. 1968. "IC Packages: Which One To Use?" *Solid State Technology,* September: 45–46.

Romander, Hugo. 1933. "The Inverted Ultraudion Amplifier." *QST,* September: 14–18.

Rosenberg, Nathan. 1963. "Technological Change in the Machine Tool Industry, 1840–1910." *Journal of Economic History* 23: 414–443.

Rotsky, George. 1988. "The 30th Anniversary of the Integrated Circuit: Thirty Who Made a Difference." *Electronic Buyers' News*, September.

Rotsky, George. 1991. "Let's Just Say Bob Widlar Had Some Non-Linear Traits." *Electronic Engineering Times*, March 11: 137–138.

Ryerson, Clifford. 1962. "The Reliability and Quality Control Field from Its Inception to the Present." *Proceedings of the IRE* 50: 1321–1338.

Saad, Ted. 1964. "The Microwave Engineer." *Microwave Journal*, August: 12.

Sah, Chih-Tang, Robert Noyce, and William Shockley. 1957. "Carrier Generation and Recombination in P-N Junctions and P-N Junction Characteristics." *Proceedings of the IRE* 45: 1228–1243.

Santo, Brian. 1988. "1K-Bit RAM." *IEEE Spectrum* 25, no. 11: 108–112.

Saxenian, AnnaLee. 1994. *Regional Advantage: Culture and Competition in Silicon Valley and Route 128*. Harvard University Press.

Scheffler H. S. 1981. The Minuteman High Reliability Component Parts Program: History and Legacy. Report C81–451/201, Rockwell International.

Schrecker, Ellen. 1998. *Many Are the Crimes: McCarthyism in America*. Little, Brown.

Scranton, Philip. 1989. *Figured Tapestry: Production, Markets, and Power in Philadelphia Textiles, 1885–1941*. Cambridge University Press.

Scranton, Philip. 1991. "Diversity in Diversity: Flexible Production and American Industrialization, 1880–1930." *Business History Review* 65: 27–90.

Scranton, Philip. 1997. *Endless Novelty: Specialty Production and American Industrialization, 1865–1925*. Princeton University Press.

Seidenberg, Philip. 1997. "From Germanium to Silicon: A History of Change in the Technology of Semiconductors." In *Facets: New Perspectives on the History of Semiconductors*, ed. A. Goldstein and W. Aspray. IEEE.

Sheets, John. 1970. "Three-State Switching Brings Wired OR to TTL." *Electronics*, September 14: 78–84.

Shima, Mastoshi, and Federico Faggin. 1974. "In Switch to N-MOS Microprocessor Gets a 2-μs Cycle Time." *Electronics*, April 18: 95–100.

Shoolery, James. 1993. "NMR Spectroscopy in the Beginnings." *Analytical Chemistry* 65: 731A-741A.

Sideris, George. 1973. "The Intel 1103: The MOS Memory That Defied Cores." *Electronics*, April 26: 108–113.

Siegel, Lenny, and John Markoff. 1985. *The High Cost of High Tech.* Harper & Row.

Sinclair, Donald. 1965. *The General Radio Company, 1915–1965.* Newcomen Society.

Skinner, Wickham, and David Rogers. 1968. *Manufacturing Policy in the Electronics Industry: A Casebook of Major Production Problems.* Irwin.

Smith, Merritt Roe, ed. 1985. *Military Enterprise and Technological Change: Perspectives on the American Experience.* MIT Press.

Smith, W. J. 1963. "Minuteman Guidance System—Evaluation of Achieved Reliability." In *Aerospace Reliability and Maintainability Conference.* AIAA.

Smits, F. M., ed. 1985. *A History of Engineering and Science in the Bell System: Electronics Technology (1925–1975).* AT&T Bell Laboratories.

Sobel, Robert. 1982. *ITT: The Management of Opportunity.* Time Books.

Sobel, Robert. 1986. *RCA.* Stein and Day.

Southworth, George. 1962. *Forty Years of Radio Research.* Gordon and Breach.

Spiegel, J. 1961. "Military System Reliability: Some Department of the Air Force Contributions." *IRE Transactions on Reliability and Quality Control* 10: 53–63.

Spiegel, J., and E. M. Bennett. 1961. "Military System Reliability: Department of Defense Contributions." *IRE Transactions on Reliability and Quality Control* 10: 1–8.

Sporck, Charles. 2001. *Spin-Off: A Personal History of the Industry That Changed the World.* Saranac Lake Publishing.

Staller, Jack. 1965. "The Packaging Revolution." *Electronics* 38, October 18: 72–87 and November 1: 75–87.

Stearns, Myrl 1963. "Thomson-Varian S.A.: Microwave Industry and the Common Market." *Microwave Journal,* June: 24–25

Stine, Jeffrey. 1992. "Scientific Instrumentation as an Element of U.S. Science Policy: National Foundation Support of Chemistry Instrumentation." In *Invisible Connection,* ed. R. Bud and S. Cozzens. SPIE Optical Engineering Press.

Stokes, John W. 1982. *70 Years of Radio Tubes and Valves.* Vestal.

Stranix, Richard. 1960. "Minuteman Reliability: Guide for Future Component Manufacturing!" *Electronics Industries,* December: 90–104.

Sturgeon, Timothy. 2000. "How Silicon Valley Came to Be." In *Understanding Silicon Valley,* ed. M. Kenney. Stanford University Press.

Taylor, A. Hoyt. 1961. *Radio Reminiscences: A Half Century.* Naval Research Laboratory.

Terman, Frederick. 1938. *Fundamentals of Radio.* McGraw-Hill.

Thiessen, Arthur. 1965. *A History of the General Radio Company.* General Radio.

Thornton, James. 1970. *Design of a Computer: The Control Data 6600.* Scott, Foresman.

Thornton, James. 1980. "The CDC 6600 Project." *Annals of the History of Computing* 2, no. 4: 338–348.

Tucker, Gardiner. 1968. "A Defense View of Microelectronics and Total Cost of Ownership." in Conference Proceedings of Impact of Microelectronics, II, Chicago.

Tyne, Gerald. 1977. *Saga of the Vacuum Tube.* Howard Sams.

Ueyama, Takahiro, and Christophe Lécuyer. In press. "Initiating Science-Based Medicine: The Medical Accelerator, Cancer Therapy, and the Formation of the Stanford University Medical Center, 1945–1978."

van Keuren, David. 1994. "The Military Context of Early American Radar, 1930–1949." In *Tracking the History of Radar,* ed. O. Blumtritt et al. IEEE.

Varian, Dorothy. 1983. *The Inventor and the Pilot.* Pacific Books.

Varian, Russell, and Sigurd Varian. 1939. "A High Frequency Oscillator and Amplifier." *Journal of Applied Physics* 10: 321–327.

Waits, Robert. 1997. "Semiconductor Manufacturing and Vacuum Technology: A Memoir." *Solid State Technology* 40, no. 5.

Walker, Gerald. 1972. "Consumer Gear Benefits from IC Design Tradeoffs." *Electronics,* August 14: 93–107.

Weber, Samuel. 1967. "LSI: The Technologies Converge." *Electronics,* February 20: 124–129.

West, W. 1962. "Achievement of High Reliability Levels for Minuteman Solid-State Devices." In *Semiconductor Reliability,* volume 2, ed. W. Von Alven. Engineering Publishers.

Westman, Harold. 1928. "28,000 Kilocycles—and How!" *QST,* August: 37–42.

Weston, J. Fred, ed. 1960. *Procurement and Profit Renegotiation.* Wadsworth.

Wilson, John. 1985. *The New Venturers: Inside the High-Stakes World of Venture Capital.* Addison-Wesley.

Winterelt. Rolf. 1961. "Minuteman System Is 'Most Reliable.'" *Missiles and Rockets,* February 27: 37–39.

Wolfe, Tom. 1983. "The Tinkerings of Robert Noyce." *Esquire,* December: 346–373.

Wolff, Michael. 1976. "The Genesis of the Integrated Circuit." *IEEE Spectrum* 13: 45–53.

Wuerth, J. M. 1976. "The Evolution of Minuteman Guidance and Control." *Navigation* 23, no. 1: 64–75.

Yew, Lee Kuan. 2000. *From Third World to First: The Singapore Story, 1965–2000.* HarperCollins.

Yu, Albert. 1998. *Creating the Digital Future.* Free Press.

Zahl, Albert. 1972. *Radar Spelled Backwards.* Vantage.

... ...

Inside Technology

Peter Keating and Alberto Cambrosio, *Biomedical Platforms: Reproducing the Normal and the Pathological in Late-Twentieth-Century Medicine*

Eda Kranakis, *Constructing a Bridge: An Exploration of Engineering Culture, Design, and Research in Nineteenth-Century France and America*

Christophe Lécuyer, *Making Silicon Valley: Innovation and the Growth of High Tech, 1930–1970*

Pamela E. Mack, *Viewing the Earth: The Social Construction of the Landsat Satellite System*

Donald MacKenzie, *Inventing Accuracy: A Historical Sociology of Nuclear Missile Guidance*

Donald MacKenzie, *Knowing Machines: Essays on Technical Change*

Donald MacKenzie, *Mechanizing Proof: Computing, Risk, and Trust*

Maggie Mort, *Building the Trident Network: A Study of the Enrolment of People, Knowledge, and Machines*

Nelly Oudshoorn and Trevor Pinch, editors, *How Users Matter: The Co-Construction of Users and Technologies*

Paul Rosen, *Framing Production: Technology, Culture, and Change in the British Bicycle Industry*

Susanne K. Schmidt and Raymund Werle, *Coordinating Technology: Studies in the International Standardization of Telecommunications*

Charis Thompson, *Making Parents: The Ontological Choreography of Reproductive Technology*

Dominique Vinck, editor, *Everyday Engineering: An Ethnography of Design and Innovation*

Index

Printed in the United States
by Baker & Taylor Publisher Services

Printed in the United States
by Baker & Taylor Publisher Services